T0271196

WHEAT ANTIOXIDANTS

BICENTENNIAL
1807
⊛WILEY
2007
BICENTENNIAL

THE WILEY BICENTENNIAL–KNOWLEDGE FOR GENERATIONS

*E*ach generation has its unique needs and aspirations. When Charles Wiley first opened his small printing shop in lower Manhattan in 1807, it was a generation of boundless potential searching for an identity. And we were there, helping to define a new American literary tradition. Over half a century later, in the midst of the Second Industrial Revolution, it was a generation focused on building the future. Once again, we were there, supplying the critical scientific, technical, and engineering knowledge that helped frame the world. Throughout the 20th Century, and into the new millennium, nations began to reach out beyond their own borders and a new international community was born. Wiley was there, expanding its operations around the world to enable a global exchange of ideas, opinions, and know-how.

For 200 years, Wiley has been an integral part of each generation's journey, enabling the flow of information and understanding necessary to meet their needs and fulfill their aspirations. Today, bold new technologies are changing the way we live and learn. Wiley will be there, providing you the must-have knowledge you need to imagine new worlds, new possibilities, and new opportunities.

Generations come and go, but you can always count on Wiley to provide you the knowledge you need, when and where you need it!

WILLIAM J. PESCE
PRESIDENT AND CHIEF EXECUTIVE OFFICER

PETER BOOTH WILEY
CHAIRMAN OF THE BOARD

WHEAT ANTIOXIDANTS

EDITED BY

LIANGLI YU
Department of Nutrition and Food Science
The University of Maryland

WILEY-INTERSCIENCE
A JOHN WILEY & SONS, INC., PUBLICATION

Published by John Wiley & Sons, Inc., Hoboken, New Jersey
Published simultaneously in Canada

For general information on our other products and services or for technical support, please contact our Customer Care Department within the United States at (800) 762-2974, outside the United States at (317) 572-3993 or fax (317) 572-4002.

Wiley publishes in a variety of print and electronic formats and by print-on-demand. Some material included with standard print versions of this book may not be included in e-books or in print-on-demand. If this book refers to media such as a CD or DVD that is not included in the version you purchased, you may download this material at http://booksupport.wiley.com. For more information about Wiley products, visit www.wiley.com.

Wiley Bicentennial Logo: Richard J. Pacifico

Library of Congress Cataloging-in-Publication Data:

Yu, Liangli.
  Wheat antioxidants / edited by Liangli Yu.
       p. cm.
  Includes index.
  ISBN 978-0-470-04259-5 (cloth)
  1. Antioxidants. 2. Wheat. I. Title.
  TX553.A73Y8 2008
  613.2'8–dc22   2007034303

CONTENTS

CONTRIBUTORS

Elizabeth A. Arndt
ConAgra Foods, Inc., 6 ConAgra Drive, Omaha, NE 68102, USA

Allan Ayella
Department of Human Nutrition, Kansas State University, Manhattan, KS 66503, USA

Trust Beta
Department of Food Science, University of Manitoba, Winnipeg, Manitoba, Canada R3T 2N2

Denys Charles
Global Analytical Science Department, Colgate-Palmolive Company, 909 River Rd, Piscataway, NJ 08855, USA

Zhihong Cheng
Balchem Corporation, 52 Sunrise Park Road, P.O. Box 600, New Hampton, NY 10958, USA
Department of Nutrition and Food Science, 0112 Skinner Building, University of Maryland, College Park, MD 20742, USA

Gang Guo
The Mennel Milling Company, 1702 S. Jefferson St, Roanoke, VA 24016, USA

Zhigang Hao
Global Analytical Science Department, Colgate-Palmolive Company, 909 River Rd, Piscataway, NJ 08855, USA

Yu Jiang
Department of Human Nutrition, Kansas State University, Manhattan, KS 66503, USA

Yutaka Konishi
Central Laboratories for Frontier Technology, Kirin Brewery Co. Ltd, 1-13-5, Fukuura, Kanazawa-ku, Yokohama-shi, Kanagawa 236-0004, Japan

Chunjian Lin
The Mennel Milling Company, 1702 S. Jefferson St, Roanoke, VA 24016, USA

Marla W. Luther
ConAgra Foods, Inc., 6 ConAgra Drive, Omaha, NE 68102, USA

Chandrika Liyana-Pathirana
Department of Biochemistry, Memorial University of Newfoundland, St. John's, Newfoundland, Canada AIB 3X9

Don L. Mennel
The Mennel Milling Company, 1702 S. Jefferson St, Roanoke, VA 24016, USA

Jeffrey Moore
Department of Nutrition and Food Science, 0112 Skinner Building, University of Maryland, College Park, MD 20742, USA

Archie Mpofu
Department of Food Science, University of Manitoba, Winnipeg, Manitoba, Canada R3T 2N2

Ping Ouyang
Department of Human Nutrition, Kansas State University, Manhattan, KS 66503, USA

John Parry
Box 9061, M.T. Carter Hall, Agricultural Research Station, Virginia State University, Petersburg, VA 23834, USA

Hongyan Qu
Department of Human Nutrition, Kansas State University, Manhattan, KS 66503, USA

Harry D. Sapirstein
Department of Food Science, University of Manitoba, Winnipeg, Manitoba, Canada R3T 2N2

Fereidoon Shahidi
Department of Biochemistry, Memorial University of Newfoundland, St. John's, Newfoundland, Canada AIB 3X9

Elaine Studer
Department of Microbiology & Immunology, Virginia Commonwealth University, MCV Campus, P.O. Box 980678, Richmond, VA 23298, USA

Rong Tsao
Food Research Program, Agriculture and Agri-Food Canada, 93 Stone Road West, Guelph, Ontario, Canada N1G 5C9

Weiqun Wang
Department of Human Nutrition, Kansas State University, Manhattan, KS 66503, USA

JunJie Yin
Department of Nutrition and Food Science, 0112 Skinner Building, University of Maryland, College Park, MD 20742, USA

Liangli (Lucy) Yu
Department of Nutrition and Food Science, 0112 Skinner Building, University of Maryland, College Park, MD 20742, USA

Liangping Yu
Balchem Corporation, 52 Sunrise Park Road, P.O. Box 600, New Hampton, NY 10958, USA
Department of Nutrition and Food Science, 0112 Skinner Building, University of Maryland, College Park, MD 20742, USA

Junjun Zhang
Department of Microbiology & Immunology, Virginia Commonwealth University, MCV Campus, P.O. Box 980678, Richmond, VA 23298, USA

Huiping Zhou
Department of Microbiology & Immunology, Virginia Commonwealth University, MCV Campus, P.O. Box 980678, Richmond, VA 23298, USA

Kequan Zhou
Department of Food Science and Technology, 101 FST Building (0418), Virginia Polytechnic Institute, Blacksburg, VA 24061, USA

CHAPTER 1

OVERVIEW AND PROSPECTIVE

Liangli (Lucy) Yu

1.1 INTRODUCTION

The pathology of a number of chronic diseases including cancer involves oxidative damage to cellular components. For instance, reactive oxygen species (ROS) capable of causing damage to DNA have been associated with carcinogenesis, coronary heart disease, and many other health problems related to advancing age. Minimizing oxidative damage may well be one of the most important approaches to the primary prevention of these aging associated diseases and health problems. Antioxidants terminate ROS attacks and appear to be of primary importance in the prevention of these diseases and health problems. It has been widely accepted that diet can significantly alter the overall health and quality of life. Development of functional foods rich in bioavailable antioxidants may play an important role in this regard. The key for developing functional foods is to provide a sufficient amount of the bioavailable safe active components, the functional additives/nutraceuticals, in the finished functional food products. Multidisciplinary approaches are required to select suitable agricultural materials containing adequate concentrations of beneficial components, to enhance and preserve the bioactives through postharvest treatments and optimized storage conditions, to understand their bioavailability and efficacy, to evaluate the potential side effects of elevated intakes of these bioactive components, to preserve the bioactive components during food formulation and processing, to evaluate the potential impact of the functional food intake on biomarkers of targeted health problems, and to promote the production and the consumption of these foods.

Recent research demonstrates that wheat grain contains significant level of natural antioxidants. Wheat is an important agricultural commodity and a primary food ingredient worldwide and contains considerable beneficial nutritional components. Wheat and wheat-based food ingredients rich in natural antioxidants can ideally serve as the basis for development of functional foods designed to improve the health of millions of consumers. Development of functional foods using wheat-based ingredients may also find value-added alternative utilization of wheat grain and fractions, thus enhancing agricultural economy.

1.2 ANTIOXIDANT PROPERTIES OF WHEAT GRAIN

Growing evidence indicates that intake of whole wheat foods may be associated with potential health benefits including the reduced risk of coronary heart diseases and certain types of cancer (1–3). These beneficial effects are attributed to the bioactive factors in wheat grain such as nondigestible carbohydrates and phytochemicals (1–4). Antioxidants are a group of small molecular weight phytochemicals present in wheat grain. These include but are not limited to carotenoids, tocopherols, lignans, and phenolic acids. These antioxidative components may prevent life important molecules such as DNA and enzymes from oxidative damages through different mechanisms. For instance, wheat antioxidants may directly react with reactive oxygen species (ROS) such as hydroxyl radicals or singlet oxygen molecules to terminate their attacks to biological molecules. Wheat antioxidants may also form chelating complexes with transition metals to reduce their availability as catalysts for free radical generation. Antioxidant properties of wheat grain and fractions and their phytochemical composition have been investigated and are summarized in Chapters 2–4. In addition to showing the significant level of natural antioxidants in wheat grain, the results from these studies suggest that antioxidants are not evenly distributed in wheat grain but are concentrated in wheat bran and aleurone fraction of bran. This finding may lead to the production and consumption of "super bran" and other wheat-based food ingredients rich in natural wheat antioxidants. The effects of genotype, growing conditions, and interaction between genotype and environmental conditions on antioxidant properties of wheat grain and fractions have also been reported by several research groups and discussed in Chapter 3. The results from these studies indicate the potential for producing wheat grain rich in natural antioxidants and other bioactive factors. These results also warrant further investigations for enhancing levels of natural antioxidants and other beneficial factors in wheat grain through wheat breeding effort, improved agricultural practices, and genetic modification.

Several research groups reported that wheat phenolic acids, a group of phenolic antioxidants in wheat grain, are predominantly present in the insoluble bound forms in wheat grain, along with a small portion in the soluble free or conjugated forms (Chapter 3), suggesting the possibility to enhance the availability of wheat antioxidant availability through improved postharvest treatments. A few studies investigated the postharvest treatments, milling practice, and storage conditions for their potential influences on the antioxidant availability in wheat bran. The results from these studies showed that postharvest enzymatic and yeast treatments, bran particle size, and storage conditions may significantly alter the antioxidant availability in wheat-based food ingredients (Chapter 6). Additional research is needed to advance our understanding of these postharvest and ingredient storage approaches and optimizing these conditions to enhance the antioxidant availability in wheat-based food ingredients.

Interestingly, recent studies showed the potential effects of food processing conditions on antioxidant availability in wheat-based food products and the antioxidant properties of selected food products (Chapters 6 and 7). It is well accepted that bioavailability is required for any bioactive food factors to have any health beneficial effects. Bioavailability depends on a number of factors including their availability in food products, and their absorption and *in vivo* delivery to target organs and tissues. The availability of bioactives in food products is determined by their concentration in food ingredients and their preservation through food formulation and processing conditions. Interaction of wheat antioxidants with other food ingredients and the effects of food processing conditions such as thermal treatment on overall antioxidant properties of wheat-based food products are not fully understood.

Wheat phenolics including the phenolic acids and lignans are known antioxidants. Phenolic acid composition and lignan contents in different varieties of wheat have been studied (Chapters 3, 4, and 16). The interaction between the selected wheat phenolic acids and free radicals and transition metals was also investigated (Chapter 5). The bioavailability of phenolic acids from a few previous studies has also been discussed in Chapter 15. It is noted that food matrix may alter the availability and bioavailability of bioactives. Thoroughly designed animal and pilot human studies are needed to investigate the bioavailability of wheat antioxidants including phenolic acids and lignans from different wheat-based food ingredients and food products. This information is very important to optimize human benefits of wheat-based functional foods rich in natural antioxidants.

A number of analytical methods have been adapted and developed for investigating the antioxidant properties of wheat grain and fractions and their phytochemical compositions. These methods are summarized in details (Chapters 8–12). The limitations of these methods are also discussed along with their advantages and disadvantages. In addition to the spectrophotometric and fluorometric methods, electron spin resonance (ESR) spectroscopy has been discussed because it directly measures the presence of free radicals and has been utilized to evaluate radical scavenging activities of wheat antioxidants and to validate the analytical methods for wheat antioxidant research (Chapter 10). These analytical methods may be applied to other cereal grains and botanicals. It needs to be pointed out that there is still a need to develop new analytical methods that may evaluate antioxidant properties of a selected sample under physiologically relevant conditions and may compare hydrophilic and lipophilic antioxidants under same experimental conditions.

1.3   OTHER BIOLOGICAL ACTIVITIES OF WHEAT ANTIOXIDANTS

It is noted that wheat antioxidants may differ greatly in their chemical structures. Lutein, α-tocopherol, secoisolariciresinol, and vanillic and ferulic acids are known antioxidative compounds present in wheat grain (Fig. 1.1). They share a conjugation system with or without phenolic substitution(s). This common structural component allows them to have strong interaction with free radicals and convert the radicals to less reactive components, showing radical scavenging capacities. Besides their capacity to interact with ROS, wheat antioxidants may have other biological activities. It is well known that α-tocopherol has vitamin E activity, whereas lutein is a carotenoid compound that may affect macular pigment optical density (5). On the contrary, secoisolariciresinol is known as a phytoestrogen that may have potential in

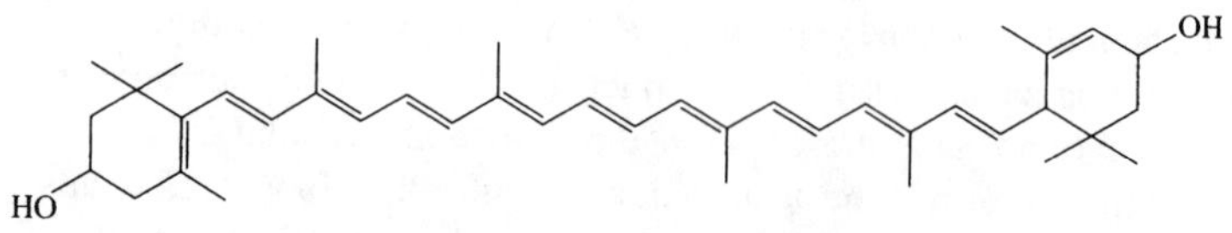

Lutein

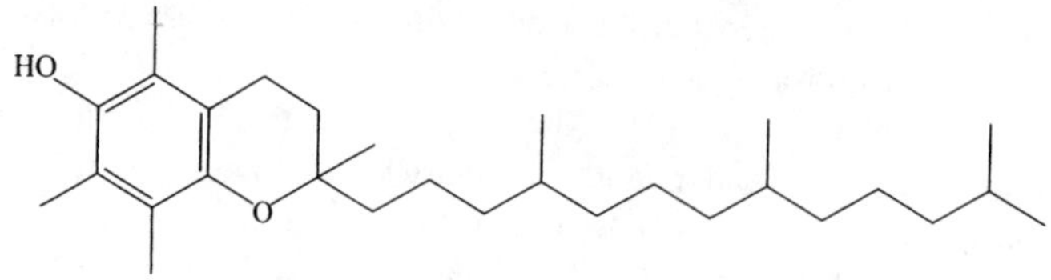

α-Tocopherol

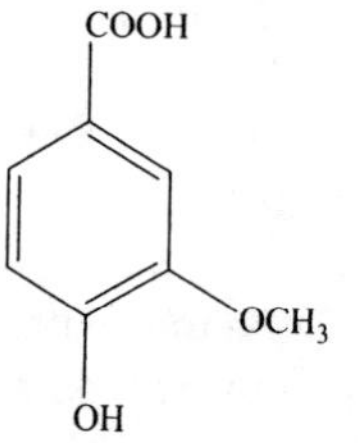

Vanillic acid

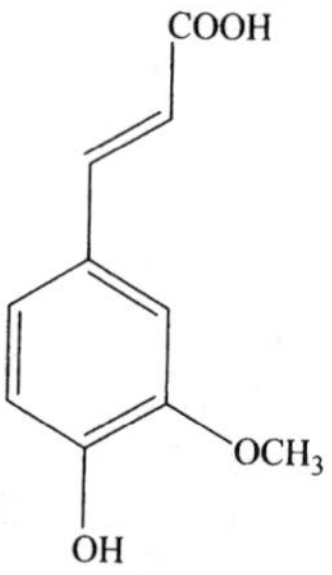

Ferulic acid

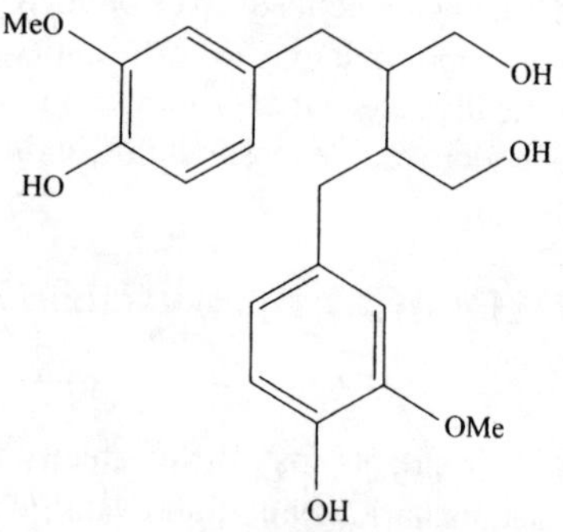

Secoisolariciresinol

Figure 1.1 Chemical structures of the phenolic acids.

chemoprevention of breast and prostate cancers, osteoporosis, and cardiovascular diseases (6,7). In addition, ferulic acid, the predominant phenolic acid in wheat grain or bran, has been shown to restore endothelial function in aortas of spontaneously hypertensive rats and to prevent trimethyltin-induced cognitive dysfunction in mice (8,9). In summary, individual antioxidant components may contribute to different health beneficial effects of whole wheat foods or wheat antioxidants because of their different chemical structures.

A recent study showed that wheat antioxidants may alter mRNA levels of 3-hydroxy-3-methylglutaryl-CoA reductase (HMG-CoA-R) and cholesterol 7α-hydroxylase (CYP7A1). These activities are also not related or mediated through their antioxidative actions. Wheat antioxidants may alter total plasma and low density lipoprotein (LDL) cholesterol levels through altering the levels of these two enzymes. Effects of wheat antioxidants on genes involved in cholesterol metabolism are discussed in Chapter 14. Chapter 15 in this book discusses the potential effect of wheat antioxidants on normal intestinal cells and nutrient absorption. These biological activities may also contribute to their overall health beneficial effects. Additional research is needed to investigate the other biological activities of individual wheat antioxidant compounds *in vitro* and *in vivo*, potential synergistic effects between wheat antioxidative components, the molecular mechanisms involved in their bioactivities, and their possible toxic effects.

1.4 WHEAT ANTIOXIDANTS: OPPORTUNITIES AND CHALLENGES

Wheat is an important agricultural commodity and a popular food ingredient worldwide. Wheat antioxidants and other beneficial phytochemicals are concentrated in the bran fraction of wheat grain. Bran is mostly used for low-value animal feed instead of human food ingredient. Research promoting the production and consumption of wheat-based food ingredients and food products rich in natural antioxidants may provide new value-adding opportunities for wheat bran, whole wheat flour, and other wheat-based food ingredients, which may benefit wheat growers, grain processing industry, food ingredient industry, and food manufacturers. The consumer desire of health beneficial functional foods also promotes research in the fields of wheat breeding, plant physiology, general plant science, food chemistry, food processing, general food science, nutrition, human health, and other biological and health sciences.

Wheat antioxidants also offer challenges for researchers in the related fields. Multidisciplinary approaches are required to advance our knowledge on wheat antioxidants. For instance, chemists and crop scientists have to work together to investigate how agricultural practices may alter antioxidant property and composition of wheat grains. It is also hard to ensure the consistency of wheat antioxidant properties in wheat grain and wheat-based food ingredients because many factors may be involved. To date, the chemical composition of wheat antioxidants is not fully understood. This makes it hard to quantify individual antioxidant compounds in the wheat-based food ingredients, which may be important for quality assurance of wheat-based food ingredients rich in antioxidants.

Food is a very complicated chemical system. Many interactions between food components may occur during food formulation, processing, and storage. These interactions may alter the status of individual antioxidants in the system, which may alter the effectiveness of analytical approaches such as antioxidant extraction. This makes it very challenging to research chemical components in food.

REFERENCES

1. Truswell, A. Cereal grains and coronary heart disease. *Eur. J. Clin. Nutr.* **2006**, *56*, 1–14.
2. Zoran, D. L.; Turner, N. D.; Taddeo, S. S.; Chapkin, R. S.; Lupton, J. R. Wheat bran diet reduces tumor incidence in a rat model of colon cancer independent of effects on distal luminal butyrate concentrations. *J. Nutr.* **1997**, *127*, 2217–2225.
3. Reddy, B. S.; Hirose, Y.; Cohen, L. A.; Simi, B.; Cooma, I.; Rao, C. V. Preventive potential of wheat bran fractions against experiemtnal colon carcinogenesis: implications for human colon cancer prevention. *Cancer Res .***2000**, *60*, 4792–4797.
4. Flight, I.; Clifton, P. Cereal grains and legumes in the prevention of coronary heart disease and stroke: a review of the literature. *Eur. J. Clin. Nutr.* **2006**, *60*, 1145–1159.
5. Mares, J. A.; LaRowe, T. L.; Snodderly, D. M.; Moeller, S. M.; Gruber, M. J.; Klein, M. L.; Wooten, B. R.; Johnson, E. J.; Chappell, R. J. Predictors of optical density of lutein and zeaxanthin in retinas of older women in the Carotenoids in Age-Related Eye Disease Study, an ancillary study of the Women's Health Initiative. *Am. J. Clin. Nutr.* **2006**, *84*, 1107–1122.
6. Begum, A. N.; Nicolle, C.; Mila, I.; Lapierre, C.; Nagano, K.; Fukushima, K.; Heinonen, S.- M.; Adlercreutz, H.; Remesy, C.; Scalbert, A. Dietary lignins are precursors of mammalian lignans in rats. *J. Nutr.* **2004**, *134*, 120–127.
7. Mazur, W.; Adlercreutz, H. Naturally occurring oestrogens in food. *Pure & Appl. Chem.* **1998**, *70*, 1759–1776.
8. Suzuki, A.; Yamamoto, M.; Jokura, H.; Fujii, A.; Tokimitsu, I.; Hase, T.; Saito, I. Ferulic acid restores endothelium-dependent vasodilation in aortas of spontaneously hypertensive rats. *Am. J. Hypertens.* **2007**, *20*, 508–513.
9. Kim, M. J.; Choi, S. J.; Lim, S. T.; Kim, H. K.; Heo, H. J.; Kim, E. K.; Jun, W. J.; Cho, H. Y.; Shin, D. H. Ferulic acid supplementation prevents trimethyltin-induced cognitive deficits in mice. *Biosci. Biotechnol. Biochem.,* **2007**, *71*, 1063–1068.

CHAPTER 2

ANTIOXIDANT PROPERTIES OF WHEAT GRAIN AND ITS FRACTIONS

Fereidoon Shahidi
Chandrika Liyana-Pathirana

There is a growing interest in phytonutrients or plant-derived bioactives that occur naturally in foods and possess curative and preventive properties above their nutritional value (1). In general, fruits and vegetables have been shown to contain compounds that contribute to human health (2). However, little attention has been paid to cereals with respect to their contribution in human health promotion and disease risk reduction, despite the fact that they serve as staple foods for most of the world's population (3). Among different cereals, wheat is the most important one in the temperate climate (3). There are two commercially important wheat types, namely, durum wheat or pasta wheat and common or bread wheat (4).

Antioxidant activity is an important biological property of many phytochemicals that protects living organisms from oxidative damage thereby preventing various deleterious events and diseases in plants and animals including human beings (5). Phenolic compounds possess antioxidant activity and these are aromatic secondary metabolites of phenylalanine, and, to a lesser extent, tyrosine that constitute one of the most diverse family of compounds found in plants (6). Simple phenols, phenylpropanoids, flavonoids, tannins (proanthocyanidins and others), and lignins are among numerous categories of plant phenolics (7). Cereals have been known to contain phenolic acids, phytoestrogens, and small quantities of flavonoids (8). The phenolic acids in cereals are benzoic and cinnamic acid derivatives; the latter being most common (9). Cereals are also a major source of dietary lignans with potent antioxidant activity (10).

In cereal grains, phenolic acids are concentrated in the cell walls of their outer layers mainly esterified to the arabinose side groups of arabinoxylans (11). On the contrary, Goupy et al. (12) have reported that phenolic acids are mainly found in the aleurone layer and endosperm of cereals. In general, ferulic acid is the major phenolic acid in many cereals that exists predominantly in the seed coat (13), while traces of ferulic acid may also be found in the starchy endosperm (14). Sosulski et al. (15) reported the presence in wheat of *trans*-ferulic, syringic, and vanillic acids, while

Wheat Antioxidants, Edited by Liangli Yu
Copyright © 2008 John Wiley & Sons, Inc.

Hatcher and Kruger (16) reported six phenolic acids, namely, sinapic, ferulic, vanillic, syringic, caffeic, and coumaric acids. Oat and corn also contain several different phenolic compounds including p-hydroxybenzoic, vanillic, protocatechuic, syringic, ferulic, caffeic, and synapic acids (15). Presence of *trans*-ferulic, syringic, p-hydroxybenzoic, and protocatechuic acids was reported in rice (15). In addition, oat (*Avena sativa* L.) has been considered as a good source of anioxidants. According to Gary et al. (17) oat contains a range of functional ingredients that are concentrated in different parts of the kernel. The main antioxidative constituents of oat include vitamin E (tocols), phytic acid, phenolic compounds, and avenantheramides; flavonoids and sterols are also present to a lesser extent. These antioxidative constituents are concentrated in the outer layers of the oat kernel (18). According to Terao et al. (19) ferulic acid is the dominant phenolic acid in rye, wheat, and barley.

The health benefits of cereal grains may possibly be attributed to the nature of their cell wall polymers and chemical architecture (20,21). Phenolic acids, which are covalently bound to the insoluble wheat bran matrix, have been shown to possess different antioxidative functions (22). Recently, phenolic acids have gained attention because of their antioxidative, antiinflammatory, antimutagenic, and anticarcinogenic properties as well as their ability to modulate some key enzymatic functions in the cell (23). However, little detailed or systematic information is known about distribution of phenolics in wheat grain. Thus, the work reported here was carried out to fill an important gap in the existing knowledge in the field and to shed light on phenolics of wheat and their potential health benefits.

2.1 SAMPLE PREPARATION

Grains were obtained from the Canadian Grain Commission (Winnipeg, MN) and processed by milling and pearling. The milling yielded bran, flour, shorts, and feed flour. During pearling, the bran layers were sequentially removed from wheat kernels by abrasion. Starting from unprocessed grain, kernels were pearled from 10% to 50% in 10% increments, and the pearled wheat and corresponding by-products were collected separately at each level. The crude phenolic compounds present in wheat milling fractions, pearled grains, and their byproducts were extracted into 80% aqueous ethanol (1 : 10, w/v) at 4°C for 16 h.

2.2 TOTAL PHENOLIC CONTENT (TPC) AND TOTAL ANTIOXIDANT CAPACITY (TAC) OF WHEAT FRACTIONS

Wheat class CWRS (Canada Western Red Spring; *Triticum aestivum* L., crop year 2001) includes common bread wheat, while CWAD (Canada Western Amber Durum; *Triticum turgidum* L. var. durum, crop year 2002) wheat class is mainly grown for production of semolina (coarse flour prepared from durum wheat). TPC and TAC of milling and pearling fractions of wheat classes CWAD and CWRS are shown in Figs. 2.1 and 2.2, respectively. The content of total phenolics was

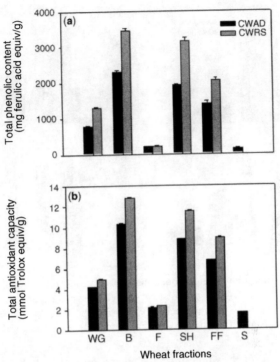

Figure 2.1   Total phenolic content (**a**) and total antioxidant capacity (**b**) of milling fractions of CWAD and CWRS wheat types. Abbreviations are WG, whole grain; B, bran; F, flour, SH, shorts; FF, feed flour; S, semolina; CWAD, Canadian Western Amber Durum; and CWRS, Canadian Western Red Spring.

determined according to a modified version of the procedure described by Singleton and Rossi (24), while TAC was determined according to the Trolox equivalent antioxidant capacity (TEAC) assay described by van den Berg et al. (25) with slight modifications. The TPC of milling fractions ranged from 140 to 2279 µg FAE/g defatted wheat and from 216 to 3437 µg FAE/g defatted wheat for CWAD and CWRS, respectively. The TAC expressed as µM Trolox equivalents (TE), was positively correlated with TPC of CWAD and CWRS. The TPC and TAC decreased in the order of bran > shorts > feed flour > whole grain > flour for both wheat classes. However, CWAD possessed an additional fraction, namely semolina, which is a product derived only from endosperm of durum wheat. Semolina possessed the lowest TPC and TAC among various fractions examined for CWAD. In both wheat classes, bran had significantly higher TPC and TAC compared to the endosperm that affords the flour (endosperm of wheat kernel) fraction. The higher TPC and TAC in shorts (a mixture of bran, endosperm, and germ) and feed flour (a mixture of bran and low grade endosperm) compared to flour can be explained by the presence of various proportions of bran (the outermost layers of the wheat kernel including aleurone layer) and germ (wheat embryo) in them. In the whole grain (unprocessed wheat grain), the endosperm dilutes the antioxidant substances present in the bran and

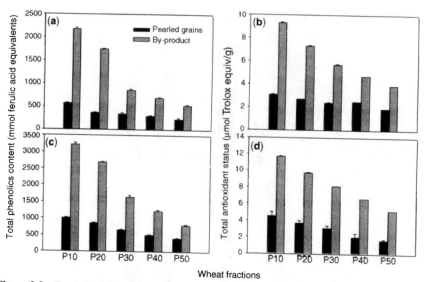

Figure 2.2 Total phenolic content and total antioxidant status of pearled grains and byproducts of CWAD (**a** and **b**) and CWRS (**c** and **d**) wheat types. Abbreviations are P10, 10% pearling; P20, 20% pearling; P30, 30% pearling; P40, 40% pearling; P50, 50% pearling; CWAD, Canadian Western Amber Durum; and CWRS, Canadian Western Red Spring.

hence TPC and TAC in the whole grain are lower compared to the bran fraction alone. In another study performed to obtain bran-rich and starch-rich fractions of wheat using milling and sieving, a higher TPC and antioxidant activity in the bran-rich fraction was noticed when compared with those in the starch-rich fraction (17). Although whole grains are known to provide benefits to humans owing to their unique phytochemical composition, the antioxidant activity of pearled wheat fractions and their by-products has not been studied to any great extent. Figure 2.2 represents the TPC and TAC of pearled wheat grains and their by-products. The whole grains and the grains pearled to remove 10% of the external layers possessed the highest TPC and TAC. The TPC and TAC decreased in the processed grain as the degree of pearling was increased. For CWRS, TPC in the 10% pearled grains and by-product, respectively, was 2.5 and 4 times higher than that of pearled grains and by-products of 50% pearling level; the corresponding values for CWAD were 2.4 and 4.1 times. Moreover, TAC decreased from 3.1 ± 0.04 to 1.9 ± 0.03 and 4.5 ± 0.04 to $1.6 \pm 0.02\,\mu mol$ TE/g, respectively, for CWAD and CWRS from 10% to 50% pearling. Significantly higher values of TPC and TAC for CWRS compared to those of CWAD may be attributed to the existing differences in genetic composition of the two wheat classes that are comprised of different species. Despite this difference, whole grains of both wheat classes possessed higher TPC and TAC compared to their pearled wheat grains. The TPC and TAC decreased with each level of pearling and the 10% pearled grains had the highest values when compared to a higher pearling level. However, the by-products always had a greater TPC and TAC compared to the respective pearled grains. Pearled samples of both CWAD and CWRS wheat classes contained remnants of both bran and germ. When the degree of pearling was increased, the existing phytochemicals were diluted as the pearled grains composed mainly of endosperm. At each level of pearling, phytochemical-rich external layers were removed leaving a greater proportion of endosperm in the pearled grain. However, the deep crease in the wheat grain did not yield uniform removal of the external layers. Hence, complete removal of external bran layers did not result in total removal of pericarp and aleurone from the crease of the pearled wheat grains. According to Dexter and Wood (26) the proportion of bran and germ in a whole grain is 14% and 1%, respectively. The by-product at 10% level of pearling predominantly includes bran layers, while at 20% level it contains a mixture of endosperm, aleurone, and bran. Zhou et al. (27) have shown that aleurone layer contributed significantly to the TPC and hence to antioxidant capacity of wheat. The same phenomenon can be attributed to the higher antioxidant capacity of by-products resulting from 20% pearling than those resulting from 30% to 50% pearling. Theoretically, beyond 20% of pearling, the by-product should be composed of endosperm alone; however, contamination always exists due to the residual crease in the pearled grain.

Phenolic compounds were concentrated in the bran fraction, while endosperm also possessed some that contributed to TAP and hence TAC (28,29). Fulcher (30) has reported that wheat phenolics are in general concentrated in cell walls of the aleurone layer. The 10% by-product fraction would be primarily pericarp with significant amounts of aleurone, while 20% fraction would be rich in aleurone. However, with the latter, the antioxidant levels are similar to 10% because of the presence of a higher amount of endosperm, which is poor in antioxidants. Martinez-Tome et al. (31) observed that oat and wheat bran were associated with high antioxidant activity. It has also been demonstrated that whole wheat and its milling fractions, especially bran, are rich sources of phenolic compounds (32–34). Peterson et al. (35) determined the antioxidant activity of oat by pearling them for 5 to 180 s. The authors observed higher antioxidant activity in the short-pearling fractions as longer pearling times removed most of the external layers leaving starchy endosperm in the pearled products. Moreover, Peterson (18) reported that antioxidative components were concentrated in the outer layers of the wheat kernel. Emmons et al. (36) have shown that extracts of oat pearling fractions possessed higher TPC and antioxidant activity than those prepared from the flour fractions. The study reported here also suggests that the degree of pearling should be kept at a minimum in order to retain the high-value bioactives in the pearled grains.

2.3   IRON(II)-CHELATING ACTIVITY OF WHEAT FRACTIONS

According to Kehrer (37), iron chelation may render important antioxidative effects by retarding metal-catalyzed oxidation. The effective iron chelators may afford protection against oxidative damage by removing iron(II) that may otherwise participate in hydroxyl radical-generating Fenton-type reactions. Minimizing iron(II) protects against oxidative damage by inhibiting production of reactive oxygen species (ROS) and lipid peroxidation. The iron(II)-chelating activity of various wheat milling

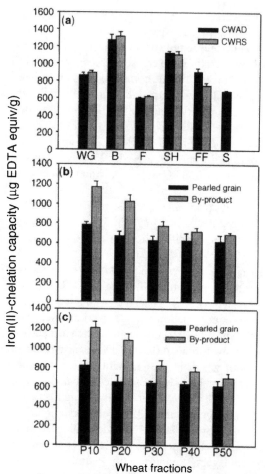

Figure 2.3   Iron(II)-chelation capacity (μg EDTA equiv/g) of milling fractions (**a**), pearled grains, and by-products of two wheat class, CWAD (**b**) and CWRS (**c**). Abbreviations are WG, whole grain; B, bran; F, flour; SH, shorts; FF, feed flour; S, semolina; P10, 10% pearling; P20, 20% pearling; P30, 30% pearling; P40, 40% pearling; P50, 50% pearling; CWAD, Canadian Western Amber Durum; and CWRS, Canadian Western Red Spring.

fractions and pearled wheat grains and by-products were determined by measuring the formation of iron–ferrozine complex. Figure 2.3 summarizes the results obtained for chelating effects of wheat fractions on iron(II). Among different milling fractions, bran demonstrated superior chelating properties over the other fractions examined. Flour and semolina demonstrated the lowest iron(II)-chelation capacity. The iron(II)-chelation capacity of pearled grains varied from 618 ± 69 to 682 ± 35 and 620 ± 47 to 723 ± 42 μg EDTA (ethylinediaminetetraacetic acid) equiv/g of defatted wheat of CWAD and CWRS, respectively. The corresponding values for the by-products were from 692 ± 21 to 1173 ± 59 and 703 ± 38 to 1206 ± 68 μg EDTA equiv/g, respectively. The whole grains always demonstrated higher chelation capacity compared to the pearled grains. Results indicated the presence of iron(II)-chelating agents such as phenolics in association with bran layers and hence the by-products are examined. Metal-chelating properties have been reported for amino acids and short peptides that can be easily extracted into an aqueous medium (38). Shahidi (39) reported that constituents such as amino acids, peptides, and proteins might play a significant role as physiological and dietary antioxidants thereby augmenting the antioxidant properties against oxidative damage. The antiradical properties exhibited by wheat extracts were also directly proportional to their iron(II)-chelating properties. It has been reported that high iron status or iron overload is positively correlated with coronary heart disease risk (40). Hence, increased intake of phenolic compounds may maintain a relatively low iron status thereby reducing the risk of iron overload (41).

2.4 OXYGEN RADICAL ABSORBANCE CAPACITY (ORAC) OF WHEAT FRACTIONS

The antioxidant activity of wheat milling fractions, pearled grains, and their by-products measured by the ORAC assay, as determined according to the method of Dávalos et al. (42), showed effective scavenging of peroxyl radical especially by the products containing bran (Fig. 2.4). The ORAC values of wheat fractions were expressed as μmol TE/g of defatted wheat. The ORAC values of wheat milling fractions ranged from 45 ± 2 to 301 ± 5 and 54 ± 2 to 310 ± 3 μmol TE/g for CWAD and CWRS cultivars, respectively. The ORAC of milling fractions was in the order of bran > shorts > feed flour > whole grain > flour. A general agreement in the rank order with regard to ORAC was noted between CWAD and CWRS. Hence, bran possessed the highest ORAC, while flour and semolina fractions had the lowest ORAC. The ORAC values for samples of CWRS were much higher than those of CWAD that may be attributed to the varietal differences. The ORAC of wheat bran was approximately 6.8 folds higher than those of flour fractions for both CWAD and CWRS. The antioxidant activity of pearled wheat grains and their by-products showed effective scavenging of peroxyl radical, especially by the by-products at 10% to 20% pearling. According to Hendelman et al. (43), the ORAC of oat varied from 2.08 to 8.13 μmol/g. These authors found that the bran and flour had similar antioxidant activity due to the mixing of the bran with the starchy endosperm. However, the aleurone layer possessed the highest ORAC value. In contrast, this study demonstrated that the by-products at 10% and 20% pearling, containing mainly bran and aleurone portions, possessed a much higher ORAC value than all pearled grains and by-products at 30 to 50% pearling; wheat demonstrated better antioxidant activity than oat in the ORAC assay (43). However, ORAC was determined using two different methods in the two studies, and hence comparison becomes rather difficult. Thus, the antioxidant capacity of wheat can primarily be attributed to the constituents present in the bran layers. Zhou et al. (27) determined the antioxidant activity of bran and aleurone layers of a Swiss red wheat variety and found that the aleurone layer exerted 7 to 8 times more antioxidant activity than that of the bran layer as determined by the ORAC assay. Thus, consumption of whole grains may provide full advantage of all antioxidative compounds present in wheat.

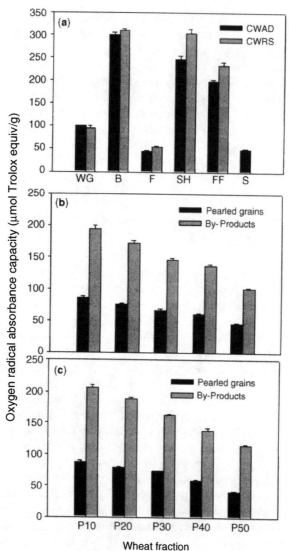

Figure 2.4 Oxygen radical absorbance capacity (ORAC, μmol Trolox equiv/g) of milling fractions (**a**), pearled grains, and by-products of two wheat classes, CWAD (**b**) and CWRS (**c**). Abbreviations are WG, whole grain; B, bran; F, flour, SH, shorts; FF, feed flour; S, semolina; P10, 10% pearling; P20, 20% pearling; P30, 30% pearling; P40, 40% pearling; P50, 50% pearling; CWAD, Canadian Western Amber Durum; and CWRS, Canadian Western Red Spring.

2.5 INHIBITION OF PHOTOCHEMILUMINESCENCE (PCL) BY WHEAT FRACTIONS

The antioxidative potential of whole grains and milling fractions and of pearled wheat grains and their by-products as measured by PCL method is shown in Figs. 2.5 and 2.6,

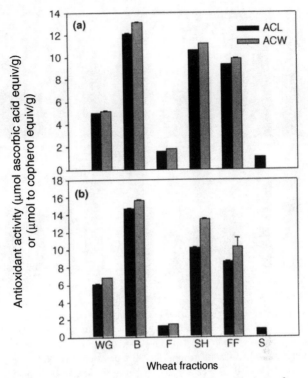

Figure 2.5   Antioxidant activity of whole grains and milling fractions of two wheat classes, CWAD and CWRS, as evaluated by photochemiluminescence in ACW (μmol ascorbic acid equiv/g) and ACL (μmol tocopherol equiv/g defatted material) systems. Abbreviations are WG, whole grain; B, bran; F, flour; SH, shorts; FF, feed flour; S, semolina; CWAD, Canadian Western Amber Durum; and CWRS, Canadian Western Red Spring.

respectively. The antioxidant capacity of water- (ACW) and lipid-soluble (ACL) compounds was assessed using a PHOTOCHEM® (Analytik Jena USA, Delaware, OH). In this method, the radical-scavenging capacity is evaluated by measuring the inhibition of photoinduced, chemiluminescent autoxidation of luminol (44).

The highest inhibition was observed with the fractions that contained bran of both wheat cultivars. The inhibition of PCL for milling fractions varied from 0.9 to 14.7 and 1.5 to 15.6 μmol α-tocopherol equiv/g of defatted wheat for CWRS and CWAD, respectively, in the ACL system. The corresponding values for the ACW system were 1.1 to 12.1 and 1.8 to 13.1 μmol ascorbic acid equiv/g of defatted wheat, respectively. The inhibition of PCL varied significantly among different milling fractions and was in the decreasing order of bran > shorts > feed flour > whole grain > flour. Moreover, in the CWAD cultivar semolina rendered the lowest inhibition. With respect to pearling, the antioxidant activity was highest for the by-products resulting from 10% pearling of both CWAD and CWRS. In the ACW system, the pearled grains of CWAD and CWRS at 10% pearling demonstrated 2 and 2.9 times stronger antioxidant activity than those pearled to 50%, respectively. The corresponding

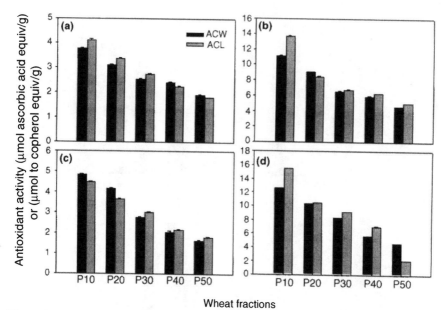

Figure 2.6 Antioxidant activity of pearled grains, and byproducts of two wheat classes CWAD (**a, b**) and CWRS (**b, c**) as evaluated by photochemiluminescence in ACW (µmol ascorbic acid equiv/g) and ACL (µmol tocopherol equiv/g) systems. Abbreviations are P10, 10% pearling; P20, 20% pearling; P30, 30% pearling; P40, 40% pearling; P50, 50% pearling; CWAD, Canadian Western Amber Durum; and CWRS, Canadian Western Red Spring.

values for the ACL system were 2.3 and 2.5 times. Moreover, the by-products were much stronger than the pearled grains. Whole grains of both CWAD and CWRS exhibited a higher inhibitory activity against PCL compared to that of pearled wheat grains.

Superoxide radical anions are produced by means of a photosensitizer and are detected by their reaction with a chemiluminogenic substance and by measuring the chemiluminescence produced. Luminol acts as the radical generator and detector (45). In the presence of an antioxidant compound, the intensity of PCL is attenuated due to scavenging of $O_2^{\cdot-}$. Hence, the antiradical properties of the additive may be quantified and expressed in equivalent concentration units of a reference compound; for instance, ascorbic acid or α-tocopherol equivalents in water- and lipid-based systems, respectively. Hence, the PCL method allows determination of the effects of both hydrophilic and hydrophobic compounds using the same system (46). The PCL method has been used to assess the antioxidant activity of plant extracts and blood plasma (46–50). In this study, we have reported on antioxidant activity of wheat milling and pearling fractions determined using the PCL method. With many fractions, the hydrophobic antioxidants demonstrated greater activity than that of hydrophilic constituents and hence values of ACL analysis were higher than those resulting from ACW analysis.

2.6   EFFECT OF MILLING AND PEARLING OF WHEAT ON INHIBITION OF LOW-DENSITY LIPOPROTEIN (LDL) OXIDATION

Dietary antioxidants that prevent LDL from oxidation are of great importance in protection against atherosclerosis (51). Cereals such as wheat (52,53), rye (54–56), and barley (57) have been shown to be rich sources of phenolic acids. Hydroxycinnamic acids and ferulic acid dehydrodimers are most common among cereal phenolic acids (52,53,57).

The inhibition of copper-induced oxidation of human LDL by whole wheat grains and milling fractions thereof as well as pearled wheat grains and their by-products are presented in Fig. 2.7. The retention capacity for CWAD milling fractions varied from 401 to 1603 µg protein/g, while in CWRS it ranged from 464 to 1673 µg protein/g of defatted wheat. In CWAD cultivar, the bran was 4 and 3.7 times more effective than the flour and semolina fractions, respectively, in inhibiting LDL against copper-induced oxidation. The bran of CWRS was 3.6 times more powerful than its flour fraction in the LDL system. The fractions that contained a higher proportion of bran exhibited a better antioxidant activity than those with a higher proportion of endosperm.

With regard to pearling, the activity of wheat samples in preventing the oxidation of human LDL was greater for by-products at 10% to 20% pearling than higher pearling levels. The inhibition of oxidation of human LDL by wheat decreased with sequential removal of the bran layers. There was a significant difference in the inhibition of LDL oxidation by the fractions, with increased degree of pearling. The highest activity of 10% to 20% by-products may be attributed to the presence of a greater proportion of bran and/or aleurone layer in these fractions than in the pearlings of 30% to 50%.

Low density lipoprotein is known to contain endogenous antioxidants such as α-tocopherol that may contribute to antioxidative activity in copper-induced LDL oxidation assay. There may be synergism between the well-known chain-breaking antioxidant α-tocopherol and the hydrogen donors (58) in the wheat samples. Thus, the phenolic hydrogen donors in wheat extracts may spare tocopherols and revert the tocopheryl radical directly to tocopherol. Moreover, phenolic antioxidants may scavenge other radicals thereby preventing them from attacking tocopherol (58). Another factor that may bring about antioxidant action in the LDL system is the interaction of wheat phenolics with LDL. Phenolic antioxidants may bind with apo-lipoprotein B, thus preventing the copper catalyst from binding to LDL. Moreover, the phenolic-protein binding may promote the access of phenolics to lipids (59). Several antioxidant mechanisms such as hydrogen donation, metal chelation, and protein binding may explain the antioxidant activity of phenolics against *in vitro* LDL oxidation (59). The ability of an antioxidant to inhibit copper-induced LDL oxidation may also be attributed to efficient removal of copper from the LDL surface (60). Thus, dietary antioxidants may play a significant role in retarding the development of atherosclerosis that may otherwise lead to coronary heart disease.

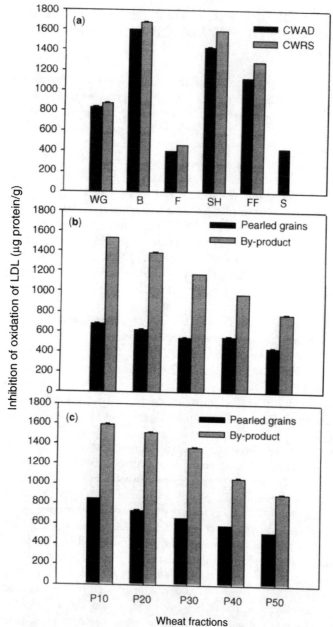

Figure 2.7   Inhibition of oxidation of LDL (μg protein/g) by milling fractions (**a**), pearled grains, and by-products (**b** and **c**) of two wheat classes, CWAD (**b**) and CWRS (**c**). Abbreviations are WG, whole grain; B, bran; F, flour, SH, shorts; FF, feed flour; S, semolina; P10, 10% pearling; P20, 20% pearling; P30, 30% pearling; P40, 40% pearling; P50, 50% pearling; CWAD, Canadian Western Amber Durum; and CWRS, Canadian Western Red Spring.

2.7 INFLUENCE OF MILLING AND PEARLING ON HOMEDIATED SUPERCOILED DNA SCISSION BY WHEAT

The supercoiled pBR DNA is converted into a nicked open circular form and a linear form upon strand breakage. It has been shown that HO$^•$ can effectively induce single

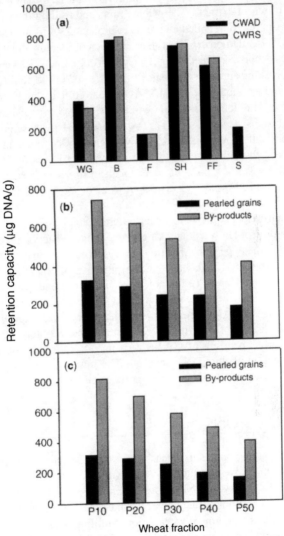

Figure 2.8 Retention capacity (μg DNA/g) of pBR 322 supercoiled DNA against hydroxyl radical mediated scission by milling fractions (**a**), pealed grains, and by-products of two wheat classes, CWAD (**b**) and CWRS (**c**). Abbreviations are WG, whole grain; B, bran; F, flour; SH, shorts; FF, feed flour; S, semolina; P10, 10% pearling; P20, 20% pearling; P30, 30% pearling; P40, 40% pearling; P50, 50% pearling; CWAD, Canadian Western Amber Durum; and CWRS, Canadian Western Red Spring.

strand breaks in DNA (61). Monitoring of single strand breaks in DNA induced by $HO^•$ may be used in the evaluation of antioxidant properties of a compound or extract of interest (62). Wheat antioxidants present in milling fractions, pearled grains, and their by-products differed in their ability to protect DNA from nicking by hydroxyl radical-mediated Fenton reaction. The $HO^•$ cleaved supercoiled plasmid pBR 322 DNA completely into nicked circular and linear DNA in the absence of any protection. Figure 2.8 summarizes the results of inhibition of DNA strand cleavage upon introduction of wheat extracts. The whole grains and milling fractions of two wheat cultivars were quite different in their ability to protect DNA from nicking by hydroxyl radical; the CWRS demonstrated higher activity compared to that of CWAD. The extracts of bran, shorts, and feed flour of the wheat cultivar CWRS were most effective against DNA nicking and their efficiency ranged from 90% to 95% (data not shown). The corresponding fractions of the CWAD wheat cultivar inhibited DNA scission by 87% to 90%. The flour extracts of CWAD and CWRS demonstrated 29–49% and 32–52% inhibitions, respectively. The extract derived from semolina fraction of CWAD possessed the lowest inhibiting power against DNA nicking at 26% to 47%. The poor capacity of flour and semolina in preventing DNA nicking indicates that Fe(II) chelation was not a possible mechanism in this assay. Bran had the highest retention capacity followed by shorts, feed flour, whole grains, and flour for both wheat cultivars.

Wheat antioxidants present in pearled grain and by-products also differed in their ability to protect DNA from nicking by the Fenton reaction mediated $HO^•$. The by-products of wheat pearling fractions were more effective than pearled-grain products in preventing $HO^•$ damage. In the presence of wheat antioxidants, DNA was converted mostly to nicked circular DNA and rarely to linear DNA and fragments. The by-products of pearling were more efficient than pearled grains in retaining DNA against $HO^•$-mediated oxidation. Moreover, the effects were significantly reduced with increased degree of pearling of the grains and resultant by-products. The by-products at 10% pearling of CWRS at 6 mg/mL concentration rendered more than 90% protection against DNA scission. The inhibitory activity of wheat anti-oxidants may be attributed to their scavenging of $HO^•$ or chelation of iron(II).

2.8   CONCLUSIONS

The antioxidative components in wheat grain were distributed asymmetrically. The concentration of bioactive constituents was greater in the external layers of the grain. Processing of cereals may thus have a significant effect on their antioxidant activity. Concentration of grain antioxidants will be drastically reduced during the refining process. Industrial debranning of wheat prior to milling would remove about 10% of the bran layers, so the residual grain would have reduced antioxidant potential. However, this is still much better compared to flour. Thus, a significant amount of the antioxidative compounds present in the external layers usually end up in the by-product following debranning or pearling. As phenolic compounds are concen-trated in the outermost layers, the bran fractions obtained as milling by-products and also by-products of wheat pearling may be used as a natural source of antioxidants and as a value-added product in the preparation of functional food ingredients and/or for enrichment of certain products.

REFERENCES

1. Balentine, D. A.; Albano, M. C.; Nair, M. G. Role of medicinal plants, herbs and spices in protecting human health. *Nutr. Rev.* **1999**, *7*, S41–S45.
2. Kurulich, A. C.; Tsau, G. J.; Brown, A.; Howard, L.; Klein, B. P.; Jeffery, E. H.; Kushad, M.; Wallig, M. A.; Juvik, J. A. Carotene, tocopherol and ascorbate contents in subspecies of Brassica oleracea. *J. Agric. Food Chem.* **1999**, *47*, 1576–1581.
3. Belderok, B. Developments in bread-making processes. *Plant Food Hum. Nutr.* **2000**, *55*, 1–14.
4. Dick, J. W.; Matsuo, R. R. Durum wheat and pasta products. In: Pomeranz, Y. (Ed). *Wheat, Chemistry and Technology*. American Association of Cereal Chemists, St. Paul, MN, **1988**. pp. 507–547.
5. Wang, M.; Jin, Y.; Ho, C.-T. Evaluation of resveratrol derivatives as potential antioxidants and identification of a reaction product of resveratrol and 2,2-diphenyl-1-picrylhydrazyl radical. *J. Agric. Food Chem.* **1999**, *47*, 3974–3977.
6. Cuppett, S. Plant production of biochemical compounds. *INFORM* **1998**, *588–590*, 592–595.
7. Brielmann, H. L., Jr. Phytochemicals: the chemical components of plants. In: Kaufman, P. B.; Cseke, L. J.;Warber, S.; Duke, J. A.; Brielmann, H. L. (Eds.). *Natural Products from Plants*. CRC Press, Boca Raton, FL, **1999**. pp. 1–36.
8. Zieliski, H.; Kozowska, H. Antioxidant activity and total phenolics in selected cereal grains and their different morphological fractions. *J. Agric. Food Chem.* **2000**, *48*, 2008–2016.
9. Natella, F.; Nardini, M.; DiFelice, M.; Scaccini, C. Benzoic and cinnamic acid derivatives as antioxidants: Structure-activity relation. *J. Agric. Food Chem.* **1999**, *47*, 1453–1459.
10. Cassidy, A. Physiological effects of phytoestrogens in relation to cancer and other human health risks. *Proc. Nutr. Soc.* **1996**, *55*, 399–418.
11. Maillard, M.-N.; Berset, C. Evolution of antioxidant activity during kilning: role of insoluble bound phenolic acids of barley and malt. *J. Agric. Food Chem.* **1995**, *43*, 1789–1793.
12. Goupy, P.; Hugues, M.; Boivin, P.; Amiot, M. J. Antioxidant composition and activity of barley (*Hordeum vulgare*) and malt extracts and of isolated phenolic compounds. *J. Sci. Food Agric.* **1999**, *79*, 1625–1634.
13. Watanabe, M.; Ohshita, Y.; Tsushida, T. Antioxidant compounds from buckwheat (*Fagopyrum esculentum* Möench) hulls. *J. Agric. Food Chem.* **1997**, *45*, 1039–1044.
14. Pussayanawin, V.; Wetzel, D. L.; Fulcher, R. G. Fluorescence detection and measurement of ferulic acid in wheat milling fractions by microscopy and HPLC. *J. Agric. Food Chem.* **1988**, *36*, 515–520.
15. Sosulski, F.; Krzysztof, K.; Lawrence, H. Free, esterified, and insoluble-bound phenolic acids. 3. Composition of phenolic acids in cereal and potato flours. *J. Agric. Food Chem.* **1982**, *30*, 337–340.
16. Hatcher, D. W.; Kruger, J. E. Simple phenolic acids in flours prepared from Canadian wheat: relationship to ash content, color, and polyphenol oxidase activity. *Cereal Chem.* **1997**, *74*, 337–343.
17. Gray, D. A.; Auerbach, S.; Hill, S.; Wang, R.; Campbell, G. M.; Webb, C.; South, B. Enrichment of oat antioxidant activity by dry milling and sieving. *J. Cereal Sci.* **2000**, *32*, 89–98.
18. Peterson, D. M. Oat antioxidants. *J. Cereal Sci.* **2001**, *33*, 115–129.
19. Terao, J.; Karasawa, H.; Arai, H.; Nagao, A.; Suzuki, T.; Takama, K. Peroxyl radical scavenging activity of caffeic acid and its related phenolic compounds in solution. *Biosci. Biotech. Biochem.* **1993**, *57*, 1204–1205.
20. Bunzel, M.; Ralph, J.; Martia, J. M.; Hatfield, R. D.; Steinhart, H. Diferulates as structural components in soluble and insoluble cereal dietary fibre. *J. Sci. Food Agric.* **2001**, *81*, 653–660.
21. Bunzel, M.; Ralph, J.; Kim, H.; Lu, F.; Ralph, S. A.; Marita, J. M.; Hatfield, R. D.; Steinhart, H. Sinapate dehydrodimers and sinapate-ferulate heterodimers in cereal dietary fiber. *J. Agric. Food Chem.* **2003**, *51*, 1427–1434.
22. Kroon, P. A.; Faulds, C. B.; Ryden, P.; Robertson, J. A.; Williamson, G. Release of covalently bound ferulic acid from fiber in the human colon. *J. Agric. Food Chem.* **1997**, *45*, 661–667.

23. Ho, C.-T .,Chen, Q.; Shi, H.; Zhang, K. Q.; Rosen, R. T. Antioxidative effects of polyphenol extract prepared from various Chinese teas. *Prev. Med.* **1992**, *21*, 520–525.
24. Singleton, V. L.; Rossi, J. A. Colorimetry of total phenolics with phosphomolybdic–phosphotungstic acid reagents. *Am. J. Enol. Viticul.* **1965**, *16*, 144–158.
25. van den Berg, R.; Haenen, G. R. M. M.; van den Berg, H.; Bast, A. Applicability of an improved Trolox equivalent antioxidant capacity (TEAC) assay for evaluation of antioxidant capacity measurements of mixtures. *Food Chem.* **1999**, *66*, 511–517.
26. Dexter, J. E.; Wood, P. J. Recent applications of debranning of wheat before milling. *Trends Food Sci. Technol.* **1996**, *7*, 35–40.
27. Zhou, K.; Laux, J. J.; Yu, L. Comparison of Swiss red wheat grain and fractions for their antioxidant properties. *J. Agric. Food Chem.* **2004**, *52*, 1118–1123.
28. Liyana-Pathirana, C. M.; Shahidi, F. Antioxidant activity of commercial soft and hard wheat (*Triticum aestivum* L.) as affected by gastric pH conditions. *J. Agric. Food Chem.* **2005**, *53*, 2433–2440.
29. Liyana-Pathirana, C. M.; Shahidi, F. Antioxidant properties of commercial soft and hard winter wheats (*Triticum aestivum* L.) and their milling fractions. *J. Sci. Food Agric.* **2006**, *86*, 477–485.
30. Fulcher, R. G. Morphological and chemical organization of the oat kernel. In:Webster, F. H. (Ed). *Oats Chemistry and Technology*. American Association of Cereal Chemists, St. Paul, MN, **1996**. pp. 47–74.
31. Martinez-Tome, M.; Murcia, M. A.; Frega, N.; Ruggieri, S.; Jimenez, A. M.; Roses, F.; Parras, P. Evaluation of antioxidant capacity of cereal beans. *J. Agric. Food Chem.* **2004**, *52*, 4690–4699.
32. Adom, K. K.; Sorrells, M. E.; Liu, R. H. Phytochemicals and antioxidant activity of milled fractions of different wheat varieties. *J. Agric. Food Chem.* **2005**, *53*, 2297–2306.
33. Beta, T.; Nam, S.; Dexter, J. M.; Sapirstein, H. D. Phenolic content and antioxidant activity of pearled wheat and roller-milled fractions. *Cereal Chem* .**2005**, *82*, 390–393.
34. Liyana-Pathirana, C. M.; Shahidi, F. Importance of insoluble-bound phenolics to antioxidant properties of wheat. *J. Agric. Food Chem.* **2006**, *54*, 1256–1264.
35. Peterson, D. M.; Emmons, C. L.; Hibbs, A. H. Phenolic antioxidants and antioxidant activityin pearling fractions of oat groats. *J. Cereal Sci.* **2001**, *33*, 97–103.
36. Emmons, C. L.; Peterson, D. M.; Paul, G. L. Antioxidant capacity of oat (*Avina sativa* L.) extracts. 2. *In vitro* antioxidant activity and contents of phenolic and tocol antioxidants. *J. Agric. Food Chem.* **1999**, *47*, 4894–4898.
37. Kehrer, J. P. The Haber-Weiss reaction and mechanisms of toxicity. *Toxicology* **2000**, *149*, 43–50.
38. Yee, J. J.; Shipe, W. F. Using enzymatic proteolysis to reduce copper–protein catalysis of lipid peroxidation. *J. Food Sci.* **1981**, *46*, 966–967.
39. Shahidi, F. Antioxidants in food and food antioxidants. *Nahrung* **2000**, *44*, 158–163.
40. Salonen, Y.; Ylä-Herttuala, S.; Yamamoto, R.; Butler, S.; Korpela, H.; Salonen, R.; Nyyssönen, K.; Palinski, W.; Witztum, J. L. Autoantibody against oxidized LDL and progression of carotid athero-sclerosis. *Lancet* **1992**, *339*, 883–887.
41. Samman, S.; Sandstrom, B.; Toft, M. B.; Bukhave, K.; Jensen, M.; Sorensen, S. S.; Hansen, M. Green tea or rosemary extract added to food reduces nonheme-iron absorption. *Am. J. Clin. Nutr.* **2001**, *73*, 607–612.
42. Dávalos, A.; Gómez-Cordovés, C.; Bartolomé, B. Extending applicability of the oxygen radical absorbance capacity (ORAC-fluorescein) assay. *J. Agric. Food Chem.* **2004**, *52*, 48–54.
43. Handelman, G.; Cao, G.; Walter, M. F.; Nightingale, Z. D.; Paul, G. L.; Prior, R. L.; Blumberg, J. B. Antioxidant capacity of oat (*Avena sativa* L.) extracts. 1. Inhibition of low-density lipoprotein oxidation and oxygen radical absorbance capacity. *J. Agric. Food Chem.* **1999**, *47*, 4888–4893.
44. Popov, I. N.; Lewin, G. Photochemiluminescent detection of antiradical activity: IV. Testing of lipid-soluble antioxidants. *J. Biochem. Biophys. Method* **1996**, *31*, 1–8.
45. Popov, I.; Lewin, G. Antioxidative homeostasis: characterization by means of chemiluminescent technique. *Method Enzymol* .**1999**, *300*, 437–456.
46. Amarowicz, R.; Raab, B.; Karamac, M. Antioxidative activity of an ethanolic extract of evening primrose. *Nahrung* **1999**, *43*, 216–217.
47. Vichi, S.; Zitterl-Eglseer, K.; Jugl, M.; Franz, C. Determination of the presence of antioxidants deriving from sage and oregano extracts added to animal fat by means of assessment of the radical scavenging capacity by photochemiluminescence analysis. *Nahrung* **2001**, *45*, 101–104.
48. Amarowicz, R.; Raab, B.; Shahidi, F. Antioxidant activity of phenolic fractions of rapeseed. *J. Food Lipid* **2003**, *10*, 51–62.

49. Gahler, S.; Otto, K.; Bohm, V. Alterations of vitamin C, total phenolics, and antioxidant capacity as affected by processing tomatoes to different products. *J. Agric. Food Chem.* **2003**, *51*, 7962–7968.

50. Lee, J.; Renita, M.; Fioritto, R. J.; Martin, S.; Schwartz, S. J.; Vodovotz, Y. Isoflavone characterization and antioxidant activity of Ohio soybeans. *J. Agric. Food Chem.* **2004**, *52*, 2647–2651.

51. Esterbauer, H. J.; Gebicki, J.; Puhl, H.; Juergens, G. The role of lipid peroxidation and antioxidants on oxidative modification of LDL. *Free Radical Biol. Med.* **1992**, *13*, 341–390.

52. Lempereur, I.; Rouau, X.; Abecassis, J. Genetic and agronomic variation in arabinoxylan and ferulic acid contents of durum wheat (*Triticum durum* L.) grain and its milling fractions. *J. Cereal Sci.* **1997**, *25*, 103–110.

53. Lempereur, I.; Surget, A.; Rouau, X. Variability in dehydrodiferulic acid composition of durum wheat (*Triticum durum* Desf.) and distribution in milling fractions. *J. Cereal Sci.* **1998**, *28*, 251–258.

54. Andreasen, M. F.; Christensen, L. P.; Meyer, A. S.; Hansen, A. Content of phenolic acids and ferulic acid dehydrodimers in 17 rye (*Secale cereale* L.) varieties. *J. Agric. Food Chem.* **2000**, *48*, 2837–2842.

55. Andreasen, M. F.; Christensen, L. P.; Meyer, A. S.; Hansen, A. Ferulic acid dehydridimers in rye (*Secale cereale* L.). *J. Cereal Sci.* **2000**, *31*, 303–308.

56. Andreasen, M. F.; Landbo, A.-K.; Christensen, L. P.; Hansen, A.; Meyer, A. S. Antioxidant effects of phenolic rye (*Secale cereale* L.) extracts monomeric hydroxycinnamates, and ferulic acid dehydrodimers on human low density lipoprotein. *J. Agric. Food Chem.* **2001**, *49*, 4090–4096.

57. Zupfer, J. M.; Churchill, K. E.; Rasmusson, D. C.; Fulcher, R. G. Variation in ferulic acid concentration among diverse barley cultivars measured by HPLC and microspectrophotometry. *J. Agric. Food Chem.* **1998**, *46*, 1350–1354.

58. Rice-Evans, C. A.; Miller, N. J.; Paganga, G. Structure-antioxidant activity relationships of flavonoids and phenolic acids. *Free Radical Bio. Med.* **1996**, *20*, 933–956.

59. Satue-Gracia, M. T.; Heinonen, M.; Frankel, EN. Anthocyanins as antioxidants on human low-density lipoprotein and lecithin-liposome systems. *J. Agric. Food Chem.* **1997**, *45*, 3362–3367.

60. Decker, E. A.; Ivanov, V.; Zhu, B-Z .,Frei, B. Inhibition of low-density lipoprotein oxidation by carnosine and histidine. *J. Agric. Food Chem.* **2001**, *49*, 511–516.

61. Breen, A. P.; Murphy, JA. Reactions of oxyradicals with DNA. *Free Radical Bio. Med.* **1995**, *18*, 1033–1078.

62. Hiramoto, K.; Ojima, N.; Sako, K.; Kikugawa, K. Effect of plant phenolics on the formation of the spin-adduct of hydroxyl radical and the DNA strand breaking by hydroxyl radical. *Biol. Pharm. Bull.* **1996**, *19*, 558–563.

CHAPTER 3

EFFECTS OF GENOTYPE, ENVIRONMENT AND GENOTYPE × ENVIRONMENT INTERACTION ON THE ANTIOXIDANT PROPERTIES OF WHEAT

Archie Mpofu
Trust Beta
Harry D. Sapirstein

3.1 INTRODUCTION

Wheat is a very diverse and widely adaptable cereal crop (1). Wheat breeding programs are primarily targeted at selecting new cultivars that have higher grain yields (2). In some countries such as Canada and Australia, end use quality is an important secondary breeding objective (2). Wheat yield has been increased by improving the following genetic traits: ear emergence time, plant stature, disease resistance, pest resistance, and stress tolerance (1). Changes in the end use properties that have enabled the use of wheat in a diverse variety of cultural foods were made possible through unconsciously changing the genetic characteristics of the grain constituents by exploiting natural variation of alleles at specific gene loci (2). The same gene pool has been exploited to adapt the crop to compositional and processing requirements (2). The biochemistry and genetic control of the major components of the starchy endosperm of wheat (protein and starch) are now relatively well understood. The embryo, grain coats, pericarp, and aleurone of wheat have complex compositions; they contain a range of constituents including antioxidant compounds whose biochemistry is still partially unknown and genetic control is poorly understood (1). Knowledge of genotype and environmental influences on the concentration of

Wheat Antioxidants, Edited by Liangli Yu
Copyright © 2008 John Wiley & Sons, Inc.

antioxidant compounds and the associated antioxidant activities likely serves as an elucidation of their genetic control and stability across different environments.

Genotype × growing environment (G × E) studies involve evaluation of genotypes for a given property in a set of environments (3). In these studies, in order to establish genotype and environment effects, it is necessary to use experimental designs where effects can be separated using analysis of variance (ANOVA) or other statistical techniques that can establish patterns in the data attributable to these effects (4). G, E, and G × E effects can only be established when genotypes are replicated within and over several locations (4). Genotype in the strict sense refers to the set of genes possessed by plants or other biological materials that is responsible for the expression of trait(s) under investigation (3). In this chapter, genotype is as well used to refer to wheat cultivars, varieties, and breeding lines. Growing environment can be defined as the set of biophysical conditions that influence the growth and development of biological material and thereby influence the expression of some trait(s). Examples of such biophysical conditions likely include water, nutrition, temperature, and disease (3). Significant differences in the performance of genotypes when they are grown in different environments indicate effects of G × E interaction (5). Absence of G × E interaction simplifies breeding because the best genotype in one environment would also be the best for all target environments. Significant G × E interaction may be either (i) a noncrossover G × E interaction where the ranking of genotypes remains constant across environments and the interaction is significant because of changes in the magnitude of the response or (ii) a crossover G × E interaction where a significant change in rank occurs from one environment to another. When selecting genotypes for wide adaptation, in the presence of significant G × E interaction, plant breeders look for a noncrossover G × E interaction (6).

Reports on the G × E variation of wheat biochemical properties have included those on bread making quality (7–12) and that on wheat color (6). Not much information is currently available on the genotype by environment variation of the antioxidant properties of wheat, with reports by Mpofu (13), Mpofu et al. (14), and Moore et al. (15) being the only ones from studies designed and precisely reported as G × E, encompassing G, E, and G × E effects on antioxidant properties of specific genotypes grown in multiple environments.

In wheat and other plants, antioxidant compounds are synthesized naturally as part of multifunctional defense systems against the detrimental effects of oxidation. These antioxidants can be added to food products in order to enhance their quality, stability, and safety. When consumed in human diets, these antioxidants may terminate free radical chain reactions and therefore reduce cell injury and death, slow down the aging process and reduce the incidence of such diseases as cancer, cardiovascular disease, and Parkinson's disease (16). Given these biological properties, it is apparent that selecting and breeding wheat genotypes that are high in antioxidant compounds will improve agronomical traits of the wheat plants, enhance the keeping quality, stability, and safety of wheat products and improve the health beneficial properties associated with wheat consumption. This chapter reviews the influences of genotype (G), the growing environment (E), and G × E interaction on the total phenolic content (TPC), phenolic acid composition and the associated antioxidant activities in wheat and wheat fractions.

Several reports on wheat antioxidant studies have indicated genotype and/or environmental variation of antioxidant properties of wheat (17–26). Most of these studies were either not designed as genotype by environment studies or probably designed as such but not reported in the manner typical of $G \times E$ studies. The relative contributions of genotype and environmental variances to variation in antioxidant activity (AOA) in terms of DPPH scavenging capacity, total phenolic content, and concentrations of six phenolic acids has been reported (14). Mpofu (13) reported G, E, and $G \times E$ variations in the superoxide scavenging capacities (SSCs) and peroxyl radical scavenging capacities (PSCs) of wheat. Similarly, Moore et al. (15) reported G, E, and $G \times E$ effects on the free radical scavenging capacities against DPPH, ABTS cation, peroxyl (ORAC), and superoxide anion radicals and chelating properties, as well as TPC and phenolic acid compositions of bran from 20 hard winter wheat varieties grown at two locations in eastern Colorado. Moore et al. (15) additionally pooled results from the reports by Yu et al. (20), Zhou and Yu (27), and Yu and Zhou (28) and analyzed them as $G \times E$. These reports (13–15) were presented in a manner typical of $G \times E$ studies, their results and implications are reviewed here and wherever possible contrasted with those from other reports on antioxidant properties of specific wheat genotype(s) grown in single or multiple environments. For comparison purposes, the parameters mainly focused on are DPPH SC, PSC, SSC, TPC, and phenolic acid concentrations as these were reported in at least two of the $G \times E$ studies on wheat antioxidants.

3.2   GENOTYPE EFFECTS

Six Canadian hard spring genotypes (AC Barrie, AC Vista, Neepawa, AC Elsa, Superb, and Snowbird) grown in the 2003 crop year differed significantly in several antioxidant properties (13,14). These genotypes had highly significant differences in TPC, DPPH SC, and concentrations of each of eight phenolic acids (14). AC Barrie, AC Elsa, Neepawa, and Superb belong to the Canada Western Red Spring (CWRS) commercial wheat class. AC Vista and AC Snowbird are white colored wheat genotypes belonging to the Canada Prairie Spring White (CPS) and Canada Western Hard White Spring (CWHS) classes, respectively.

Moore et al. (15) reported significant differences in the antioxidant properties of bran samples from 20 hard winter wheat genotypes. The genotypes investigated by Moore et al. (15) included Prowers 99, Co 99534, Verango, Stanton, Alliance, CO980607, CO980630, CO99508, CO980376, Akron, Enhancer, CO980719, Kalvesta, Wichita, and Halt, which are all red colored. The rest of the genotypes Moore et al. (15) investigated are white colored and are namely Trego, Avalanche, Lakin, Gmt10002, and Intrada. Moore et al. (15) also reported significant genotype effects on the DPPH scavenging capacity, TPC, and PSC of bran samples of three hard winter wheat varieties grown at five locations in eastern Colorado based on analysis of the pooled results from the reports by Yu et al. (20), Zhou and Yu (27), and Yu and Zhou (28).

The variations in the antioxidant properties of these genotypes are discussed in this section and contrasted with variations of other wheat genotypes. These variations likely indicate significance of the genotype effects on the antioxidant properties of whole wheat and wheat fractions including bran.

3.2.1   Total Phenolic Content

According to a review by Prior et al. (29), TPC has been expressed in terms of ferulic acid equivalents, gallic acid equivalents, catechin equivalents, tannic acid equivalents, chlorogenic acid equivalents, caffeic acid equivalents, protocatechuic acid equivalents, and vanillic acid equivalents. Use of different standards in TPC measurement makes comparison of levels impossible. In the $G \times E$ study by Mpofu et al. (14), TPCs were reported using ferulic acid as a standard and expressed in ferulic acid equivalents (FAE), whereas in the study by Moore et al. (15), gallic acid was used as the TPC standard and results accordingly expressed in gallic acid equivalents (GAE). On this basis, TPC levels in the genotypes from these two studies (14,15) cannot be compared, however, common to both studies was the conclusion of significant genotype effects on TPC. TPC results of these and several non-$G \times E$ studies are reviewed below.

Highly significant differences ($p < 0.0001$) were detected among the TPCs of AC Barrie, AC Vista, Neepawa, AC Elsa, Superb, and Snowbird (14). AC Elsa and Neepawa, with TPCs of 1990 and 1985 µg ferulic acid equivalents (FAE)/g, respectively, had the highest ranked TPCs, while AC Snowbird had the lowest TPC of 1709 µg FAE/g averaged across locations. These genotypes differed significantly ($p < 0.05$) in TPC at each of the four western Canada locations in which they were grown (Melfort, Regina, Swift Current, and Winnipeg). The TPCs reported by Mpofu et al. (14) (1709–1990 µg FAE/g) are about twofold higher than those reported by Li et al. (24) for four Chinese wheat samples (Chinese black-grained wheat (BGW), Dongjian purple-grained wheat (DPGW), Wu blue-grained wheat (WBGW), and Dongjian white-grained wheat (DWGW)) possibly due to genotype differences between samples from these studies. Beta et al. (23) reported TPCs with a greater degree of variation. These TPCs ranged between 1420 and 5300 µg FAE/g and were from pearling fractions of six wheat genotypes (AC Barrie, AC Superb, AC Crystal, AC Vista, AC Corinne, and AC Snowbird), two of which (AC Superb and AC Snowbird) were replicated in two growing environments. The larger TPC range reported by Beta et al. (23) compared to those reported by Mpofu et al. (14) and Li et al. (24) for whole-wheat samples reflects the increase in phenolic content of wheat from the inner to the outer parts of the grain as also reflected in the report by Adom et al. (18).

The TPCs of wheat bran from 20 hard winter wheat genotypes: Prowers 99, Co 99534, Verango, Stanton, Alliance, CO980607, CO980630, CO99508, CO980376, Akron, Enhancer, CO980719, Kalvesta, Wichita, and Halt, Trego, Avalanche, Lakin, Gmt10002, and Intrada grown at Walsh and Burlington testing locations differed significantly ($p < 0.05$) at both locations (15). TPCs of the bran from the 20 genotypes ranged between 2700 and 3500 µg GAE/g.

Adom et al. (17) reported significant differences between the TPCs of 11 wheat genotypes: W7985, Jennah Khetifa, Stoa, Cham1, Clark's Cream, NY6432-18, Opata, Caledonia, Sinton, Superior, and Roane. Significant differences were also reported between the TPCs of bran/germ fractions as well as between endosperm fractions of CayugaNY, RoaneNY, and CaledoniaNY harvested from plots grown near Ithaca, NY, USA in 2002, and Caledonia MI/OH and RoaneMI/OH composites of seeds harvested from plots near East Lansing, MI, USA and Wooster, OH, USA grown in the same year (18). TPCs of flour samples prepared from three hard winter wheat varieties (Akron, Trego, and Platte) grown at the five locations in Colorado differed significantly (21). Likewise, TPCs of eight wheat genotypes (Choptank, MV5-46, McCormick, Sisson, VA97W-024, SS560, Vigoro Tribute, and Roane) grown in Maryland, USA were significantly different (25). Significant TPC differences were also reported by Gélinas et al. (26) for wheat samples from three genotypes (AC Barrie, AC Brio, Celtic) harvested from six growing locations in Quebec, Canada. Table 3.1 summarizes these and other results on the antioxidant properties of wheat genotypes.

Among the TPCs reported in gallic acid equivalents (Table 3.1), bran/germ samples (18) had the highest TPCs whereas flour samples (21) had the least TPCs. This was not surprising since phenolic compounds are concentrated in the bran and germ fractions of wheat that are removed during the milling of wheat into flour. Among the whole-wheat samples with TPCs reported in GAE, those investigated by Adom et al. (17) had the highest TPCs. While genotype differences likely contributed to the TPC differences between the whole-wheat samples reported by Adom et al. (17), Gélinas et al. (26), and Moore et al. (25), however, differences in extraction procedures most likely also played a role.

Measurement of total phenolic content approximates the amount of phenolic compounds present without distinguishing between phenolic structures. The phenolic compounds present in wheat include phenolic acids, alkylresorcinols, flavanoids, ferulates, tannins, suberin, and lignin (30). The genotype differences reported in total phenolic content of wheat or wheat fractions likely indicate that genotype has a significant influence on the biosynthesis and accumulation of one or more of these phenolic compounds. Phenolic compounds have potent antioxidant activity (31), TPC has as such been found to be significantly correlated with different measures of antioxidant activity including DPPH scavenging capacity (14,23,24).

3.2.2   Phenolic Acid Composition

The concentrations of each of six phenolic acids in AC Barrie, AC Vista, Neepawa, AC Elsa, Superb, and Snowbird determined using HPLC differed significantly ($p < 0.05$) averaged across growing environments and at each location when results were evaluated by location (14). The patterns of variation for phenolic acid concentrations were dissimilar among genotypes. Representing approximately 49.7–64.8% of the total amount of measured phenolic acids, ferulic acid was the predominant phenolic acid (14). The overall mean concentration of ferulic acid in the wheat genotypes ranged from 371 to 441 µg/g. As with the results for TPC, AC Elsa and Neepawa had the highest ferulic acid concentrations and AC Snowbird had the lowest concentration that was in this case not significantly different from that of Superb (14). Contrary to the TPC results, Superb had the highest concentration of *o*-coumaric acid (OCA) (229 µg/g). OCA was the second most prominent phenolic acid among the six

TABLE 3.1 Ranges and Significance of Differences Between Antioxidant Properties of Wheat Genotypes

Parameter	Range	Sample	# of genotypes	Genotype effects	References
Total phenolic content	1709–1990 µg FAE/g	Whole-wheat meal	6	Significant	Mpofu et al. 14
	706–1108 µg FAE/g	Whole-wheat meal	4	Significant	Li et al. 24
	1420–5300 µg FAE/g	Wheat pearling fractions	6	Significant	Beta et al. 23
	1335–1618 µg GAE/g	Whole-wheat meal	11	Significant	Adom et al. 17
	631–735 µg GAE/g	Whole-wheat meal	3	Significant	Gélinas et al. 26
	400–800 µg GAE/g	Whole-wheat meal	8	Significant	Moore et al. 25
	~177–257 µg GAE/g	Wheat flour	3	Significant	Yu et al. 21
	5394–5870 µg GAE/g	Bran/germ fractions	3	Significant	Adom et al. 18
	331–367 µg GAE/g	Endosperm fractions	3	Significant	Moore et al. 15
	2700–3 500 µg GAE/g	Bran	20	Significant	Moore et al. 15
Ferulic acid concentration	371–441 µg/g	Whole-wheat meal	6	Significant	Mpofu et al. 14
	1550–1849 µg/g	Wheat bran	4	Significant	Li et al. 24
	130–144 µg/g	Wheat bran	2	Significant	Zhou et al. 32
	455–621 µg/g	Whole-wheat meal	8	Significant	Moore et al. 25
	101.14–136.29 µg/g	Wheat bran	2	Significant	Yu et al. 22
	287–588 µg/g	Whole-wheat meal	11	Significant	Adom et al. 17
	1950–2192 µg/g	Bran/germ fractions	3	Significant	Adom et al. 18
	29.1–40.74 µg/g	Endosperm fractions	3	Significant	Adom et al. 18
	89.4–193.9 µg/g	Bran	20	Significant	Moore et al. 15
DPPH radical scavenging capacity	13.21–14.22% DPPH discoloration	Whole-wheat meal	6	Significant	Mpofu 13
	23.66–33.51% DPPH discoloration	Whole-wheat meal	4	Significant	Li et al. 24
	2.5–26.0% DPPH discoloration	Wheat pearling fractions	6	Significant	Beta et al. 23
	ED_{50}=0.95–7.10 mg/ml	Whole-wheat meal	3	Significant	Yu et al. 19
	ED_{50}=23.2–27.42 mg/ml	Whole-wheat meal	8	Significant	Moore et al. 25
	64.4–70.6% DPPH remaining	Wheat flour	3	Significant	Yu et al. 21
	16.6–33.8% DPPH remaining	Bran	20	Significant	Moore et al. 15
Superoxide radical scavenging capacity	SSCW = 307–497 µg AAE/g	Whole-wheat meal	6	Significant	Mpofu 13
	SSCL = 332–412 µg TE/g	Whole-wheat meal	6	Significant	Mpofu 13
Peroxyl radical scavenging capacity	H-ORA C = 32.99–40.98 µmol TE/g	Whole-wheat meal	6	Not significant	Mpofu 13
	L-ORAC = 32.99–40.98 µmol TE/g	Whole-wheat meal	6	Not significant	
	T-ORAC = 32.99–40.98 µmol TE/g	Whole-wheat meal	6	Not significant	

measured, it comprised approximately 25% of the total content of individual phenolic acids averaged over genotypes and environments. Vanillic acid, caffeic acid, syringic acid, and p-coumaric acid combined, on average comprised about 13% of the total amount of phenolic acids quantified (14).

Significant differences ($p < 0.05$) were detected in concentrations of each of the following phenolic acids: p-hydroxybenzoic acid, coumaric acid, vanillic acid, syringic acid, and ferulic acid in bran from 20 wheat genotypes (15). Ferulic acid, ranging between 89.4 and 193.9 µg/g was the predominant phenolic acid (15). Concentrations of p-hydroxybenzoic, vanillic, syringic, and coumaric acids had the following concentration ranges: 7.2–34.1, 7.1–34.2, 10.1–57.1, and 2.5–17.0 µg/g, respectively (15).

The predominant concentration of ferulic acid among the phenolic acids and significant differences in its concentration among different wheat genotypes has as well been reported by Li et al. (24), Moore et al. (25), Yu et al. (22), and Zhou et al. (32). Adom et al. (17,18) did not compare concentration of ferulic acid to that of other phenolic acids, but also reported significant genotype effects on ferulic acid concentration. Results of these studies are summarized in Table 3.1. The concentrations of ferulic acid in whole wheat reported by Moore et al. (25) were higher but somehow comparable to those reported by Adom et al. (17) and Mpofu et al. (14) (Table 3.1), likely indicating genotype differences between the whole-wheat samples investigated by these authors. Such a conclusion can, however, only be reached through use of the same extraction solvents, protocols, and conditions, which was not the case for these studies.

Moore et al. (15) reported relatively low ferulic acid concentrations in wheat bran (89.4–193.9 µg/g), a range comparable to those obtained by Zhou et al. (32) and Yu et al. (22) that were 130–144 and 101–136 µg/g, respectively. The concentrations of ferulic acid in wheat bran reported by Moore et al. (15), Yu et al. (22), and Zhou et al. (32) are about 10 fold lower than those reported by Li et al. (24). This large difference is likely attributable to the differences in extraction procedures since Li et al. (24) hydrolyzed bran samples prior to liquid–liquid extraction, while Moore et al. (15), Yu et al. (22), and Zhou et al. (32) first extracted their bran samples using 50% acetone then hydrolyzed the extracts thereby likely extracting less phenolic acids.

Significant variation in the FA concentration of wheat genotypes has been found to be correlated with disease resistance. Three Fusarium head blight susceptible genotypes (Roblin, Wheaton, and Oslo) had a considerable lag in ferulic acid synthesis in the first 7–10 days after anthesis compared to the resistant genotypes (Sumai, BacUp, and 2375). At maturity, concentrations of ferulic acid in the grain were, however, similar (33). Abdel-Aal et al. (34), based on GLC results, reported on FA concentrations between 570 and 705 µg/g in four wheat genotypes: Arin, Katepwa, AC Foremost, and Roblin and three breeding lines: 94M-011, 94M-014, and 94M-025FA grown at four locations in Saskatchewan, Canada. Arin, a genotype that is highly resistant to orange blossom midge had the highest FA content (705 µg/g) compared to only 570 µg/g in the relatively susceptible Katepwa (34).

3.2.3   DPPH Scavenging Capacity

Ranging between 13.21% and 14.22% DPPH discoloration, the DPPH scavenging capacities of AC Barrie, AC Vista, Neepawa, AC Elsa, Superb, and Snowbird averaged across all locations were significantly different (14). As with the ranking of genotypes according to their TPC and ferulic acid concentrations, AC Elsa had the highest DPPH scavenging capacity, which was significantly different from those of all the other genotypes except Neepawa. Similar to TPC and ferulic acid concentration results as well, AC Snowbird had the lowest DPPH scavenging capacity and DPPH scavenging capacities of the genotypes were also significantly different at each of the four locations (14).

Moore et al. (15) expressed DPPH scavenging capacity results in terms of % DPPH remaining after 10 min of reaction. Bran samples from the genotypes they investigated differed significantly in DPPH scavenging capacity. The DPPH values ranged from 16.6% to 33.8% with the lower percentages indicating greater scavenging capacities (15).

Significant genotype effects have been detected or can be inferred from several non-$G \times E$ studies including those by Yu et al. (19), Yu et al. (21), Beta et al. (23), and Moore et al. (25) (Table 3.1). Differences in the extraction protocols and units of measure used however prevent comparison of values.

3.2.4   Superoxide Scavenging Capacities

AC Barrie, AC Vista, Neepawa, AC Elsa, Superb, and Snowbird had highly significant differences ($p < 0.0001$) in superoxide scavenging capacities of water (SSCW) and lipid (SSCL) soluble substances measured using a photochemiluminescence (PCL) assay (13). AC Barrie (497 µg ascorbic acid equivalents (AAE)/g) had the highest SSCW averaged across locations, whereas AC Snowbird (307 µg AAE/g) had the lowest. The SSCL of the genotypes followed a different trend, it ranged between 332 µg trolox equivalents (TE)/g for AC Elsa and 412 µg TE/g for Neepawa (13).

Other reports on the superoxide scavenging capacities (SSCs) of wheat genotypes have been based on different methods of extraction and determination and did not distinguish between SSCs of water and lipid soluble substances. In their $G \times E$ study, Moore et al. (15), using the hypoxanthine (HPX)/xantine oxidase (XOD) method detected significant genotype effects on the SSC of wheat bran from 20 genotypes. The SCCs reported by Moore et al. (15) were expressed in terms of percent $O_2^{\cdot -}$ remaining and ranged between 56.3% and 69.3%. Yu et al. (22) also used the HPX/XOD method and they detected no significant differences between the SSCs of Akron and Trego bran from wheat harvested at each of the following locations: Julesburg, Walsh, and Fort Collins. SSCs of the Akron and Trego bran were on average 27.48% and 28.97% $O_2^{\cdot -}$ quenched, respectively (22). Zhou et al. (32) reported a significant difference between the SSCs of two genotypes, Alliance (12.9 superoxide dismutase (SOD) equivalents) and Wichita (10.01 SOD equivalents) based on an electron spin resonance (ESR) spectrometry method.

3.2.5   Peroxyl Radical Scavenging Capacities

Peroxyl radical scavenging capacities (PSCs) measured using the hydrophilic (H-ORAC) and lipophilic (L-ORAC) oxygen radical absorbance capacity assays did not differ significantly among genotypes (13). Similarly, total ORAC (T-ORAC) (i.e., the sum of H-ORAC and L-ORAC values) did not vary significantly among the genotypes. Genotype effects were as well not significant at each of the test locations (Winnipeg, Swift Current, Melfort, and Regina) (13). Contrary to the observations by Mpofu (13), Moore et al. (15) reported significant genotype effects on the PSCs of wheat bran samples. Whole-wheat samples were used in the former.

Genotype effects were likely significant in the PSCs of Maryland grown soft wheat (25). The bran from Akron and Trego varieties differed significantly in PSCs when the wheat was grown at Julesburg and Fort Collins, but not at Walsh (22). No significant differences were detected between the PSCs of bran from Alliance and Wichita varieties grown at Walsh or Burlington (32). The inconsistency in effects of genotype on PSCs might be reflective of varietal differences, but differences in extraction and determination methods might also have a major role in addition to the use of whole-grain versus wheat bran samples.

Mpofu (13) reported on H-ORAC values between 28.1 and 36.5 µmol TE/g of ground whole-wheat grain. Moore et al. (15) determined PSCs of Maryland soft wheat varieties based on the predominantly hydrophilic traditional ORAC method, which does not distinguish between hydrophilic and lipophilic PSCs. The range of values reported by Moore et al. (25) (32.9–47.7 µmol of TE/g) was comparable to that obtained by Mpofu (13) (Table 3.1).

Moore et al. (15) detected PSCs between 45.0 and 78.0 µmol of TE/g using 50% acetone extracts of wheat bran. A comparable range (50.59–65.94 µmol of TE/g of bran) was reported by Zhou et al. (32) who used a similar extraction protocol. Much lower ranges 3.41–6.25 and 4.46–5.60 µmol of TE/g of bran were detected in ethanol extracts of Trego and Akron wheat brans, respectively (22). These differences were likely due to the effects of extraction solvent on PSC estimation and likely indicate that 50% acetone is a better extraction solvent than ethanol for PSC determination.

3.3   ENVIRONMENT EFFECTS

Growing environment had significant ($p < 0.05$) effects on the TPC, DPPH scavenging capacities, and concentrations of six phenolic acids (14). The PSCs and SSCs of the same wheat were also significantly affected by growing environment (13). Moore et al. (15) detected significant environment effects on DPPH scavenging capacity, TPC, and concentrations of PHBA, VA, SA, coumaric acid, and FA in 20 wheat genotypes grown at two locations. Environment effects were not significant for PSC (15). Analysis of the pooled results from the reports by Yu et al. (20), Zhou et al. (27), and Yu and Zhou (28) by Moore and others (15) revealed significant environment effects on the DPPH scavenging capacity, PSC and TPC, but not on the SSC of bran samples of three hard winter wheat varieties grown at five locations in eastern Colorado.

The mean concentrations of OCA in wheat from each location was negatively correlated ($r = -0.970, p < 0.05$) to average temperature during grain filling (14). No other significant relationship was detected between concentrations of the other phenolic acids, TPC, or any of the AOA parameters (DPPH, peroxyl, or superoxide scavenging capacities) and either total rainfall or average temperature during the kernel development period (13,14). Given that only four growing environments were investigated in this study and that only one harvest was involved thereby limiting the number of site year results, the relationship between concentration of OCA and average temperature during grain filling must be interpreted with caution.

Examination and comparison of the TPC and free radical scavenging properties of bran from Akron wheat grown at four nonirrigated and one irrigated testing locations in Colorado showed potential influences of growing conditions on antioxidant properties (20). A significant negative correlation was detected between the Fe^{2+} chelating activities of the bran samples from the four nonirrigated locations and total solar or daily average solar radiation ($r = -0.999$ and $p = 0.001$) (20). The bran from Trego wheat grown at the same testing locations had significantly different radical scavenging activities, chelating capacities, and total phenolic contents also suggesting that growing conditions likely influence the antioxidant properties of wheat (28). Positive correlations were detected between the DPPH$^{\bullet}$ scavenging activity and either total solar radiation (TSR) ($r = 0.97, p = 0.03$) or average daily solar radiation ($r = 0.97, p = 0.03$) (28). Unlike in the Akron wheat study (20), chelating activities of the Trego bran samples were not significantly correlated with any recorded environmental conditions including the hours exceeding 32 °C, total solar radiation, or daily average radiation showing possible genotype-dependent differential effects of environmental factors.

Analysis of the pooled results from the reports by Yu et al. (20), Zhou and Yu (27), and Yu and Zhou (28) by Moore et al. (15) revealed significant ($p < 0.05$) negative correlations between DPPH scavenging capacity and TSR ($r = -0.443$), ABTS SC and TSR ($r = -0.392$), and between TPC and hours exceeding 32 °C ($r = -0.417$). The results from the analysis of pooled results from the reports by Yu et al. (20), Zhou and Yu (27), and Yu and Zhou (28) by Moore et al. (15) are likely more accurate than those from the individual reports in elucidating the correlations between the antioxidant properties and environmental factors. Given that these results were from one growing season, they have to be interpreted with caution.

3.4   GENOTYPE BY ENVIRONMENT INTERACTION EFFECTS

Significance of interaction effects was not reported in most of the wheat antioxidant studies reviewed, and as this can only be determined statistically, only the $G \times E$ interaction results from Mpofu (13), Mpofu et al. (14), and Moore et al. (15) are reviewed in this section.

Genotype by environment interaction effects were significant for TPC and not significant for concentration of phenolic acids (ferulic acid (FA), o-coumaric acid (OCA), p-coumaric acid (PCA), syringic acid (SA), caffeic acid (CA), and vanillic acid (VA), DPPH radical scavenging capacities, or peroxyl radical scavenging capacities (14). Significant $G \times E$ interaction effects were as well detected for SSCW and SSCL but not any of the PSC parameters (H-ORAC, L-ORAC or T-ORAC) (14). Moore et al. (15) detected significant $G \times E$ interaction effects for the DPPH SC, PSC, TPC, ABTS$^{\cdot+}$ SC, and Fe^{2+} chelating capacity of the bran of 20 genotypes grown at two locations. Analysis of pooled results from the reports by Yu et al. (20), Zhou and Yu (27), and Yu and Zhou (28) revealed significant $G \times E$ interaction effects on the DPPH SC, PSC, TPC, ABTS$^{\cdot+}$ SC, and Fe^{2+} chelating capacity of the bran of three hard winter wheat varieties grown at five locations in eastern Colorado (15). $G \times E$ interaction effects become significant when genotypes perform differently when they are grown in different environments (5). These interactions can either be crossover or noncrossover as defined by Matus-Cadiz et al. (6).

Table 3.2 shows the rankings of genotypes at each growing location for TPC, SSCW, and SSCL. Ranking of genotypes at each location for each of these parameters was inconsistent likely indicating significant crossover interaction. Taking TPC as an example, Neepawa was in the highest Duncan grouping at all growing environments but Winnipeg, where it was second highest. Superb was in the second highest grouping

TABLE 3.2 Rankings of Genotypes in Different Locations for Parameters Significantly Influenced by $G \times E$ Interaction

Genotype	Location			
	Melfort	Regina	Swift current	Winnipeg
Total phenolic content				
AC Barrie	1862a	1837b	1844b	2030b
AC Elsa	1902a	1793b	2039a	2227a
Neepawa	1884a	1987a	2032a	2037b
AC Snowbird	1630c	1505d	1765b	1936cd
Superb	1736b	1677c	1821b	1873d
AC Vista	1747b	1680c	1830b	1956c
Superoxide scavenging capacities of water soluble substances (μg AAE/g)				
AC Barrie	558a	425a	428a	578a
AC Elsa	269f	268c	305d	424c
Neepawa	360d	332b	386c	525b
AC Snowbird	320e	343b	302d	261e
Superb	418b	254c	406b	438c
AC Vista	401c	345b	315d	398d
Superoxide scavenging capacities of lipid soluble substances (μg TE/g)				
AC Barrie	392b	375a	357c	360d
AC Elsa	316d	284d	325d	405c
Neepawa	413a	351ab	451a	433a
AC Snowbird	333c	323bc	407b	418b
Superb	409a	309cd	429ab	347d
AC Vista	295e	339bc	364c	429ab

For each parameter, values within the same column with different letters are significantly different at $p < 0.05$. (Adapted from Mpofu (13), Mpofu et al. (14).)

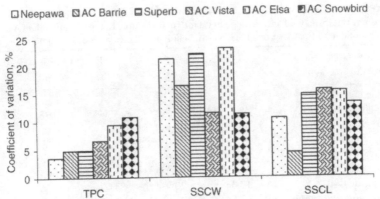

Figure 3.1   Stability of genotypes across environments for total phenolic content (TPC), superoxide scavenging capacity for water soluble substances (SSCW), and superoxide scavenging capacity for lipid soluble substances (SSCL). (Adapted from Mpofu 13, Mpofu et al. 14).

for Melfort and Swift Current wheat, whereas for Regina and Winnipeg wheat it was in the lowest grouping. AC Barrie, AC Elsa, AC Snowbird, and AC Vista also changed ranks (Table 3.2).

The relative stability of genotypes in expressing TPC, SSCW and SSCL across the growing locations can be visualized in Fig. 3.1. Although they were inconsistently ranked in wheat from different locations, Neepawa, AC Barrie, and Superb were relatively stable for TPC (CV < 5%) with Neepawa being the most stable. AC Snowbird with a CV of 10.81 across growing environments was the least stable for TPC. AC Barrie (CV = 4.33%) was relatively stable for SSCL, otherwise all genotypes were relatively unstable for SSCW and SSCL (CVs > 10).

3.5   RELATIVE CONTRIBUTION OF G, E, AND G × E EFFECTS TO TOTAL VARIATION

Table 3.3 shows the relative contribution of genotype and growing environment variances to total variance in the reports by Mpofu (13) and Mpofu et al. (14). Effects of growing environment largely dominated the total variance of antioxidant parameters (Table 3.3). Variance due to growing environment effects was at least 13.62% more than that due to genotype for TPC, DPPH SC, concentrations of VA, SA, and FA, SSCL, H-ORAC, L-ORAC, and T-ORAC. Genotype effects contributed more to the total variance of SSCW and concentrations of CA, PCA, and OCA than did environmental effects (Table 3.3). The contribution of genotype × environment interaction to total variances was relatively small. For the three variables that were significantly influenced by G × E interaction (TPC, SSCW and SSCL), its contribution to total variation was between 3.34% and 14.30%. When combined, G and E effects accounted for 57.93–95.98% of the total variation (when all variables in Table 3.3 are considered together), indicating the importance of these in influencing antioxidant properties.

TABLE 3.3   Variance Components (% of Total Mean Squares) for Genotype (G), Growing Environment (E), and G × E Interaction Effects for Total Phenolic Content, Antioxidant Activity, and Phenolic Acid Composition of Six Wheat Genotypes Grown at Four Locations

Variance component	G^a	E^b	$G \times E^a$	$G \times Block/E$	Block/E
Degrees of freedom	5	3	15	40	8
Parameter					
Total phenolics (µg FAE/g)	38.12***	57.86***	3.34***	0.38	0.29
DPPH SC (% DPPH discoloration)	36.99***	50.61***	6.71	4.00	1.68
Vanillic acid (µg/g)	24.42***	71.09***	0.79	1.43	2.28
Caffeic acid (µg/g)	60.89***	29.29***	1.93	2.03	5.87
Syringic acid (µg/g)	17.34***	70.91***	3.63	1.66	6.45
p-Coumaric acid (µg/g)	63.54***	30.61***	1.17	0.70	3.98
Ferulic acid (µg/g)	36.86***	56.57***	2.34	0.95	3.29
o-Coumaric acid (µg/g)	46.74***	41.44***	2.76	2.45	6.61
SSCW (µg AAE/g)	52.50*	36.42*	9.12*	0.76	1.20
SSCL (µg TE/g)	25.31*	53.68*	14.30*	2.05	4.67
H-ORAC$_{FL}$ (µmol TE/g)	13.16	52.33*	8.36	14.35	11.8
L-ORAC$_{FL}$ (µmol TE/g)	11.94	45.99*	14.08	13.17	14.82
T-ORAC$_{FL}$ (µmol TE/g)	14.76	48.93	8.30	16.16	11.84

Adapted from Mpofu (13), Mpofu et al. (14).
[a]Significance is based on genotype × block/E mean square.
[b]Significance is based on block/E mean square. * Significant ($p < 0.05$). ** Significant ($p < 0.01$). *** Highly significant ($p < 0.0001$).

Tables 3.4 and 3.5 respectively show the relative contribution of G, E, and G × E interaction variances to total variance in the antioxidant parameters of bran samples of 20 hard winter wheat varieties grown at two locations in eastern Colorado and bran samples of three hard winter wheat varieties grown at five locations in eastern

TABLE 3.4   Variance Components (% of Total Mean Squares) for G, E, and G × E Interaction Effects for DPPH SC, PSC, SSC, and TPC of Bran Samples of 20 Wheat Genotypes Grown at Two Locations

Variance component	G	E	G × E
Degrees of freedom	19	1	19
Parameter			
DPPH SC (% DPPH remaining)	85.78*	7.35*	5.88*
ORAC (µmol of TE/g)	56.13*	0.53	29.16*
SSC (% remaining)	26.60*	62.90*	10.26*
TPC (mg of GAE/g)	21.61*	68.32*	8.02*
p-Hydroxybenzoic acid (µg/g)	12.38*	72.46*	13.50*
Vanillic acid (µg/g)	9.28*	76.95*	10.63*
Syringic acid (µg/g)	17.94*	74.14*	7.87*
Coumaric acid (µg/g)	12.59*	79.20*	7.95*
Ferulic acid (µg/g)	31.66*	56.99*	11.09*

Adapted from Moore et al. (15).* Highly significant ($p < 0.0001$).

TABLE 3.5   Variance Components (% of Total Mean Squares) for G, E, and G×E Interaction Effects for DPPH SC, PSC, SSC and TPC of Bran Samples of Three Wheat Genotypes Grown at Five Locations

Variance component	G	E	G×E
Degrees of freedom	2	4	8
Parameter			
DPPH SC (% DPPH remaining)	88.58**	7.84**	3.57**
ORAC (μmol of TE/g)	25.63**	51.75**	22.23**
SSC (% remaining)	2.82	27.83	54.22*
TPC (mg of GAE/g)	5.35**	79.54**	15.00**

Adapted from Moore et al. (15).
* Significant ($p < 0.05$).
** Significant ($p < 0.01$).

Colorado. G and E, and G×E interaction were significant for all antioxidant properties except PSC (ORAC) which was significantly influenced by G and G×E interaction but not E (Table 3.4). E contributed the highest proportion of total variance (56–79%) for SSC, TPC and concentrations of PHBA, VA, SA, coumaric acid, and FA. For DPPH SC and ORAC, G contributed the highest proportion of total variance (56–86%) (Table 3.4) on the contrary to the observations by Mpofu (13) and Mpofu et al. (14) (Table 3.3). Similar to the observation by Mpofu (13) and Mpofu et al. (14) (Table 3.3), G×E interaction was not dominant in influencing any of the parameters presented in Table 3.4, it contributed between 7% and 29% to total variation of these parameters.

The relative contribution of G, E, and G×E interaction to total variation in phenolic composition and antioxidant properties of three Maryland hard wheat varieties grown in two eastern Colorado locations is presented in Table 3.5. In agreement with the observations by Mpofu (13) and Mpofu et al. (14) (Table 3.3) as well as Moore et al. (15) (Table 3.4), TPC was influenced most by E (Table 3.5). Similar to the observations of Moore et al. (15) for bran samples of 20 wheat genotypes grown at two locations (Table 3.4), analysis of pooled data for the bran samples of three Maryland wheat genotypes grown in five eastern Colorado locations revealed domination of G in influencing DPPH SC (Table 3.5). G×E interaction had the greatest influence on the SSCs of the three Maryland wheat genotypes grown in five eastern Colorado locations (Table 3.5) on the contrary to variation in any of the parameters measured by Mpofu (13), Mpofu et al. (14), and Moore et al. (15).

3.6   CONCLUDING REMARKS

Conclusions that can be reached from G×E results on a given antioxidant parameter are likely limited to the sample, solvents, and experimental methods and conditions used. Results from the G×E study by Moore et al. (15) and analysis of pooled results (15) from the studies by Yu et al. (20), Zhou and Yu (27), and Yu and Zhou (28) contradict each other and/or the reports by Mpofu (13) and Mpofu et al. (14) on significance and relative influence of G, E and/or G × E interaction to several phenolic and antioxidant properties. For example Moore et al. (15) reported significant G, lack of significant E, and significant E effects on the PSCs of the 20 wheat genotypes grown in two eastern Colorado locations and significant G, E, and G × E interaction on PSCs of the bran of three wheat genotypes grown in five eastern Colorado locations. Mpofu (13) on the contrary reported lack of significant E and G × E interaction and significant G effects on PSCs of six wheat genotypes grown in four western Canada locations. Differences in extraction solvents and protocols likely played a major role in these differences. Standardization of the methods for antioxidant determination as proposed by Prior et al. (29) will, if accomplished, knock off the influence of solvents and extraction conditions thereby improving validity of interpretations of G × E results from different studies.

Apart from the G × E reports by Mpofu (13), Mpofu et al. (14), and Moore (15), other reports on significant genotype and/or growing environment effects on wheat antioxidant properties have included those by Adom et al. (17,18), Yu et al. (19,20,22), Beta et al. (23), Li et al. (24), Moore et al. (25), and Gélinas et al. (26). Some of the major stumbling blocks in comparing results from different antioxidant studies as identified throughout this chapter include use of different extraction methods, solvents or standards. Different wheat components (whole wheat, bran, wheat pearling fractions, etc.) have also been used further complicating some of the possible comparisons. It is however important to note that the majority of the reports reviewed indicated significant genotype and/or growing environment effects on antioxidant properties.

Lack of significant genotype effects has been reported for some antioxidant parameters. Mpofu (13) detected lack of significant genotype effects on the following PSC measures: H-ORAC, L-ORAC, and T-ORAC in six wheat genotypes. Genotype effects were also not significant on the PSCs of the bran of Alliance and Wichita varieties (32) and PSCs of the bran of three hard winter wheat varieties (15). The lack of G effects on the PSCs reported by Mpofu (13), Zhou et al. (32), and from analysis of pooled results from the studies by Yu et al. (20), Zhou and Yu (27), and Yu and Zhou (28,15) indicates that this antioxidant parameter is likely not significantly influenced by G. Moore et al. (15), however, detected significant G effects on the PSCs of the bran of 20 wheat varieties indicating the need for further, more extensive studies to elucidate the role of G in influencing this parameter. Yu et al. (22) reported lack of significant differences between the SSCs of Akron and Trego bran from wheat harvested at each of three Colorado locations: Julesburg, Walsh and Fort Collins. Similarly, Moore et al. (15) based on analysis of pooled results from the studies by Yu et al. (20), Zhou an Yu (27), and Yu and Zhou (28) reported lack of significant genotype effects on the PSCs of the wheat bran of three Maryland wheat genotypes. Contrary to these results, Mpofu (13) and Moore et al. (15) reported significant G effects on PSCs of six wheat genotypes and 20 wheat genotypes, respectively.

Significant environment effects on phenolic and antioxidant properties likely reflects the role of environmental factors in synthesis of phenolic and other antioxidant compounds. Phenolic compounds are synthesized via the phenylpropanoid pathway, which readily responds to various plant stressors including pathogens, UV-B radiation, and heavy metals (35). The environmental factors influencing antioxidant properties were not identified in the reports by Mpofu (13) and Mpofu et al. (14), save for the possible relationship between mean OCA concentration in wheat from each location and average temperature during grain filling. Yu et al. (20) detected significant negative correlations between the chelating activities of Akron bran samples and total solar or daily average solar radiation, while Zhou and Yu (28) detected positive correlations between the DPPH˙ scavenging activity of Trego wheat and either total solar radiation or average daily solar radiation. Analysis of pooled results from the studies by Yu et al. (20), Zhou and Yu (27), and Yu and Zhou (28) revealed the existence of significant ($p < 0.05$) negative correlations between DPPH scavenging capacity and TSR ($r = -0.443$), ABTS SC and TSR ($r = -0.392$), and between TPC and hours exceeding $32\,°C$ ($r = -0.417$) (15) indicating possible relations between these parameters and environmental factors. In order to conclude existence of such relationships, repeated testing over years will be required.

Significant genotype effects for TPC, DPPH SC, and concentrations of FA, SA, and VA (14,15) indicate that it would likely be possible to select for these quantitative traits in a genotype development program. Successful breeding of wheat with high phenolic antioxidant properties and associated agronomical and human health benefits will, however, most likely depend on identification of some genotype(s) stable to influences of the environment and interactions between genotype and environmental factors. No such genotype(s) has yet been identified.

Apart from breeding, G × E effects also have a bearing on sampling, particularly, sample identification. Significant genotype and/or growing environment effects identified in studies on wheat antioxidants as well as studies on other biochemical properties of wheat highlight the need for proper sample identification when reporting these properties. Proper identification would entail specification of the name of the genotype, where and probably as well in which year it was grown.

Future studies on genotype × environment effects on antioxidant properties should cover a more diverse range of genotypes and growing environments and be conducted over several years to determine the stability, adaptability, and heritability of the antioxidant properties. Apart from phenolic acids, antioxidant compounds present in wheat include alkylresorcinols, flavanoids, ferulates, tannins, suberin, and lignin (30). Future research should additionally investigate G × E variation in concentrations of these as well as tocopherols and carotenoids. Including *in vivo* antioxidant tests in future G × E studies on wheat antioxidant properties would also be desirable as these would elucidate the actual physiological effects of wheat antioxidants and G, E as well as G × E effects on these.

REFERENCES

1. Worland, A. J.; Snape, J. W. Genetic basis of worldwide wheat varietal improvement. In:Bonjean, A. P.; Angus, W. J. (Eds.).The World Wheat Book: A History of Wheat Breeding. Lavoisier Publishing, Paris, **2001**. pp. 59–100.
2. Bushuk, W. Wheat breeding for end-product use. *Euphytica* **1998**, *100*, 137–145.
3. Basford, K. E.; Cooper, M. Genotype × environment interactions and some considerations of their implications for wheat breeding in Australia. *Aust. J. Agric. Res.* **1998**, *49*, 153–174.
4. O' Brien, L. Genotype and environment effects on feed grain quality. *Aust. J. Agric. Res.* **1999**, *50*, 703–720.
5. Crossa, J. Statistical analyses of multilocation trials. *Adv. Agron.* **1990**, *44*, 55–85.

6. Matus-Cadiz, M. A.; Hucl, P.; Perron, C. E.; Tyler, R. T. Genotype × environment interaction for grain color in hard white spring wheat. *Crop Sci.* **2003**, *43*, 219–226.
7. Baenziger, P. S.; Clements, R. L.; McIntosh, M. S.; Yamazaki, W. T.; Starling, T. M.; Sammons, D. J.; Johnson, J. W.;Effects of cultivar, environment and their interaction and stability analyses on milling and baking quality of soft red winter wheat. *Crop Sci.* **1985**, *2*, 5–8.
8. Graybosch, R. A.; Peterson, C. J.; Shelton, D. R.; Baenziger, P. S.;Genotypic and environmental modification of wheat flour protein composition in relation to end-use quality. *Crop Sci.* **1996**, *36*, 296–300.
9. Mikhaylenko, G. G.; Czuchajowska, Z.; Baik, B. K.; Kidwell, K. K. Environmental influences on flour composition, dough rheology, and baking quality of spring wheat. *Cereal Chem.* **2000**, *77*, 507–511.
10. Peterson, C. J.; Graybosch, R. A.; Baenziger, P. S.; Grombacher, A. W. Genotype and environment effects on quality characteristics of hard red winter wheat. *Crop Sci.* **1992**, *32*, 98–103.
11. Peterson, C. J.; Graybosch, R. A.; Shelton, D. R.; Baenziger, P. S. Baking quality of hard winter wheat: responses of cultivars to environment in the great plains. *Euphytica* **1998**, *100*, 157–162.
12. Collaku, A.; Harrison, S. A.; Finney, P. L.; Van Sanford, D. A. Clustering of environments of southern soft red winter wheat regions for milling and baking quality attributes. *Crop Sci.* **2002**, *42*, 58–63.
13. Mpofu, A. Genotype and environment variation in the phenolic antioxidant properties of hard spring wheat, [MSc thesis] University of Manitoba, Winnipeg (MB), Canada, **2006**. pp.56–92.
14. Mpofu, A.; Sapirstein, H. D.; Beta, T. Genotype and environmental variation in phenolic content, phenolic acid composition, and antioxidant activity of hard spring wheat. *J. Agric. Food Chem.* **2006**, *54*, 1265–1270
15. Moore, J.; Liu, J. G.; Zhou, K.; Yu, L. Effects of genotype and environment on the antioxidant properties of hard winter wheat bran. *J. Agric. Food Chem.* **2006**, *54*, 5313–5322.
16. Halliwell, B. Oxidative stress, nutrition and health. Experimental strategies for optimization of nutritional antioxidant intake in humans. *Free Radic. Res.* **1996**, *25*, 57–74.
17. Adom, K. K.; Sorrells, M. E.; Liu, R. H. Phytochemical profiles and antioxidant activity of wheat varieties. *J. Agric. Food Chem.* **2003**, *51*, 7825–7834.
18. Adom, K. K.; Sorrells, M. E.; Liu, R. H. Phytochemicals and antioxidant activity of milled fractions of different wheat varieties. *J. Agric. Food Chem.* **2005**, *53*, 2297–2306.
19. Yu, L.; Haley, S.; Perret, J.; Harris, M. Antioxidant properties of hard winter wheat extracts. *Food Chem.* **2002**, *78*, 457–461.
20. Yu, L.; Perret, J.; Harris, M.; Wilson, J.; Haley, S. Antioxidant properties of bran extracts from "Akron" wheat grown at different locations. *J. Agric. Food Chem.* **2003**, *51*, 1566–1570.
21. Yu, L.; Haley, S.; Perret, J.; Harris, M. Comparison of wheat flours grown at different locations for their antioxidant properties. *Food Chem.* **2004**, *86*, 11–16.
22. Yu, L.; Zhou, K.; Parry, J. W. Inhibitory effects of wheat bran extracts on human LDL oxidation and free radicals. *Lebensm.-Wiss.-Technol.* **2005**, *38*, 463–470.
23. Beta, T.; Nam, S.; Dexter, J. E.; Sapirstein, H. D. Phenolic content and antioxidant activity of pearled wheat and roller-milled fractions. *Cereal Chem.* **2005**, *82*, 390–393.
24. Li, W.; Shan, F.; Sun, S.; Corke, H.; Beta, T. Free radical scavenging properties and phenolic content of Chinese black-grained wheat. *J. Agric. Food Chem.* **2005**, *53*, 8533–8536.
25. Moore, J.; Hao, Z.; Zhou, K.; Luther, M.; Costa, J.; Yu, L. Carotenoid, tocopherol, phenolic acid, and antioxidant properties of Maryland-grown soft wheat. *J. Agric. Food Chem.* **2005**, *53*, 6649–6657.
26. Gélinas, P.; McKinnon, C. M. Effect of wheat variety, farming site, and bread-baking on total phenolics. *Int. J. Food Sci. Technol.* **2006**, *41*, 329–332.
27. Zhou, K.; Yu, L. Antioxidant properties of bran extracts from Trego wheat grown at different locations. *J. Agric. Food Chem.* **2004**, *52*, 1112–1117.
28. Yu, L.; Zhou, K. Antioxidant properties of bran extracts from 'Platte' wheat grown at different locations. *Food Chem.* **2004**, *90*, 311–316
29. Prior, R. L.; Wu, X.; Schaich, K. Standardized methods for the determination of antioxidant capacity and phenolics in foods and dietary supplements. *J. Agric. Food Chem.* **2005**, *53*, 4290–4302.
30. Shahidi, F.; Naczk, M. Phenolics in Food and Nutraceuticals. CRC Press, Boca Raton, FL, **2004**. pp.50–53.
31. Decker, E. A. Antioxidant mechanisms. In:Akoh, C. C.;Min, D. B. (Eds.).Food Lipids: Chemistry, Nutrition and Biotechnology. Marcel Dekker, New York, **1988**. pp.397–421.

32. Zhou, K.; Yin, J. J.; Yu, L. Phenolic acid, tocopherol and carotenoid compositions, and antioxidant functions of hard red winter wheat bran. *J. Agric. Food Chem.* **2005**, *53*, 3916–3922.

33. McKeehen, J. D.; Busch, R. H.; Fulcher, R. G. Evaluation of wheat (*Triticum aestivum* L.) phenolic acids during grain development and their contribution to *Fusarium* resistance. *J. Agric. Food Chem.* **1999**, *47*, 1476–1482.

34. Abdel-Aal, E.-S. M.; Hucl, P.; Sosulski, F. W.; Graf, R.; Gillot, C.; Pietrzak, L. Screening spring wheat for midge resistance in relation to ferulic acid content. *J. Agric. Food Chem.* **2001**, *49*, 3559–3566.

35. Dixon, R. A.; Paiva, N. L. Stress induced phenylpropanoid metabolism. *Plant Cell* **1995**, *7*, 1085–1097.

CHAPTER 4

CAROTENOID, TOCOPHEROL, LIGNAN, FLAVONOID, AND PHYTOSTEROL COMPOSITIONS OF WHEAT GRAIN AND ITS FRACTIONS

Rong Tsao

4.1 INTRODUCTION

Consumption of whole grain and whole-grain products has been associated with reduced incidence of chronic diseases such as cardiovascular disease (1–3), diabetes (3,4), and cancer (5). Whole-grain diet has also been found to inhibit inflammation (6). These health benefits have been often attributed to dietary fibers, however, as the evidence of the health benefits of phytochemicals and antioxidants in fruits and vegetables accumulates, roles of the phytochemicals in whole grains including whole wheat have become another focus of many studies. Phytochemicals together with many other micronutrients are often found in the germ and the bran of wheat. Foodstuffs made of processed wheat flour, therefore, lack these natural bioactive components, consequently may have lowered health benefits compared to those contain whole wheat. Whole wheat is not only a good source of B vitamins, vitamin E, magnesium, iron and fiber, it also contains other valuable phytochemicals not found in some fruits and vegetables. Recent research has shown that these microcomponents at least in part contribute to the health benefits of whole wheat. These findings have generated great consumer interest in whole-grain foods or food supplements.

The finding of the bioactivities of phytochemicals in fruits and vegetables also instigated researchers to examine the full phytochemical spectrum of wheat. Systematic approaches to the phytochemical profile of wheat in recent years have shown that wheat contains many health beneficial compounds such as phenolic acids, flavonoids, lignans, carotenoids, tocopherols/tocotrienols, and phytosterols/phytostanols. This chapter is therefore intended to summarize the most recent findings in the phytochemical studies of wheat and the health protective roles of the major components. Detailed discussion on phenolic acids will not be part of this chapter and readers are referred to Chapter 5 of this book for in-depth discussions on the composition and their role in maintaining human health.

4.2   PHYTOCHEMICAL COMPOSITION OF WHEAT

As in all plant foods, the phytochemical composition of wheat is greatly affected by genetics, environmental factors, and processing conditions. The total phenolic content of the bran that contains mainly the phenolic acids and polyphenols is 3362–3967 µg/g gallic acid equivalent, in which only 15% was from the methanol/water extractable fraction, i.e., 85% of the total phenolic content in wheat bran was in bound form (7). Bound phenolic acids and other polyphenols may be more important in terms of the health benefits (7–9).

4.2.1   Polyphenols (Lignans, Flavonoids)

Despite the large volume of studies on the phenolic acids in wheat, information on other polyphenol content of wheat has been lacking. However, available literature indicates lignans and flavonoids are found in wheat and they may have contributed to the many health benefits of wheat.

4.2.1.1   Lignans   Lignans are a group of polyphenolic compounds in plants formed primarily from oxidative coupling of two *p*-propylphenol moieties (Fig. 4.1). The best food source of plant lignans is perhaps flax (10), however, wheat has also been reported to contain the same. In wheat, lignans were found in the bran layer, and the major lignan in wheat bran is secoisolariciresinol diglycoside (SDG) (11). Wheat bran lignans, like other dietary lignans such as secoisolariciresinol and matairesinol, are precursors of the mammalian lignans enterodiol and enterolactone, produced by the intestinal microflora from dietary precursors, and commonly found in animal and human plasma and urine (12–15) (Fig. 4.1). Most of the lignans in wheat are found in the bran. Six lignan aglycones have been identified in wheat bran with a total concentration of 2774 µg/100 g (16). Syringaresinol was the predominant lignan aglycone (1953 µg/100 g) in wheat bran, followed by isolariciresinol, lariciresinol, secoisolariciresinol, pinoresinol, and matairesinol at 297, 257, 142, 106, and 9.4 µg/100 g, respectively (16). A much lower set of data was reported by Penalvo et al. with a total of only 507 µg/100 g of all the above lignans (17). Lignans such as secoisolariciresinol and matairesinol of wheat bran are converted by the colonic microflora in mammals to enterolactone and enterodiol. Enterolactone urinary excretion in mice fed with wheat bran (15%) for 5 d was found to be significantly reduced from 30.5 to 6.2 nmol/d ($p < 0.001$) when they were removed from wheat bran. These results suggest that lignins, embedded in the cell wall and retained in the bran during solvent extraction, account for 26–32% of the enterolactone formed from cereal bran (16). In terms of the kinetics in human, more research to determine the bioavailability has been suggested (16). Increasing *in vitro* and animal

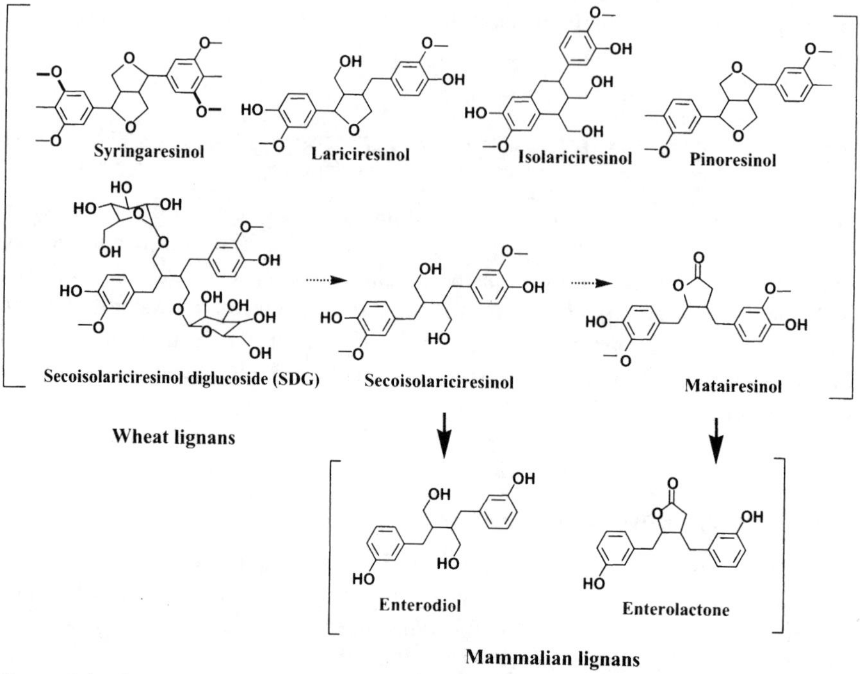

Figure 4.1 Structures of wheat lignans and their mammalian metabolites.

studies have shown that plant lignans are phytoestrogens, and mammalian lignans may provide a protective role against several chronic diseases such as breast, prostate, and colon cancers (18–20). Mammalian lignans may also reduce risk of cardiovascular and coronary heart disease (CHD) and diabetes (21–25). Vanharanta et al. investigated the association of serum enterolactone concentration with the risk of acute coronary events in a prospective nested case control study in middle-aged men from eastern Finland. They found that men in the highest serum enterolactone quarter had a 65.3% lower risk of acute coronary events than men in the lowest quarter (21).

Dietary lignans have been implicated to play an important role in reduced risk for CHD and some cancers (26). Serum enterolactone concentrations increased rapidly in the first 2 weeks and were higher when eating the whole grain than refined-grain diet by 6.2 nmol/L (26). An animal study by Drankhan et al. (27) demonstrated that even when the wheat fiber contents were equal in the diets, the antitumor activities of wheat bran from various wheat cultivars were significantly different. Qu et al. hypothesized that phytochemical lignans in wheat bran may account for the differences among wheat cultivars in cancer prevention (28). Results from their study indeed suggested that lignans may contribute, at least in part, to the cancer prevention by wheat bran observed in APC-Min mice. They further discovered that inhibition of cancer cell growth by lignan metabolites seemed to be mediated by cytostatic and apoptotic mechanisms. Lignans are involved in the antitumor activity of wheat bran in colon cancer SW480 cells. Lignans such as SDG in flax seed were found to withstand normal baking temperatures in all bakery products (29). SDG was also a relatively stable compound during storage (1 week at room temperature, 1 and 2 months at $-25°C$) (29).

It needs to be pointed out that, although ample studies have shown the health benefit of lignans, epidemiological evidence to date is conflicting. Prospective large scale studies have therefore been recommended to assess the consumption of lignans, antibiotics, and dietary habits during adolescence in order to obtain definitive conclusions (30,31).

4.2.1.2 Flavonoids

Although grains in general contain higher concentrations of phenolic acids than fruits and vegetables, they contain lower concentrations of flavonoids. On the contrary, grains often have unique phytochemical contents and profiles that complement those found in fruits and vegetables (32). In wheat, the phytochemical profiles, particularly the flavonoid profiles, have only been thoroughly studied in recent years. Several different flavones including apigenin, luteolin, chrysoeriol, and tricin were identified in the leaf tissue of wheat in their monoglycosides, diglycosides, triglycosides, and tetraglycoside forms (33), however, the flavonoid profile in the seed is not as diverse. In wheat grain, flavonoids are mainly in glycosidic and acylglycosidic forms; two apigenin-C-diglycosides and their Wessely–Moser isomers, together with their sinapic acid esters, were isolated and identified (34) (Fig. 4.2). Colorometric method has also been used for rapid identification and estimation of total flavonoid content of wheat (8).

In some special varieties of wheat such as blue and purple wheat grains, anthocyanins can be the major flavonoid component. Studies have shown that these colored wheat grains have a distinct anthocyanin profile (35). Like other polyphenols, the majority of the anthocyanin pigments have been found in the bran of red, blue, and purple wheat with total concentrations at 10, 235, and 452 $\mu g/g$, respectively (35). Cyanidin and peonidin glycosides (Fig. 4.2) were the major anthocyanins found in the blue and purple wheat, however, other unidentified anthocyanins were also important, particularly in blue wheat (35). The predominant anthocyanin in blue wheat, making up approximately 41% of the total anthocyanin content, remains to be structurally unidentified (35). Blue wheat anthocyanins are thermally stable at pH 1, but slow degradation occurs at higher pH 3–5. Higher temperature (65–95 °C) also increased degradation of blue wheat anthocyanins. SO_2 was found to stabilize the anthocyanin pigments during heating of blue wheat and its extract (35).

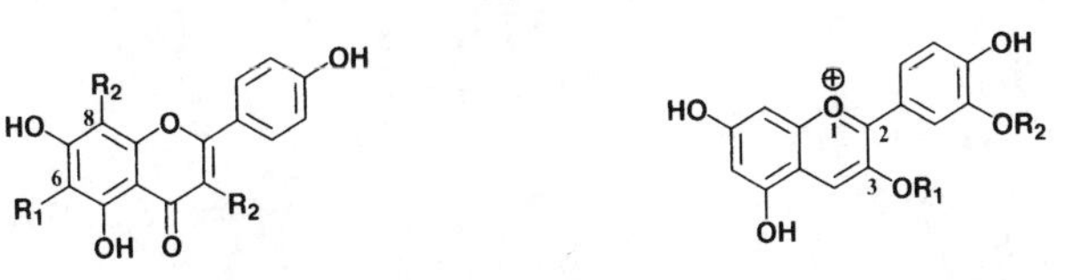

Apigenin-6C-arabinoside-8C-glucoside: R_1=arabinose, R_2=glucose
Apigenin-6C-glucoside-8C-arabinoside: R_1=glucose, R_2=arabinose
Apigenin-6C-arabinoside-8C-galactoside: R_1=arabinose, R_2=galactose
Apigenin-6C-galactoside-8C-arabinoside: R_1=galactose, R_2=arabinose

Cyanidin-3-glucoside: R_1=glucose, R_2=H
Cyanidin-3-galactoside: R_1=galactose, R_2=H
Peonidin-3-glucoside: R_1=glucose, R_2=Me

Figure 4.2   Major flavonoids found in wheat are flavonols (apigenin-C-diglycosides) and anthocyanins (cyanidin-3-glycoside and peonidin-3-glycoside).

The potential contributions of flavonoids to human health have been well studied in recent years (32,34). The antioxidant activities and health benefits of these and other phytochemicals will be discussed later in this chapter.

4.2.2   Carotenoids

Carotenoids, another group of phytochemicals contributing to the pigments, have been known to exist in different wheat, however, recent findings in the roles of carotenoids in human health have made this group of compounds the focus of many studies. The carotenoid or xanthophyll composition of wheat has been mainly analyzed using high performance liquid chromatography (HPLC), however, when specific carotenoid compounds are not of primary focus, spectrophotometric determination of carotenoid concentration following extraction of flour or meal with water-saturated butanol has been used (36). Such spectrophotometric measurement of carotenoids in wheat has been found to correlate well with the HPLC analysis, providing a fast alternative measurement to the slow and costly HPLC method (37,38). However, HPLC (both normal and reversed-phase) not only gives quantitative data but also provides identification of the carotenoids separated, particularly when coupled with photodiode array detector and mass spectrometer. The total carotenoid concentration in wheat is typically at 2–5 µg/g DM (dry matter), with lutein as the single most predominant carotenoid, followed by zeaxanthin and β-carotene (38) (Fig. 4.3).

Lutein was the primary carotenoid present in the soft red winter wheat genotypes at 0.82–1.14 µg/g along with significant amounts of zeaxanthin and β-carotene (39). Similar results were found in hard red winter wheat bran (40). Lutein was the leading carotenoid in the bran of seven wheat from four different countries, followed by significant amount of zeaxanthin and cryptoxanthin (41) (Fig. 4.3). Canadian durum and Australia general purpose wheat bran had the same highest total carotenoid content at 0.68 µmol/100 g bran, but the US Akron wheat had the highest concentration of lutein at 1.8 µg/g bran. Lutein concentration varied significantly among the 11 wheat varieties and experimental lines, ranging from 26 to 143 µg/100 g of grain, a 5.4-fold difference ($p < 0.05$). Zeaxanthin concentration was lower than lutein in all varieties examined (0.54–27 µg/100 g of grain). β-Cryptoxanthin was at even lower concentration, however, the lutein content varied the most with a 12-fold difference between the highest and lowest concentrations (8). A recent study on the content and composition of free lipids and carotenoids in spring and winter classes of wheat flour showed that, on average, spring wheat flour had higher carotenoid content (352 µg/100 g) than winter wheat (242 µg/100 g), and in both wheat, lutein was the predominant carotenoid (42). However, the total lutein content was higher in winter wheat (81 µg/100 g) than in spring wheat (76 µg/100 g) (42). Hidalgo et al. recently surveyed the carotenoid content of 54 accessions of einkorn wheat (*Triticum monococcum* ssp. *monococcum*) originating from different ecogeographical areas in Europe, and compare the carotenoid content with common durum (*T. turgidum* ssp. *durum*) and bread (*T. aestivum* ssp. *aestivum*) wheats, and found that carotenoids, mostly lutein, averaged 8.41 µg/g DM in einkorn wheat, with a maximum of 13.4 µg/g DM, 2–4 times more

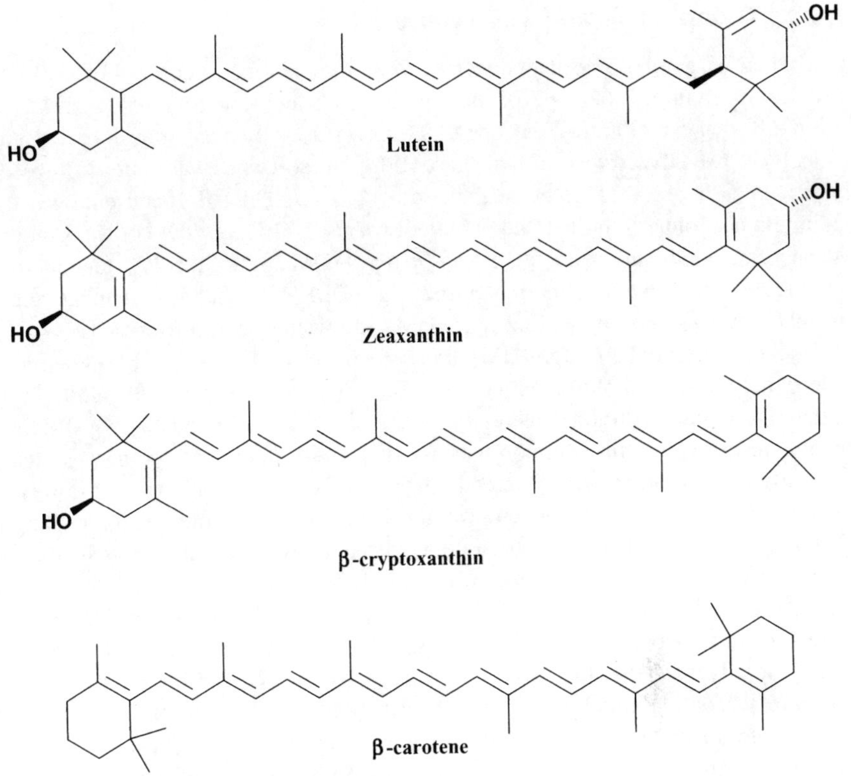

Figure 4.3   Carotenoids commonly found in wheat.

than common wheats. The carotenoid was also affected by geographic location (43). Genetic variation of lutein and zeaxanthin concentrations were also found to be significantly different among cultivated diploid (einkorn, *T. monococcum L.*), tetraploid (durum wheat, *T. turgidum L.* subsp. *durum* (Desf.) Husn.), and hexaploid (bread wheat, *Triticum aestivum L.*) wheat species and bread wheat varieties. The highest concentrations, both lutein and total carotenoids, were found in einkorn (44). Storage conditions also affect the stability of lutein in wheat seeds. Pinzino et al. studied wheat seeds stored over 36 years and found that lutein content decreases significantly as the seed ages during storage (45).

Lutein and zeaxanthin are found to exist in different isomeric forms. In a study on the levels of lutein and zeaxanthin the (*E/Z*)-geometrical isomers in wheat and pasta products, lutein was found to have higher concentration than zeaxanthin (46). The ratio of lutein to zeaxanthin in these foods ranged from 2.5 to 12. The *all-E*-isomers of lutein and zeaxanthin in wheat and pasta products were also higher than their corresponding Z-isomers, except for lasagne in which *all-E*- and 9Z-zeaxanthin were present at the same level (46). They also found that concentrations of lutein and zeaxanthin in Freekeh, a green-harvested wheat imported from Australia, were much higher than the two North American bread wheat, Catoctin and Pioneer (46).

4.2.3 Tocopherols and Tocotrienols

Wheat is also a good source of tocopherols (vitamin E) and tocotrienols (43). The eight vitamers of vitamin E (α-, β-, δ-, and γ-tocopherols and -tocotrienols) (Fig. 4.4) in seven different cereal grains (oat, spelt, durum wheat, soft wheat, maize, barley, and triticale) were analyzed with normal phase HPLC and soft wheat was found to have the highest total tocol level together with barley (~75 μg/g DM). β-Tocotrienol was the main vitamer found in hulled and dehulled wheats (33–43 μg/g DM) (47). Einkorn wheat contained an average of 77.96 μg/g DM total tocols (tocopherols + tocotrienols), with a maximum of 115.85 μg/g DM (43). The most abundant tocol in einkorn was β-tocotrienol (48.22 μg/g DM), followed by α-tocotrienol (12.77 μg/g DM), α-tocopherol (12.18 μg/g DM), and β-tocopherol (4.79 μg/g DM). The mean tocotrienol/tocopherol ratio in einkorn was 3.68. These values were all significantly higher than common durum and bread wheats that contained 53 and 63 μg/g DM total tocols, respectively. In common wheats, the most dominant tocol was again β-tocotrienol (34 and 35 μg/g DM, for durum and bread wheat, respectively); however, in both of these wheats, α-tocopherol was the distant second averaging 10 and 14 μg/g DM, respectively (43). The tocotrienol/tocopherol ratio was significantly smaller as well, averaging at 2.97 for the durum wheat and 1.79 for the bread wheat (43). A study on eight soft red winter wheat genotypes grown in Maryland, USA, showed significant difference in α-tocopherol content, ranging from 3.4 to 10.1 μg/g (39). These values were generally lower than that reported in European grown *T. aestivum* ssp. wheat (43). Geographic location, genotype, and possibly fractionation of the wheat seemed to have significant impact on the total and individual tocopherol content, as in another study bran of the hard red winter wheat (*T. aestivum* ssp.) only contained 4.1–6.5 and 3.7–5.8 μg/g, respectively, for the two predominant tocopherols, α- and γ-tocopherols (40). From the above studies, we can also conclude that not only the concentration of tocopherols may differ among different genotypes and growing environment, the tocol profile (including tocotrienols) may be affected, too. Wheat bran of seven varieties from four different countries were examined and compared for their tocopherol composition, and was found to differ significantly in the concentrations of α-, δ- and

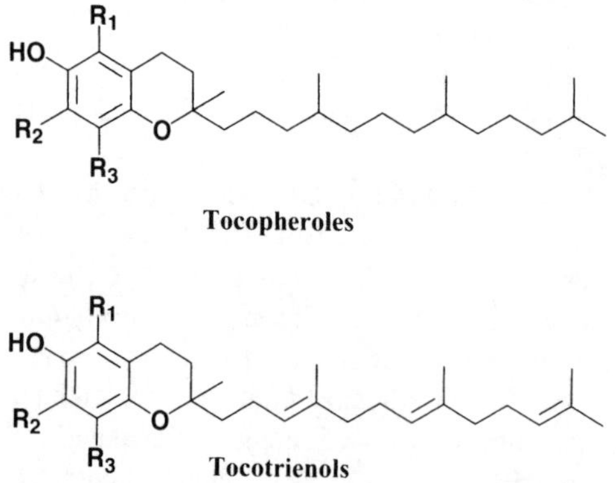

α-Tocopherol: R_1=R_2=R_3=CH_3
β-Tocopherol: R_1=R_3=CH_3; R_2=H
γ-Tocopherol: R_1=H; R_2=R_3=CH_3
δ-Tocopherol: R_1=R_2=H; R_3=CH_3

α-Tocotrienol: R_1=R_2=R_3=CH_3
β-Tocotrienol: R_1=R_3=CH_3; R_2=H
γ-Tocotrienol: R_1=H; R_2=R_3=CH_3
δ-Tocotrienol: R_1=R_2=H; R_3=CH_3

Figure 4.4 Tocopherols and tocotrienols found in wheat and wheat-based food products.

γ-tocopherols, all with a wide range 1.28–21.29, 0.23–7.0, and 0.92–6.90 µg/g, respectively (41). Canadian durum wheat bran topped the list in α-tocopherol and total tocopherols in the same study (41).

Tocopherols were found in higher concentrations in wheat germ. The concentration and the composition of different tocopherol analog varied significantly among different extraction methods (48). All α-, β-, δ- and γ-tocopherols were extracted by supercritical fluid extraction (SFE), whereas only α-, and β-tocopherols were found in hexane and chloroform/methanol extracts. However, β-tocopherol seemed to be better extracted by the latter two solvent extraction systems. The total tocopherols were 2179, 2154, 1874 mg/100 g wheat germ in SFE, hexane, chloroform/methanol extracts, respectively (48).

4.2.4 Phytosterols and Phytostanols

In addition to the above discussed phytochemicals, wheat also contains certain amount of other bioactive compounds. Phytosterols and Phytostanols, for example, a group of phytochemicals known for cholesterol reduction in human, were also found in wheat, although at very low concentrations as compared to other food sources (49,50) (Fig. 4.5). Most of the phytosterols are found in wheat germs at 413 mg/100 g in total; β-sitosterol was the predominant known phytosterol in wheat germ at 228.6 mg/100 g, followed by campesterol, stigmasterol, Δ^5-avenasterol, sitostanol, campestanol at 78.7, 3.9, 16.1, 6.9, and 12.7 mg/100 g (49) (Fig. 4.5). A study focused on phytosterol quantification by LC/APCIMS^2 in saponified wholemeal extracts of 16 dehulled spelt and 5 winter wheat (*Triticum aestivum L.*) varieties grown in Belgium during 2001–2002 at the same location, also showed similar results, but slightly higher concentrations, possibly due to the inclusion of both free and glycosylated phytosterols (50). A recent study indicated that certain amount of phytosterols in wheat may be conjugated with ferulic acid as steryl ferulates (51,52).

4.2.5 Antioxidant Activities and Health Benefits

Increasing evidence points to the health benefits of whole-grain consumption (1–6). In a most recent study, dietary intake of whole grain has been shown to significantly inversely relate to the metabolic syndrome ($p = 0.005$) and mortality from cardiovascular disease ($p = 0.04$), independent of demographic, lifestyle, and dietary factors (3). Whole-grain intake also reduced fasting glucose concentrations and body mass index across increasing quartile categories (p for trend $= 0.01$ and 0.03, respectively), independent of confounders; however, intake of refined grain increased fasting glucose concentrations (p for trend $= 0.04$) and led to a higher prevalence of the metabolic syndrome (p for trend $= 0.01$). The authors of the study concluded that whole-grain intake is a modifiable dietary risk factor, and older and young adults should be encouraged to increase their daily intake to ≥ 3 servings/d (3). Wheat as the most consumed staple food contains various phytochemicals in addition to the basic and essential nutrients such as dietary fibers and vitamins. Wheat bran in particular is a rich source of dietary fibers that have structures and compositions, which indicate that

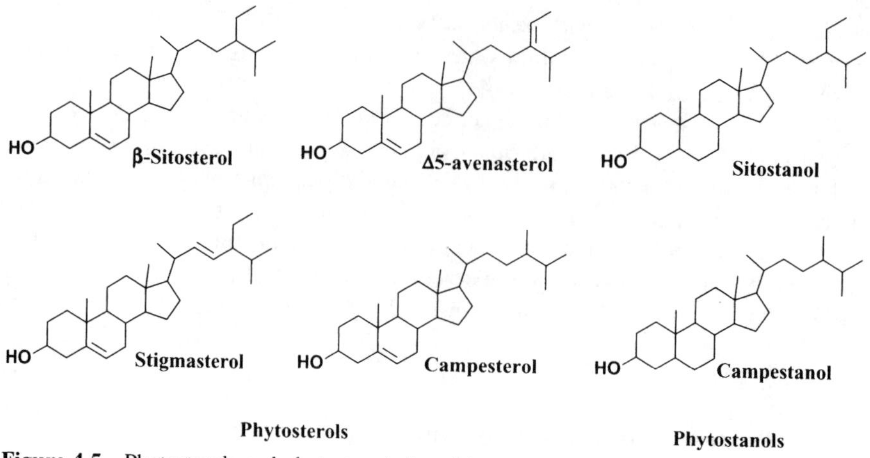

Phytosterols   **Phytostanols**

Figure 4.5 Phytosterols and phytostanols found in wheat.

they may protect against cancer and cardiovascular diseases. Nevertheless, dietary fiber makes up less than half of wheat bran. Other nutrients and phytochemicals present in wheat bran may therefore play equally important roles in the reduction of human chronic diseases, if not more. Phytochemicals such as phenolic acids, flavonoids, tocopherols and tocotrienols, and carotenoids have been attributed to the antioxidant activities of other food crops (7,53,54). These phytochemical antioxidants neutralize excess free radicals such as reactive oxygen species (ROS) that are the cause of many chronic diseases. More evidence has shown that, it is the phytochemicals such as polyphenols and carotenoids, rather than the essential vitamins, that provide the majority of the total antioxidant activities of foods (55). In wheat, most of these antioxidant phytochemicals are found in the bran fraction that has been the target of research in recent years. For this reason, determining the roles of different phyto-chemicals in wheat bran in the prevention of various chronic diseases has been proposed (56). As a result, the phytochemistry of wheat bran has been systemically investigated and the phytochemicals responsible for the strong antioxidant activity of wheat bran have been identified (7,32,41).

Phytochemicals, particularly phenolic acids, lignans, flavonoids (including anthocyanins), carotenoids (mainly xanthophylls such as lutein) and the tocols (tocopherols and tocotrienols) have been thoroughly studied (7,9,32,41). Wheat ranked the second among the four commonly consumed grains (corn, wheat, oat, and rice) for their total phenolic and flavonoid contents, and the majority of these phytochemicals (75% of the total phenolics, 93% of the flavonoids) were in bound forms, contributing 90% to the total antioxidant activity (57).

It needs to be pointed out that although many studies have shown strong antioxidant activities of wheat and wheat bran fractions, most of the antioxidant activities have been measured using *in vitro* models. These model systems are mostly limited to one mechanism, thus do not reflect the actual antioxidant potential in the human body. However, when different methods all point to the strong antioxidant activity of a wheat extract or a phytochemical from wheat, it certainly indicates a great potential. It is therefore highly recommended that at least two different methods be used in any evaluation of the antioxidant capacity (54). Different *in vitro* models have been developed or modified for the estimation of the antioxidant capacity of wheat extracts or the phytochemicals therein. Extracts from wheat or wheat bran have been evaluated for their ability to inhibit lipid peroxidation, to quench or scavenge free radicals, and for their reducing power or potency in chelating with transition metals (7,32,58,59). All of these methods showed strong antioxidant capacity of wheat or wheat bran.

The antioxidant activity of wheat or wheat-based food can be affected by different factors. Postharvest storage conditions, for example, can have major impact on the stability of the antioxidant activity of wheat (45,60). Pinzino et al. studied the changes of lutein content and antioxidant activity of wheat seeds stored over 36 years (45). Although no particular correlation study was attempted, the study did indicate that lutein might be responsible for the depression of free radicals and increased antioxidant activities (45). Postharvest treatment and heat stress were found to have major impacts on the availability of wheat antioxidants. When wheat grain, bran, and 40-mesh bran samples of both Ankor and Trego wheat were kept at 25, 60, and 100°C for 9 d, the antioxidant activities were significantly reduced, and the reduction seemed to be different among different sample types. The ORAC (oxygen radical absorbance capacity) values of Ankor bran and corresponding 40-mesh bran samples kept at 100°C for 9 days were reduced to 61% and 40%, respectively, whereas no significant change of ORAC was observed in grains. The overall loss of radical scavenging capacity was similar. It was suggested that although smaller particle size may help the release of phytochemical antioxidants, whole grain as opposed to its fractions is a preferred form of long-term storage for better preserving natural antioxidants and reduced particle size may accelerate the loss of natural antioxidants in wheat bran during storage and thermal processing (60).

For the phytochemicals of wheat to be effective *in vivo*, it is important that they are bioavailable after being ingested. The antioxidant potential of several wheat samples were evaluated using different methods, and found to be significantly increased following gastrointestinal tract-simulated pH changes (61). The authors attributed the increased antioxidant activity to the enhanced release of phenolic acids, particularly, vanillic, and ferulic acids. However, other polyphenols, such as flavonoids and lignans, may also be released upon hydrolysis (57).

The strong antioxidant activities reported for the various wheat varieties, and their phytochemical compositions only suggest that these individual groups of phytochemicals or individual compounds may contribute to the antioxidant activity. However, in actual wheat and wheat-based food, all phytochemicals are in a mixture. No one single specific group of phytochemicals or a specific individual compound can be accredited for the total antioxidant activity. It is generally considered that all of these compounds synergistically or additively contribute to the antioxidant activities and to the health beneficial effects of wheat found in epidemiological studies.

4.3 CONCLUSION

Wheat grain, particularly, its bran fraction contains several classes of phytochemicals. Among them, phenolic acids, polyphenols (flavonoids and lignans), carotenoids, tocopherols/tocotrienols, and phytosterols/phytosterols have been characterized and linked to many bioactivities related to human health. Most of these phytochemicals have shown strong antioxidant activities in both pure and mixed forms and have been implicated to play a protective role against chronic diseases such as cancer, cardiovascular diseases, and diabetes. There is strong evidence that high intake of whole wheat or wheat bran-based food or food supplement helps alleviate these chronic diseases, and recent studies, with the help of modern analytical techniques, indeed indicate the existence of the diverse phytochemical profile in wheat. However, further studies are needed to understand better how these different classes contribute to the maintenance of human health, particularly, their stability and bioavailability in human, and the synergistic or additive effect among the different classes of phytochemicals.

REFERENCES

1. Jacobs, D. R. Jr.; Meyer, K. A.; Kushi, L. H.; Folsom, A. R. *Am. J. Clin. Nutr.* **1998**, *68*, 248–257.
2. Thompson, L. U. *Crit. Rev. Food Sci. Nutr.* **1994**, *34*, 473–497.
3. Sahyoun, N. R.; Jacques, P. F.; Zhang, X. L.; Juan, W.; McKeown, N. M. *Am. J. Clin. Nutr.* **2006**, *83*, 124–131.
4. Meyer, K. A.; Kushi, L. H.; Jacobs, D. R. Jr.; Slavin, J.; Sellers, T. A.; Folsom, A. R. *Am. J. Clin. Nutr.* **2000**, *71*, 921–930.
5. Jacobs, D. R.; Marquart, L.; Slavin, J.; Kushi, L. H. *Nutr. Cancer* **1998**, *30*, 85–96.
6. Esposito, K.; Giugliano, D. *Am. J. Clin. Nutr.* **2006**, *83*, 1440–1441.
7. Kim, K.-H.; Tsao, R.; Yang, R.; Cui, S. W. *Food Chem* .**2006**, *95*, 466–473.
8. Adom, K. K.; Sorrells, M. E.; Liu, R. H. *J. Agric. Food Chem.* **2003**, *51*, 7825–7834.
9. Liyana-Pathirana, C. M.; Shahidi, F. *J. Agric. Food Chem.* **2006**, *54*, 1256–1264.
10. Thompson, L. U.; Robb, P.; Serraino, M.; Cheung, F. *Nutr. Cancer* **1991**, *16*, 43–52.
11. Mazur, W. M. *Bailliere's Clin. Endocrinol. Metab.* **1998**, *12*, 729–742.
12. Setchell, K. D. R. In:Cunnane, S. C.;Thompson, L. U. (Eds.). *Flaxseed in Human Nutrition*. AOCS Press, Champaign, IL, **1995**. pp.82–98.
13. Heinonen, S.; Nurmi, T.; Liukkonen, K.; Poutanen, K.; Wähälä, K.; Deyama, T.; Nishibe, S.; Adlercreutz, H. *J. Agric. Food Chem.* **2001**, *49*, 3178–3186.
14. Bowey, E.; Adlercreutz, H.; Rowland, I. *Food Chem. Toxicol.* **2003**, *41*, 631–636.
15. Nicolle, C.; Claudine, M.; Christine, M.; Witold, M.; Herman, A.; Christian, R.; Augustin, S. *J. Agric. Food Chem.* **2002**, *50*, 6222–6226.
16. Begum, A. N.; Nicolle, C.; Mila, I.; Lapierre, C.; Nagano, K.; Fukushima, K.; Heinonen, S. M.; Adlercreutz, H.; Remesy, C.; Scalbert, A. *J. Nutr.* **2004**, *134*, 120–127.
17. Penalvo, J. L.; Haajanen, K. M.; Botting, N.; Adlercreutz, H. *J. Agric. Food Chem.* **2005**, *53*, 9342–9347.
18. Bowen, P. E. *Nutr. Today* **2001**, *36*, 144–158.
19. Demark-Wahnefried, W.; Price, D. T.; Polascik, T. J.; Robertson, C. N.; Anderson, E. E.; Paulson, D. F.; Walther, P. J.; Gannon, M.; Vollmer, R. T. *Urology* **2001**, *58*, 47–52.
20. Yang, C. S.; Landau, J. M.; Huang, M.-T.; Newmark, H. L. *Annu. Rev. Nutr.* **2001**, *21*, 381–406.
21. Vanharanta, M.; Voutilainen, S.; Lakka, T. A.; van der Lee, M.; Adlercreutz, H.; Salonen, J. T. *Lancet* **1999**, *354*, 2112–2115.
22. Prasad, K. *Circulation* **1999**, *99*, 1355–1362.

23. Prasad, K. *Mol. Cell. Biochem.* **2000**, *209*, 89–96.

24. Prasad, K. *J. Lab. Clin. Med.* **2001**, *138*, 32–39.

25. Prasad, K.; Mantha, S. V.; Muir, A. D.; Westcott, N. D. *Mol. Cell. Biochem.* **2000**, *206*, 141–149.

26. Jacobs, D. R., Jr.; Pereira, M. A.; Stumpf, K.; Pins, J. J.; Adlercreutz, H. *Br. J. Nutr.* **2002**, *88*, 111–116.

27. Drankhan, K.; Carter, J.; Madl, R.; Klopfenstein, C.; Padula, F.; Lu, Y.; Warren, T.; Schmitz, N.; Takemoto, D. J. *Nutr. Cancer* **2003**, *47*, 188–194.

28. Qu, H.; Madl, R. L.; Takemoto, D. J.; Baybutt, R. C.; Wang, W. *J. Nutr.* **2005**, *135*, 598–602.

29. Hyvarinen, H. K.; Pihlava, J.-M.; Hiidenhovi, J. A.; Hietaniemi, V.; Korhonen, H. J. T.; Ryhanen, E.-L. *J. Agric. Food Chem.* **2006**;*54*, 48–53.

30. Boccardo, F.; Puntoni, M.; Guglielmini P.; Rubagotti, A. *Clin. Chim. Acta* **2006**, *365*, 58–67.

31. Ganry, O. *Preve. Med.* **2005**, *41*, 1–6.

32. Adom, K.; Sorrells, M.; Liu, H. *J. Agric. Food Chem.* **2005**, *53*, 2297–2306.

33. Cavaliere, C.; Foglia, P.; Pastorini, E.; Samperi, R.; Lagana, A. *Rapid. Commun. Mass Spectrom.* **2005**, *19*, 3143–3158.

34. Asenstorfer, R. E.; Wang, Y.; Mares, D. J. *J. Cereal Sci.* **2006**, *43*, 108–119.

35. Abdel-Aal, E.-S. M.; Hucl, P. *J. Agric. Food Chem.* **2003**, *51*, 2174–2180.

36. American Association of Cereal Chemists, 1983. Approved methods of the AACC. Method 8-01, eighth ed., **1983**.

37. Humphries, J. M.; Graham, R. D.; Mares, D. J. *J. Cereal Sci.* **2004**, *40*, 151–159.

38. Fratianni, A.; Mario, I.; Gianfranco, P.; Rita, A. *J. Agric. Food Chem.* **2005**, *53*, 2373–2378.

39. Moore, J.; Hao, Z.; Zhou, K.; Luther, M.; Costa, J.; Yu, L. L. *J. Agric. Food Chem.* **2005**, *53*, 6649–6657.

40. Zhou, K.; Yin, J.-J.; Yu, L. *J. Agric. Food Chem.* **2005**, *53*, 3916–3922.

41. Zhou, K.; Su, L.; Yu, L. *J. Agric. Food Chem.* **2004**, *52*, 6108–6114.

42. Konopka, I.; Czaplicki, S.; Rotkiewicz, D. *Food Chem.* **2006**, *95*, 290–300.

43. Hidalgo, A.; Brandolini, A.; Pompei, C.; Piscozzi, R. J. *Cereal Sci.* **2006**, *44*, 182–193.

44. Leenhardt, F.; Lyan, B.; Rock, E.; Boussard, A.; Potus, J.; Chanliaud, E.; Remesy, C. *Eu. J. Agronomy* **2006**, *25*, 170–176.

45. Pinzino, C.; Capocchi, A.; Galleschi, L.; Saviozzi, F.; Nanni B.; Zandomeneghi, M. *J. Agric. Food Chem.* **1999**, *47*, 1333–1339.

46. Humphries, J. M.; Khachik, F. *J. Agric. Food Chem.* **2003**, *51*, 1322–1327.

47. Panfili, G.; Fratianni, A.; Irano, M. *J. Agric. Food Chem.* **2003**, *51*, 3940–3944.

48. Ge, Y.; Yan, H.; Hui, B.; Ni, Y.; Wang, S.; Cai, T. *J. Agric. Food Chem.* **2002**, *50*, 685–689.

49. Phillips, K. M.; Ruggio, D. M.; Ashraf-Khorassani, M. *J. Agric. Food Chem.* **2005**, *53*, 9436–9445.

50. Ruibal-Mendieta, N. L.; Rozenberg, R.; Delacroix, D. L.; Petitjean, G.; Dekeyser, A.; Baccelli, C.; Marques, C.; Delzenne, N. M.; Meurens, M.; Habib-Jiwan, J.-L.; Quetin-Leclercq, J. *J. Agric. Food Chem.* **2004**, *52*, 4802–4807.

51. Nystrom, L.; Makinen, M.; Lampi, A.-M.; Piironen, V. *J. Agric. Food Chem.* **2005**, *53*, 2503–2510.

52. Iwatsuki, K.; Akihisa, T.; Tokuda, H.; Ukiya, M.; Higashihara, H.; Mukainaka, T.; Iizuka, M.; Hayashi, Y.; Kimura, Y.; Nishino, H. *J. Agric. Food Chem.* **2003**, *51*, 6683–6688.

53. Tsao, R.; Yang, R. *J. Chromatogr. A* **2006**, *1112*, 202–208.

54. Tsao, R.; Yang, R.; Xie, S.; Sockovie, E.; Khanizadeh, S. *J. Agric. Food Chem.* **2005**, *53*, 4989–4995.

55. Boyer, J.; Liu, R. H. *Nutr. J.* **2004**, *3*, 5.

56. Ferguson, L. R.; Harris, P. J. *Eur. J. Cancer Prev.* **1999**, *8*, 17–25.

57. Adom, K. K.; Liu, R. H. *J. Agric. Food Chem.* **2002**, *50*, 6182–6187.

58. Yu, L.; Haley, S ,Perret, J.; Harris, M. *Food Chem.* **2002**, *78*, 457–461.

59. Pérez-Jiménez, J.; Saura-Calixto, F. *J. Agric. Food Chem.* **2005**, *53*, 5036–5040.

60. Cheng, Z.; Su, L.; Moore, J.; Zhou, K.; Luther, M.; Yin, J.-J.; Yu, L. *J. Agric. Food Chem.* **2006**, *54*, 5623–5629.

61. Liyana-Pathirana, C. M.; Shahidi, F. *J. Agric. Food Chem.* **2005**, *53*, 2433–2440.

CHAPTER 5

ANTIOXIDANT PROPERTIES OF WHEAT PHENOLIC ACIDS

Liangping Yu
Zhihong Cheng

5.1 INTRODUCTION

Phenolic acids are widespread in plants. Several phenolic acids have been detected in wheat grain and fractions, including ferulic, vanillic, syringic, coumaric acid, caffeic, protocatechuic, gentistic, chlorogenic, and *p*-hydroxybenzoic acids (1–10). In general, these phenolic acids are either benzoic or cinnamic acid derivatives with different numbers of hydroxyl and methoxyl groups at different positions on the phenyl ring (Fig. 5.1). Wheat and wheat-based ingredients and food products may differ in their phenolic acid composition and availability/releasability. These phenolic acids are believed to contribute to the overall antioxidant properties of wheat and wheat-based food ingredients and food products. Therefore, it is important to advance our understanding of the phenolic acid contents in wheat-based materials, the effects of genotype, growing condition(s) and postharvest treatments on their availability, and the antioxidant properties of these phenolic acids.

5.2 PHENOLIC ACID CONTENTS IN WHEAT GRAIN AND FRACTIONS

A number of experiments have been conducted to determine the phenolic acid profiles in various grain and bran samples of wheat, distinguished by such characters as color and hardness, grown at diverse locations during different seasons. In 1992, bran of durum wheat (*Triticum durum*) was examined for its phenolic acid profile (1). The bran was defatted and extracted with 95% ethanol. After removing ethanol, the aqueous extract was autoclaved followed by freeze-drying, and the residue was redissolved in water and examined for phenolic acid composition using reverse-phase-HPLC. Ferulic acid was the major phenolic acid in the extract of the durum wheat bran with the highest level of 764 mg/100 g of the extract, and followed by vanillic, *p*-coumaric, protocatechuic, syringic, *p*-hydroxybenzoic, caffeic, gentistic, and chlorogenic acids with a concentration range of 637–684 mg/100 g of the extract (1).

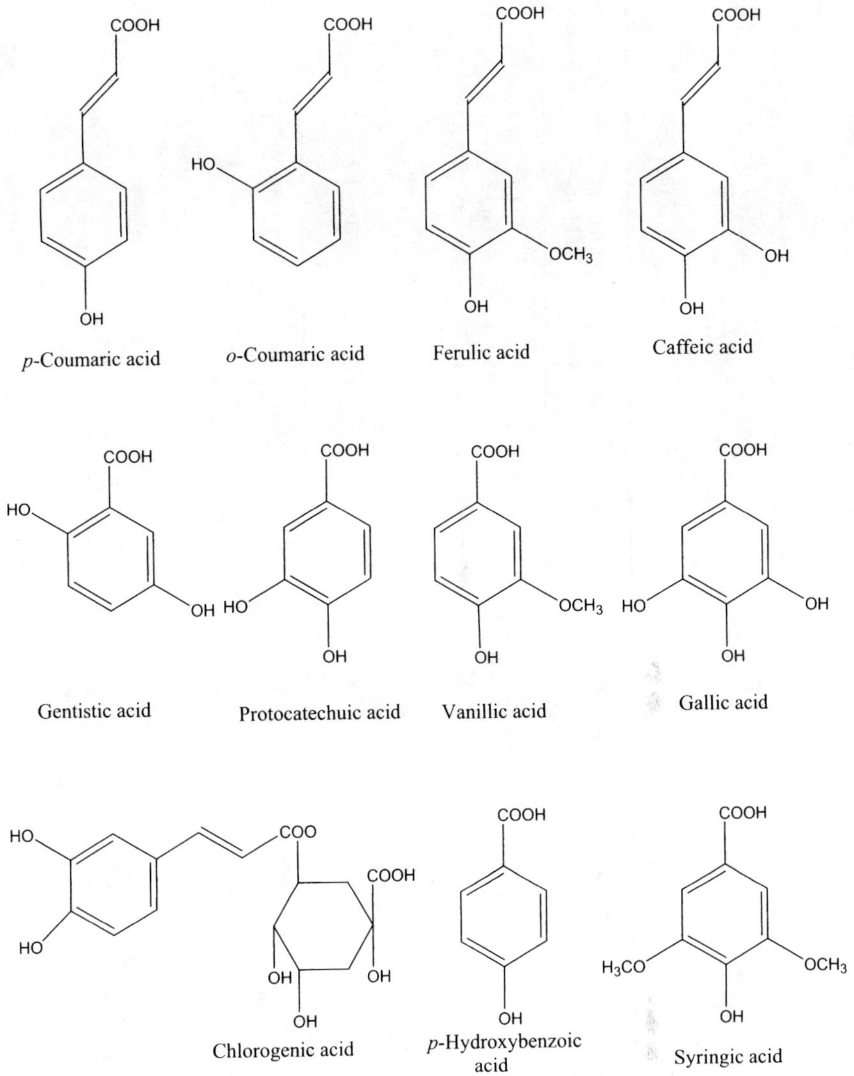

Figure 5.1 Chemical structures of the phenolic acids.

In 2003, grain samples of eleven wheat varieties and experimental lines were analyzed for their soluble free, soluble conjugated, and insoluble bound ferulic acid concentrations (2). Cham1, a white spring durum wheat, had the highest total ferulic acid level of 303 μmoles/100 g of grain, whereas Jennah Khetifa grain, a red spring durum wheat had the lowest total ferulic acid level of 148 μmoles/100 g, which were 588.4 and 287.4 μg/g grain, respectively (Table 5.1). In addition to these two varieties, hard red spring, hard white winter, soft white winter, soft red winter, and synthetic wheat were included in this study. The total ferulic acid concentration is comparable to that of wheat grain reported by other research groups. Moore and others (8) examined

TABLE 5.1 Total Phenolic Acid Composition in Wheat Grain and Fractions[a]

	Ferulic acid, μg/g	Vanillic acid, μg/g	Syringic acid, μg/g	p-Coumaric acid, μg/g	o-Coumaric acid, μg/g	Caffeic acid, μg/g	p-OH benzoic acid, μg/g	Reference
WG-US	287.4–588.4	na	na	na	na	na	na	2
SWG-US	455.9–621.5	8.4–12.7	8.9–17.8	10.4–14.1	nd	nd	nd	8
HSWG-CA	371.0–441.0	7.8–9.7	11.6–16.1	23.9–37.2	146.1–229.2	7.6–12.9	nd	9
CBGWB-CN	1550.1–2119.4	23.8–119.0	34.9–57.6	88.8–145.1	71.9–598.2	28.0–48.4	17.4–20.4	7
HRWB-AACC	1918	82.6	125	35.0	nd	0.8	39.2	10
HRWB-CA	2020	78.0	182	46.6	nd	1.78	28.3	10
SWWB-CA	1992	80.8	145	38.8	nd	nd	45.5	10
SWWB-AACC	1376	45.5	71.5	38.1	nd	nd	24.6	10

nd, not detected and na, not appliable. WG, wheat grain; SWG, soft wheat grain; HSWG, hard spring wheat grain; CBGWB, Chinese black-grained wheat grain; HRWB, hard red wheat bran; SWWB, soft white wheat bran; US, United States; CA, Canada; CN, China.
[a]Phenolic acid content was expressed on a per bran weight basis.

eight Maryland grown soft wheat varieties and the total ferulic acid level was found in the range of 456–621 µg/g grain (Table 5.1). Vanillic, syringic, and p-coumaric acids were also present in the wheat grain at total levels of 8.4–12.7, 8.9–17.8, and 10.4–14.1 µg/g, respectively, in the soft wheat grain samples (Table 5.1), but no p-hydroxybenzoic or o-coumaric acid was detected. Another study reported the effect of genotype and environmental conditions on phenolic acid composition in six western Canadian wheat genotypes grown at four locations (9). The total ferulic acid was determined at a level ranging 371–441 µg/g grain (Table 5.1). Interestingly, o-coumaric acid was the second major phenolic acid in all the grain samples, with a concentration of 145–229 µg/g grain, along with significant levels of vanillic, syringic, caffeic, and p-coumaric acids. These previous studies demonstrated the significant levels of phenolic acid contents in wheat grain, suggesting that wheat grain and grain-based food ingredients/products may contribute to total dietary phenolic acids especially ferulic acid. These previous studies also indicated the potential influence of genotype and growing conditions on phenolic acid compositions in wheat grain. It needs to be pointed out that different extraction procedures and hydrolyzing conditions have been used in these previous studies, and that may also contribute to the different phenolic acid compositions.

In addition, Moore and others (8) examined the soluble free, soluble conjugated, and insoluble bound phenolic acid profiles of eight Maryland grown soft wheat varieties and experimental lines. The soluble free and conjugated phenolic acids were extracted from the ground wheat sample using acetone-methanol-water (7 : 7 : 6, v/v/v). The solid residue was subjected to alkaline catalyzed hydrolysis, followed by acidification, ethyl ether-ethyl acetate reextraction of the freed phenolic acids, and subjected to HPLC analysis to determine the composition of the insoluble bound phenolic acids in the grain samples. The acetone-methanol-water extract was acidified and extracted with ethyl ethyl-ethyl acetate (1 : 1, v/v). The resulted ethyl ether-ethyl acetate extract was further treated and used to determine the soluble free phenolic acid composition, whereas the aqueous phase was hydrolyzed and used for analysis of soluble conjugated phenolic acids. The level of soluble free ferulic, vanillic, syringic, and p-coumaric acid was 0.6–2.3, 0.8–2.0, 0–0.9, and 0–0.2 µg/g grain, respectively. The soluble conjugated ferulic, vanillic, syringic, and p-coumaric acid was 32.0–47.2, 4.4–6.1, 6.4–13.0, and 0.9–1.3 µg/g grain, respectively. The insoluble bound ferulic, vanillic, syringic, and p-coumaric acid ranged 406.7–587.7, 2.9–5.0, 2.7–4.8, and 9.5–13.1 µg/g grain, respectively. Ferulic acid is predominant in total, insoluble, and total soluble phenolic acids, but not necessarily the primary phenolic acid in the soluble free phenolic acid fraction. It was also noted that the majority of the phenolic acids are present in the insoluble bound form and that may limit their release from matrix and reduce their bioavailability. The soluble phenolic acid content was reported for Swiss red wheat grain by the same group earlier (3). The combined soluble free and conjugated phenolic acids were determined for the 50% acetone extract of the ground grain samples at 33.7, 13.7, 5.0, 4.9, and 1.9 µg/g grain for ferulic, syringic, p-hydroxybenzoic, vanillic, and p-coumaric acids, respectively. The levels of total soluble ferulic acid was higher than that of 2.4–8.6 µg/g grain detected in the eleven wheat grains extracted with chilled 80% ethanol (Table 5.2) (2).

TABLE 5.2 Soluble Phenolic Acid Composition in Wheat Grain and Fractions[a]

	Ferulic acid, µg/g	Vanillic acid, µg/g	Syringic acid, µg/g	p-Coumaric acid, µg/g	Caffeic acid, µg/g	p-OH Benzoic acid, µg/g	References
Grain-US	2.4–8.6	na	na	na	na	na	2
TregoB-US	90.9–111.4	13.1–15.2	32.5–33.2	3.7–6.4	nd	11.1–21.9	5
IWB-In	98.5–230.5	15.0–26.5	29.3–85.0	3.7–16.2	nd	10.5–29.5	4
HRWWB-US	130.0–146.4	14.5–33.1	36.4–55.7	5.8–8.6	nd	8.9–19.0	6
HRWB-AACC	13.4	23.4	45.2	1.9	nd	5.62	10
HRWB-CA	44.7	9.6	47.5	1.9	nd	7.2	10
SWWB-CA	47.0	10.0	16.2	1.8	nd	6.9	10
SWWB-AACC	17.0	8.8	24.5	1.3	nd	4.0	10

nd, not detected and na, not appliable. TregoB, Trego wheat bran; IWB, international wheat bran; HRWWB, hard red winter wheat bran; CBGWB, Chinese black-grained wheat bran; HRWB, hard red wheat bran; SWWB, soft white wheat bran; US, United States; CN, China; and CA, Canada.
[a]Phenolic acid content was expressed on a per bran weight basis.

It is widely accepted that phenolics are concentrated in the bran fraction of wheat grain. A number of studies were conducted to examine the phenolic acid composition in wheat bran and bran-derived samples (3–7,10). Ferulic acid was the predominant phenolic acid in the 50% acetone extracts of bran samples of seven wheat varieties from four different countries along with significant levels of vanillic, syringic, p-coumaric, and p-hydroxybenzoic acids (Table 5.2) (4). The seven wheat bran samples differed in their phenolic acid compositions, suggesting the potential effects of genotype and growing conditions on phenolic acid profile in wheat bran. The potential effect of wheat variety and growing conditions on bran phenolic acid composition was observed in a later study using the bran samples of Alliance and Wichita wheat grown at two locations and 50% acetone as the extraction solvent (Table 5.2) (6). The potential effect of growing conditions on phenolic acid compositions in the 50% acetone extracts of wheat bran has also been observed for Trego wheat bran from five locations (Table 5.2) (5). In these three studies, antioxidants were extracted with 50% acetone. After removing the acetone, residues were hydrolyzed and the free phenolic acids were re-extracted and subjected to HPLC analysis. Therefore, the reported phenolic acid compositions are considered those "extractable" or "soluble".

In contrast, "total" phenolic acid composition was also determined for several wheat bran samples by alkaline hydrolysis of wheat samples followed by solvent extraction of the released free phenolic acids. Bran samples of five Chinese black-grained wheat varieties were hydrolyzed under basic condition, acidified, re-extracted with ethyl acetate, and the solutes in ethyl acetate was dissolved in methanol and quantified by HPLC (7). In addition to ferulic acid, vanillic, syringic, p-coumaric, o-coumaric, caffeic, and p-hydroxybenzoic, gallic, and gentistic acids were also detected (Table 5.1). Ferulic acid was the primary phenolic acid and had a level ranging 1550–2119 µg/g bran (7). Recently, a hard red and a soft white wheat bran samples from Canada were compared with the corresponding AACC bran samples for their extractable, bound, and total phenolic acid compositions (10). Total ferulic acid contents of these four wheat bran samples ranged 1376–2020 µg/g, which is comparable to that in the Chinese black-grained wheat bran samples (Table 5.1). No o-coumaric acid was detected in any of these Canadian or AACC bran samples. In addition, this study showed that the majority of the phenolic acids are present in the insoluble bound form in wheat bran and the hydrolysis conditions may significantly alter the phenolic acid composition estimation (10).

5.3 FREE RADICAL SCAVENGING CAPACITY OF WHEAT PHENOLIC ACIDS

A number of researches have investigated the free radical scavenging properties of these wheat phenolic acids against peroxyl (ORAC), peroxide anion ($O_2^{\bullet-}$), hydroxyl ($HO^{\bullet}$), cation ABTS ($ABTS^{\bullet+}$), and stable DPPH ($DPPH^{\bullet}$) radicals using spectrophotometric, fluorometric, electrochemical, and electron spin resonance (ESR) spectroscopic methods (11–17). Early in 1991, ferulic acid was reported to dose-dependently inhibit $O_2^{\bullet-}$ generation using the xanthine oxidase system, and exhibited 67% inhibition at an initial concentration of 10 mM (11). This inhibition was stronger

than that of 16% inhibition by 10 mM α-tocopherol measured using a spectrophotometric method under the same experimental conditions. Caffeic, p-coumaric, and
cinnamic acids exhibited no detectable inhibitory effect at 10 mM in this study (11).
The dose-dependent $O_2^{\cdot-}$ scavenging capacities of ferulic, caffeic, gentistic,
p-coumaric, gallic, vanillic acids were also investigated using the phenazine meth-
osulphate-NADH method (15). The $O_2^{\cdot-}$ scavenging capacity was measured as the
ability of a selected phenolic acid to prevent NBT reduction induced by $O_2^{\cdot-}$ generated
in the assay mixtures. The results showed that only caffeic and gallic acids were able to
inhibit NBT reduction, suggesting their capacity to act as $O_2^{\cdot-}$ scavengers, but none of
ferulic, cinnamic, vanillic, and gentistic acids exhibited detectable NBT reduction
ability (15). Recently, ferulic, p-coumaric, syringic, vanillic, and p-hydroxybenzoic
acids were compared for their $O_2^{\cdot-}$ scavenging capacities using the xanthine/xanthine
oxidase system and ESR determination (17). Syringic acid had the highest superoxide
dismutase (SOD) equivalents, which corresponds to the strongest $O_2^{\cdot-}$ scavenging
capacity among the tested phenolic acids (Fig. 5.2). SOD is a known $O_2^{\cdot-}$ scavenging
component. Ferulic acid had the stronger $O_2^{\cdot-}$ scavenging capacity than p-coumaric

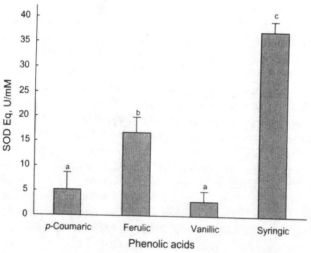

Figure 5.2   Comparison of $O_2^{\cdot-}$ scavenging capacities of the phenolic acids. $O_2^{\cdot-}$ scavenging
capacities of phenolic acids were determined by ESR method and expressed as superoxide
dismutase (SOD) equivalent (U) per mM phenolic acid). Coumaric, ferulic, syringic, and
vanillic stand for p-coumaric, ferulic, syringic, and vanillic acids, respectively, while the control
represents the control reaction containing no antioxidant. p-OH benzoic acid had no $O_2^{\cdot-}$
scavenging activity under the experimental conditions. The antioxidant-$O_2^{\cdot-}$ reaction was
initiated by adding xanthine oxidase and BMPO was the trapping agent for ESR determination.
ESR spectra were recorded at 2 min of each reaction at ambient temperature with 10 mM
incident microwave power and 100 kHz field modulation of 1 G. The final concentrations were
2 mM for xanthine, 100 mM for BMPO, 2 mM for phenolic acid, 0.2 mM for DTPA, and
0.08 U/mL for xanthine oxidase. All tests were conducted in triplicate and the means are used.
The vertical bars represent the standard deviation of each data point ($n = 3$). Values marked by
the same latter are not significantly different ($p < 0.05$). (Redrawn from Reference 17.)

acid, and p-hydroxybenzoic acid exhibited no $O_2^{\cdot-}$ scavenging capacity under the testing conditions, suggesting that the presence of the o-methoxyl group on the phenyl ring may enhance the $O_2^{\cdot-}$ scavenging capacity of phenolic acids (Fig. 5.2) (17). Taken together, these previous studies suggested the influence of testing method on $O_2^{\cdot-}$ scavenging capacity estimation.

Radical scavenging capacity against $ABTS^{\cdot+}$ was evaluated for several wheat phenolic acids using the spectrophotometric assay, known as the trolox equivalent antioxidant capacity (TEAC) or the total antioxidant activity (TAA) assay, and the ESR methods (12,14,17). The TEAC value is measured using the spectrophotometric method and represents the ability of a selected antioxidative compound to donate a hydrogen atom to $ABTS^{\cdot+}$ and convert it to a nonradical compound. The TEAC value is positively associated with the $ABTS^{\cdot+}$ scavenging capacity. The TEAC values were 0.08–3.01 mM TE/mM (12) and about 1.6–4.2 mM TE/1000 ppm (14) for p-hydroxybenzoic, o-coumaric, caffeic, syringic, vanillic, ferulic, p-coumaric, and gallic acids (Table 5.3). TE stands for trolox equivalent, and trolox is a standard antioxidant compound. Interestingly, p-hydroxybenzoic acid showed no $ABTS^{\cdot+}$ scavenging capacity in a later study using either ESR determination or the spectrophotometric method previously used in Yeh and Yen's study (17). This later study also indicated no synergistic effect between syringic, vanillic, ferulic, and p-coumaric acids in their scavenging abilities against $ABTS^{\cdot+}$. As shown in Table 5.3, the order of $ABTS^{\cdot+}$ scavenging capacity varied for these phenolic acids in the three previous studies, which may be explained by the potential influences of testing system on radical scavenging capacity estimation.

DPPH$^{\cdot}$ scavenging capacity was investigated for ferulic, p-coumaric, caffeic, syringic, vanillic, and p-hydroxybenzoic acids using spectrophotometric and ESR methods (16–19). Neither p-hydroxybenzoic nor vanillic acid had detectable DPPH$^{\cdot}$ scavenging capacity under the ESR determination conditions (17). As shown in Table 5.3, these previous studies indicated that gallic acid had stronger DPPH$^{\cdot}$ scavenging capacity than caffeic acid, and followed by ferulic acid and p-coumaric acid. These wheat phenolic acids were able to directly react with and quench stable (16–19). Also noted in Table 5.3 was that syringic acid had stronger DPPH$^{\cdot}$ scavenging capacity than ferulic acid (17). In addition, the presence of additional o-methoxyl group on the phenyl ring and the extension of the conjugated system may enhance the DPPH$^{\cdot}$ scavenging capacity of a selected phenolic acid (Table 5.3, Fig. 5.1), suggesting the potential relationship between radical scavenging capacity and chemical structure.

Peroxyl radical scavenging capacity (ORAC) of several wheat phenolic acids has been evaluated using the fluorescence method with β-phycoerythnin as the fluorescent probe and AAPH as the peroxyl radical generator (14). As shown in Fig. 5.3, all tested wheat phenolic acids, including ferulic, p-coumaric, o-coumaric, chlorogenic, caffeic, syringic, vanillic, gallic, gentistic, protocatechuic, and p-hydroxybenzoic acids, had significant ORAC. Gallic acid had the highest ORAC value of 14 μM trolox equivalents (TE) among the benzoic acid derivatives, and p-coumaric acid exhibited the greatest ORAC of 12 μM TE (Fig. 5.3). It was also reported that the ORAC of gallic, p-hydroxybenzoic, gentistic, and coumaric acids was dose dependent (14). In addition, vanillic acid had lower ORAC than p-hydroxybenzoic acid, and ferulic acid had weaker ORAC than p-coumaric acid,

TABLE 5.3 Radical Scavenging Capacities of Phenolic Acids[a]

	ABTS$^{\bullet+}$, TEAC[b]	ABTS$^{\bullet+}$, mM/mM	ABTS$^{\bullet+}$, TEAC[c]	DPPH$^{\bullet}$, TE mM/g[d]	% DPPH$^{\bullet}$ scavenged[e]	% DPPH$^{\bullet}$ scavenged[f]
Ferulic	1.90 ± 0.02	1.66 ± 0.01	3.1	1.00 ± 0.05	27.3 ± 0.8	52
p-Coumaric	2.22 ± 0.06	1.49 ± 0.03	2.3	—	7.0 ± 0.8	2
o-Coumaric	0.99 ± 0.15	—	1.8	—	—	—
Caffeic	1.26 ± 0.01	—	1.6	6.49 ± 0.30	44.0 - 49.6	—
Syringic	1.36 ± 0.01	1.22 ± 0.01	1.8	—	—	85
Vanillic	1.43 ± 0.05	1.12 ± 0.07	1.9	—	—	nd
Gallic	3.01 ± 0.05	—	4.2	9.79 ± 2.27	75 ± 2	—
p-OH benzoic	0.08 ± 0.01	nd	2.1	—	—	nd
References	12	17	14	19	16,18	17

"—" Stands for not tested, and "nd" means not detectable; ABTS$^{\bullet+}$ and DPPH$^{\bullet}$ stand for ABTS$^{\bullet+}$ and DPPH$^{\bullet}$ scavenging capacities, respectively.

[b]The TEAC value was reported as the concentration of trolox solution with equivalent antioxidant potential to a 1 mM concentration of the compound under the experimental conditions (12).

[c]The TEAC value was reported as the mM of trolox solution to obtain th antioxidant equivalent to a 1000 ppm solution of the phenolic acid (14).

[d]TE stands for trolox equivalent. Trolox is an antioxidant standard. Initial DPPH radical concentration was 104 μM. A_{517} nm of reaction mixture was measured every minute for 40 min and used for calculation (19).

[e]Initial DPPH radical concentration was 100 μM (18), and about 36.1 μg/mL (16), and phenolic acid concentration was 20 μM (18), and about 2.8 36.1 μg/mL (16), respectively. A_{517} nm was measured at 30 min (18) and 1 min (16) of reaction and used to calculate the % DPPH$^{\bullet}$ scavenged (16,18).

[f]Initial DPPH radical concentration was 250 μM, and phenolic acid concentration was 50 μM. ESR signals were measured at 25 min of reaction (16).

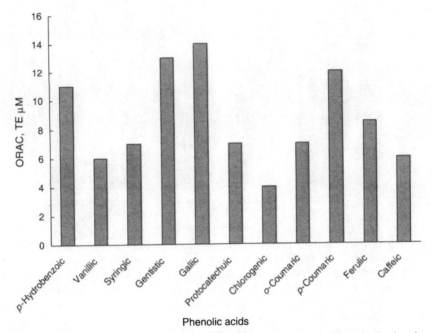

Figure 5.3   ORAC values of the phenolic acids. ORAC values were determined using the fluorescence method with β-phycoerythnin as the fluorescent probe and AAPH as the peroxyl radical generator. Final concentrations were 16.7 for β-phycoerythnin, 40 mM for AAPH. Trolox was used as the antioxidant standard at a final concentration of 1 µM. (Redrawn from Reference 14.)

suggesting that the addition of the *o*-methoxyl group on the phenyl ring may reduce the ORAC of both benzoic and cinnamic acid derivatives. In this study, the order of ORAC value in µM of TE was *p*-coumaric > *p*-benzoic > ferulic > syringic > caffeic acids (14). In contrast, the order was *p*-coumaric > ferulic > caffeic > *p*-benzoic > syringic acids for their ORAC values expressed in µM of TE per µM phenolic acid measured using a fluorometric assay with fluorescein as the fluorescent probe and AAPH as the peroxyl radical generator (20). This order was highly correlated with that for the hydroxyl radical scavenging capacities of these five phenolic acids with a correlation coefficient value of 0.954 and a P value of 0.05 (Fig. 5.4).

Hydroxyl radical (HO$^{\bullet}$) scavenging capacity (HOSC) was measured for *p*-coumaric, ferulic, syringic, vanillic, and *p*-hydroxybenzoic acids using ESR method with *N*-oxide pyrroline (DMPO) as the trapping agent (17). All the tested phenolic acids except *p*-hydroxybenzoic acid were able to scavenge HO$^{\bullet}$ generated by Fenton reaction, and *p*-coumaric acid was more effective than the other tested phenolic acids, although the ESR measurement was not quantitatively conducted. In addition, the scavenging activity seems time dependent (17). Recently, a novel fluorometric assay was developed for estimating the HOSC against pure HO$^{\bullet}$, and HOSC values determined using this new method ranged from 1.9 to 4.0 µM of TE per µM phenolic

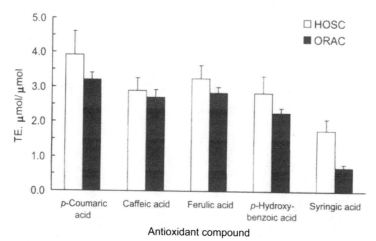

Figure 5.4   HOSC and ORAC values of five phenolic acids. Initial reaction concentrations for HOSC assay were 5.26×10^{-8} M FL, 42.5 mM sodium phosphate buffer, 26.53 mM H_2O_2, 0.686 mM $FeCl_3$, and individual phenolic acids at 2.5 μM. Initial reaction concentrations for ORAC assay were 0.067 μM FL, 53.6 mM AAPH, and individual phenolic acids at 2.7 μM. Results were reported as μM trolox equivalents (TE) per μM of phenolic acid. All tests were conducted in triplicates with mean values reported. The vertical bars represent the standard deviation of each data point ($n = 3$). (Redrawn from Reference 20.)

acid for p-coumaric, ferulic, caffeic, p-benzoic, and syringic acids (Fig. 5.4) (20). This new method uses a Fenton-like Fe^{3+}/H_2O_2 reaction to generate HO˙ under physiological pH and fluorescein as the fluorescence probe (20).

5.4   INHIBITORY EFFECT OF WHEAT PHENOLIC ACIDS ON LIPID PEROXIDATION

Lipid peroxidation in biological system is a free radical mediated oxidative chain reaction, which shares the same chemical mechanism with the lipid peroxidation in biological systems such as cell membrane. Antioxidants capable of inhibiting lipid peroxidation in model systems may suppress lipid peroxidation in the biological systems and prevent life important molecules such as DNA and membrane lipids from oxidative damages, and reduce the risk of several aging-associated health problems such as cancer and heart diseases. In 1991, caffeic, ferulic, cinnamic, p-coumaric acids were evaluated for their potential inhibitory effects on lipid peroxidation induced by peroxide anion (11). Aqueous linoleic acid system was used for the determination and the peroxide anion was generated using the xanthine/xanthine oxidase system. Caffeic and ferulic acids were shown to dose-dependently suppress the lipid peroxidation with inhibitory ratios comparable to that obtained by α-tocopherol and superoxide dismutase (SOD) (11). However, cinnamic and p-coumaric acids failed to prevent the peroxide anion induced lipid peroxidation under the same experimental conditions.

In 1992, antioxidant properties of ferulic acid and its cinnamic acid derivatives were reviewed (21). These phenolic acids were able to significantly reduce the peroxide formation in Ghee and brain tissue, and the order of inhibitory capacity was caffeic > p-coumaric > ferulic > cinnamic acids. This order was supported by an earlier observation that caffeic acid was about 1000-fold more effective than ferulic acid in suppressing lipid peroxidation of rat brain homogenates *in vitro*, and cinnamic acid had no detectable inhibitory capacity (21). In this study, ferulic acid at a final concentration of 5 μM provided 71% suppression of lipid peroxidation in rat brain homogenates. This review article also discussed the UV absorbing properties of ferulic acid and other cinnamic acid derivatives as well as the possible role of their UV absorption in their antioxidative functions. For instance, 5 mM ferulic acid significantly inhibited diene formation during the UV irradiation induced peroxidation of linoleic acid at 30°C (21). Several previous studies also showed the potential of ferulic acid in prevention of lipid peroxidation or autoxidation in food products including linseed oil and biscuits (21).

Later in 1992, nine wheat phenolic acids were tested and compared for their capacities to suppress peroxide formation in an emulsion system measured by the active oxygen method (1). All tested phenolic acids significantly reduced the peroxide formation. Protocatechuic acid had the strongest inhibitory activity with 70% reduction of peroxide formation and followed by chlorogenic > caffeic > p-hydroxybenzoic > gentistic > ferulic > vanillic > syringic > p-coumaric acids. Similar order of inhibitory activity was observed for caffeic, ferulic, and p-coumaric acids in bulk methyl linoleate in another study (18). This study also reported that caffeic acid was most effective in suppressing AAPH-induced lipid peroxidation of egg yolk phospholipid liposomes at a final testing concentration of 50 μM, and followed by ferulic and p-coumaric acids (18). In addition, all the three phenolic acids exhibited stronger inhibition than L-ascorbic acid and α-tocopherol on a per molarity basis under the same testing conditions.

Interestingly, ferulic acid dose-dependently suppressed hydroxyl-radical-induced lipid peroxidation in the synaptosomes isolated from the fresh brain samples of three-month old male Mongolian Gerbils, but neither vanillic nor coumaric acid was able to significantly prevent the lipid oxidation (13). Synaptosomes are sealed synaptic vehicles that behave in a similar manner as neurons and may be used to investigate neuron oxidation and its influence on neuronal functions. This observation was explained by either the insufficient antioxidant activity of vanillic and coumaric acids or their poor capacity to be incorporated into cell membrane (13). The inhibitory activity of ferulic acid on lipid peroxidation was also observed in isolated rat liver microsomal membranes (22). The lipid peroxidation was measured as total diene and malondialdehyde (MDA) at different incubation time periods. Ferulic acid dose-dependently inhibited both diene and MDA formations induced by AAPH or *tert*-butyl hydroperoxide (*tert*-BOOH) in the microsomal membrane suspensions (22). In the rat liver microsomal membrane system, ferulic acid showed synergistic effect with ascorbic acid in suppressing AAPH-induced MDA formation, but had no synergistic effect with either α-tocopherol or β-carotene under the experimental conditions (22). In addition, ferulic acid was more effective to prevent *tert*-BOOH-induced lipid peroxidation than that induced by AAPH, suggesting the possible alteration of the oxidative reaction mechanism on the antioxidant potency of phenolic acids.

In contrast, these phenolic acids, including p-hydroxybenzoic, protocatechuic, vanillic, syringic, p-coumaric, caffeic, and ferulic acids, failed to prevent conjugated diene formation in human LDL induced by AAPH (23). Conjugated diene is the secondary product of lipid peroxidation, and the level of conjugated diene is positively associated with the degree of lipid peroxidation. These phenolic acids were also tested for their potential in suppressing lipid oxidation in the LDL induced by $5\,\mu M\ Cu^{2+}$. Protocatechuic and caffeic acids significantly inhibited conjugated diene formation, whereas syringic acid might have prooxidant activity under the experimental conditions (23). Vanillic, p-coumaric, caffeic, and ferulic acids exhibited no inhibitory activity on lipid peroxidation induced by Cu^{2+}. These precious studies suggest that the capacity of a selected phenolic acid in suppressing lipid peroxidation may also depend on the lipid system, such as LDL or rat liver microsomal membrane.

5.5   CHELATING PROPERTIES OF WHEAT PHENOLIC ACIDS

Free radical mediated oxidative chain reaction generally involves three phases: initiation, propagation, and termination. Initiation is the rate-limiting step for lipid peroxidation chain reaction. First few free radicals are generated through several different chemical mechanisms during the initiation phase. Transition metals catalyze the formation of the first few radicals and accelerate the initiation step and the overall peroxidation reactions. Chelating agents may reduce the availability of the transition metals as the catalysts and may suppress overall degree of lipid peroxidation. Ferulic, p-coumaric, syringic, and vanillic, p-hydroxybenzoic acids were examined for their potential chelating activities against Fe^{2+} according to the spectrophotometric method or against Cu^{2+} using the ESR method (17). Details of these methods are discussed in the other chapters of this book. No Fe^{2+} chelating activity was observed for these phenolic acids under the experimental conditions. As demonstrated by the ESR measurements (Fig. 5.5), ferulic and vanillic acids may be able to alter the Cu^{2+} ESR spectrum, which reflects their interaction with Cu^{2+}, but syringic and p-coumaric acids had no Cu^{2+} chelating capacity (17). The interaction between phenolic acid and Cu^{2+} may reduce its availability as catalyst and suppress lipid peroxidation.

5.6   OTHER ANTIOXIDANT ACTIVITIES OF WHEAT PHENOLIC ACIDS

Some protein molecules such as enzymes may play critical roles in human health. Oxidative damage of these proteins may cause cell injury and lead to aging-associated health problems. Wheat phenolic acids, capable of protecting oxidative modification of cellular proteins, may reduce the risk of these health problems and have potential health beneficial effects. In 2002, Kanski and coworkers (13) reported that ferulic acid

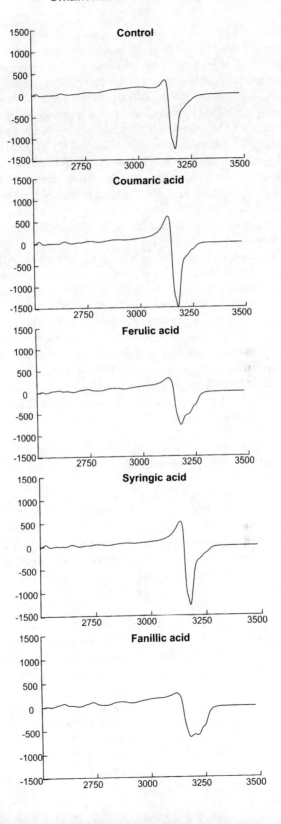

at final concentrations of 250 and 500 µM was able to almost completely suppress the protein oxidation induced by hydroxyl radicals in the synaptosomes isolated from the fresh brain samples of three-month old male Mongolian Gerbils. Protein oxidation was measured as the level of protein carbonyls using the slot-blot technique. This study also found that ferulic acid was much more effective than vanillic, coumaric, and cinnamic acids in protecting hydroxyl-radical-induced protein peroxidation in the synaptosomes (13). In addition, ESR measurement using the protein specific spin label 2,2,6,6-tetra-methyl-4-maleimidopiperidin-1-oxyl (MAL-6) showed that presence of hydroxyl and peroxyl radicals might induce the conformational alterations of the synaptosomal membrane proteins possibly due to the crosslinking reactions (13). Ferulic acid at a final concentration of 250 µM significantly prevented the conformational changes induced by either hydroxyl or peroxyl radicals (13). These data demonstrated the potential protective capacity of phenolic acids against cellular protein peroxidation.

It is well accepted that reactive oxygen species (ROS) may directly attack life important cellular components including DNA, proteins, and membrane lipids and elevate cellular oxidative stress (13). Cellular ROS level may be determined using a fluorometric assay using dichlorofluorescin diacetate (DCFH-DA) as the fluorescence probe (13). DCFH-DA is hydrolyzed by cellular esterases and release DCFH, which is converted to the highly fluorescent dichlorofluorescein (DCF) during ROS mediated oxidation. The fluorescent intensity is positively correlated with the cellular ROS level. This assay was employed to investigate the potential effect of ferulic, coumaric, vanillic, and cinnamic acids on cellular ROS level using the brain synaptosomes of Mongolian Gerbils (13). Ferulic acid at a concentration of 250 µM was able to suppress about 63% AAPH-induced fluorescence formation, whereas vanillic acid also significantly reduced the fluorescent intensity (about 15% reduction), suggesting their capacity to quench cellular ROS (13). Cinnamic and coumaric acids had no protection against AAPH-induced ROS generation in synaptosomes under the same experimental conditions (13).

In 2001, caffeic, ferulic, p-coumaric, cinnamic, gallic, vanillic, and gentistic acids were examined for their antioxidant properties using the electrochemical measurements including the potentiometry and cyclic voltammetry assays (15). Cyclic voltammetry assay determines the oxidation potential of phenolic acids, while the potentiogram represents the potential difference measured in the supporting electrolyte and in the solution containing a selected phenolic acid. The lowest oxidation potentials (E_{pa}) were 142, 410, 736, 113, 494, and 154 mV for caffeic, ferulic, p-coumaric, gallic, vanillic, and gentistic acids, respectively, but that was not detectable for cinnamic acid under the experimental conditions (15). The E_{pa} value was negatively correlated to the peroxide anion scavenging capacity of these phenolic acids, and gallic acid with the smallest E_{pa} value exhibited strongest peroxide anion scavenging activity. The E_{pa} value was also negatively correlated to the changes in

---

Figure 5.5   Interaction between Cu^{2+} and individual phenolic acids measured by ESR. The final concentrations were 5 mM for each phenolic acid and 1 mM for copper chloride ($CuCl_2$). The ESR spectrum was recorded at 1 min of reaction with 40 mW incident microwave power and 100 kHz field modulation of 5 G at 77 K. (Adapted from Reference 17.)

potential (ΔE) recorded at 180 S (15). Gallic acid had the highest ΔE value of 177 mV at 180 S, and followed by that of 148, 93, and 34 for caffeic, gentistic, and ferulic acids, respectively. The ΔE was not detectable for *p*-coumaric, cinnamic, or vanillic acids under the testing conditions.

5.7   STRUCTURE–ACTIVITY RELATIONSHIP OF WHEAT PHENOLIC ACIDS

It is widely accepted that physical and chemical properties of a selected compound are determined by its chemical structure and may be altered by a number of other factors. Structure–antioxidant activity relationships of wheat phenolic acids have been discussed in several previous publications (12,13,17,21,23). Wheat phenolic acids are generally either benzoic acid or cinnamic acid derivatives with different number of hydroxyl and/or methoxyl substituents at different positions of the phenyl ring (Fig. 5.1). The difference between the benzoic and cinnamic acid skeletons is the extension of the conjugated side chain due to presence of an additional conjugated double bond in cinnamic acid structure. When the hydroxyl derivatives of both cinnamic and benzoic acids act as free radical scavengers, they donate a proton to the radical and terminate the radical chain reactions, and form corresponding phenoxyl radicals. As shown in Fig. 5.6, the phenoxyl radicals from cinnamic acid derivatives may form more possible resonance structures than that of the corresponding hydroxyl benzoic acid derivatives, explaining why hydroxyl cinnamic acid derivatives act as stronger radical scavengers than the corresponding hydroxyl benzoic acid derivatives. This conclusion is supported by the observation that p-coumaric acid had stronger ABTS$^{\bullet+}$ and O$_2^{\bullet+}$ scavenging activity than p-hydroxybenzoic acid, and ferulic acid had stronger ABTS$^{\bullet+}$, O$_2^{\bullet+}$, and DPPH$^{\bullet}$ scavenging activity than vanillic acid (12,17). This conclusion is also supported by the previous observation that cinnamic acid derivatives had higher reactivity against the AAPH-generated peroxyl radicals and more effective in suppressing lipid peroxidation in human LDL than their benzoic counterparts (23).

Presence of additional hydroxyl and methoxyl substituents on the phenyl ring may enhance the overall antioxidant properties, including radical scavenging and chelating capacities, of the phenolic acids. Caffeic and protocatechuic acids exhibited stronger inhibitory effects against lipid peroxidation in human LDL induced by either

---

Figure 5.6   Phenoxyl radical formation and possible resonance structures for cinnamic and benzoic acid derivatives. Phenoxyl radical is initially formed at the p-OH group by abstraction of the hydroxyl H atom, regardless of the carbon skeleton of the molecule. (**a**) represents the formation and resonance structures of p-OH cinnamic acid derivatives including ferulic and p-coumaric acids. For resonance structure III, the unshared pair electrons in the p-orbital of the o-OCH$_3$ stabilizes the unpaired electron on C3 of the phenyl ring. For resonance structure VI, the unshared pair electrons in the p-orbital of the hydroxyl group stabilizes the unpaired electron on O-atom of the carboxylic group. (**b**) represents the formation and resonance structures of p-OH benzoic acid derivatives including p-OH benzoic, vanillic, and syringic acids. For resonance structure III, the unshared pair electrons in the p-orbital of the o-OCH$_3$ stabilizes the unpaired electron on C3 of the phenyl ring. For resonance structure V, the unshared pair electrons in the p-orbital of the hydroxyl group stabilizes the unpaired electron on O-atom of the carboxylic group. Cinnamic acid derivatives may have six possible resonance structures for the phenoxyl radical, whereas benzoic acid derivatives have five possible resonance structures. (Adapted from Reference 17.)

Cu^{2+} or AAPH than p-coumaric and p-hydroxybenzoic acids, respectively, indicating that the additional o-hydroxyl groups enhanced the antioxidant property of phenolic acids (23). Caffeic acid had higher $ABTS^{\cdot+}$ and $O_2^{\cdot+}$ scavenging activity, and stronger inhibitory against Cu^{2+} and AAPH-induced lipid peroxidation in human LDL than ferulic acid, suggesting that o-hydroxyl derivative of cinnamic acid may have greater antioxidant activity than their o-methoxyl counterpart (12,15,23). In contrast, vanillic acid (o-methoxyl p-hydroxybenzoic acid) had stronger $ABTS^{\cdot+}$ scavenging activity than protocatechuic acid (o-hydroxyl p-hydroxybenzoic acid). In addition, ESR measurements showed that ferulic and vanillic acids had significant capacity to interact and form chelating complex with Cu^{2+}, but syringic, p-coumaric, and p-benzoic acids could not (17). These data indicated that presence of the first o-methoxyl substituent is critical to chelating capacity against Cu^{2+} under the experimental conditions. In contrast, addition of the second o-methoxyl substituent on the phenyl ring, syringic versus vanillic acid, may eliminate the chelating capacity of the phenolic acid because of the interruption of the plane conjugation system due to steric exclusion of the substituents (17). Taking together, these previous studies suggest that presence of o-methoxyl and o-hydroxyl may enhance overall antioxidant activity of a selected phenolic acid because of their electron donating potentials.

The other factors that may alter the antioxidant activity estimation for phenolic acids may include but are not limited to (1) type of radical systems used, (2) physicochemical properties of the testing system, (3) physicochemical interactions between chemical compounds present in the testing reaction, and (4) the determination method. It has been widely recognized that type of radical systems such as the stable pure $DPPH^{\cdot}$, $ABTS^{\cdot+}$ generated by a chemical reaction outside of the testing system, $HO^{\cdot}$ generated by a chemical reaction in the testing system, or $O_2^{\cdot-}$ generated using an enzymatic reaction in the testing system may significantly alter the radical scavenging property estimation. This may be explained by the fact that individual testing systems may involve different reaction mechanisms and may interact with a selected phenolic acid differently. It is recommended that two or more radical systems be employed in evaluating the radical scavenging property of a selected phenolic acid.

The physicochemical properties of the testing system may significantly alter the antioxidant activity estimation results. For instance, caffeic acid was a stronger antioxidant than ferulic acid in the bulk methyl linoleate system to inhibit hydroperoxide formation, but ferulic acid was more effective to prevent lipid peroxidation in the ethanol-buffer system (18). This observation is in good agreement with the "polar paradox," which is the fact that polar antioxidant is generally more effective in bulk oil (low polarity) system and nonpolar antioxidant is more active in polar system (emulsion). This is because of the partition or distribution of the antioxidant in the testing system, and the antioxidant molecules at oil-air or oil-water interface may provide better prevention of lipid oxidation. It is understandable that physicochemical interactions between sample components, reagents, and contaminations may also alter the antioxidant activity estimation. For instance, presence of transition metal and chelating agents may alter the activity of a selected antioxidant in suppressing lipid peroxidation. Detailed discussion on the structure–antioxidant activity relationships for wheat phenolic acids may be defined in the previous published manuscripts (12,13,17,18,23).

REFERENCES

1. Onyeneho, S. N.; Hettiarachchy, N. S. Antioxidant activity of durum wheat bran. *J. Agric. Food Chem.* **1992**, *40*, 1496–1500.
2. Adom, K. K.; Sorrells, M. E.; Liu, R. H. Phytochemical profiles and antioxidant activity of wheat varieties. *J. Agric. Food Chem.* **2003**, *51*, 7825–7834.
3. Zhou, K.; Laux, J. J.; Yu, L. Comparison of Swiss red wheat grain and fractions for their antioxidant properties. *J. Agric. Food Chem.* **2004**, *52*, 1118–1123.
4. Zhou, K.; Lan, S.; Yu, L. Phytochemicals and antioxidant properties of wheat bran. *J. Agric. Food Chem.* **2004**, *52*, 6108–6114.
5. Zhou, K.; Yu, L. Antioxidant properties of bran extracts from Trego wheat grown at different locations. *J. Agric. Food Chem.* **2004**, *52*, 1112–1117.
6. Zhou, K.; Yin, J.; Yu, L. Phenolic acid tocopherol and carotenoid compositions, and antioxidant functions of hard red winter wheat bran. *J. Agric. Food Chem.* **2005**, *53*, 3916–3922.
7. Li, W.; Shan, F.; Sun, S.; Corke, H.; Beta, T. Free radical scavenging properties and phenolic content of Chinese black-grained wheat. *J. Agric. Food Chem.* **2005**, *53*, 8533–8536.
8. Moore, J.; Hao, Z.; Zhou, K.; Luther, M.; Costa, J.; Yu, L. Carotenoid tocopherol, phenolic acid, and antioxidant properties of Maryland-grown soft wheat. *J. Agric. Food Chem.* **2005**, *53*, 6649–6657.
9. Mpofu, A.; Sapirstein, H. D.; Beta, T. Genotype and environmental variation in phenolic content, phenolic acid composition, and antioxidant activity of hard spring wheat. *J. Agric. Food Chem.* **2006**, *54*, 1265–1270.
10. Kim, K.; Tsao, R.; Yang, R.; Cui, S. W. Phenolic acid profiles and antioxidant activities of wheat bran extracts and the effect of hydrolysis conditions. *Food Chem.* **2006**, *95*, 466–473.
11. Toda, S.; Kumura, M.; Ohnishi, M. Effects of phenolcarboxylic acids on superoxide anion and lipid peroxidation induced by superoxide anion. *Planta Med* .**1991**, *57*, 8–10.
12. Rice-Evans, C. A.; Miller, N. J.; Paganga, G. Structure–antioxidant activity relationships of flavonoids and phenolic acids. *Free Rad. Biol. Med.* **1996**, *20*, 933–956.
13. Kanski, J.; Aksenova, M.; Stoyanova, A.; Butterfield, A. Ferulic acid antioxidant protection against hydroxyl and peroxyl radical oxidation in synaptosomal and neuronal cell culture systems *in vitro*: structure-activity studies. *J. Nutr. Biochem.* **2002**, *13*, 273–281.
14. Yeh, C.; Yen, G. Effects of phenolic Acids on human phenolsulfotransferases in relation to their antioxidant activity. *J. Agric. Food Chem.* **2003**, *51*, 1474–1479.
15. Galato, D; Ckless, K.; Susin, M. F.; Giacomelli, C.; Ribeiro-do-Valle, R. M.; Spinelli, A. Antioxidant capacity of phenolic and related compounds: correlation among electrochemical, visible spectroscopy methods, and structure-antioxidant activity. *Redox Rep.* **2001**, *6*, 243–250.
16. Sroka, Z.; Cisowski, W. Hydrogen peroxide scavenging, antioxidant and anti-radical activity of some phenolic acids. *Food Chem. Toxicol.* **2003**, *41*, 753–758.
17. Zhou, K.; Yin, J.; Yu, L. ESR determination of the reactions between selected phenolic acids and free radicals or transition metals. *Food Chem* .**2006**, *95*, 446–457.
18. Kikuzaki, H.; Hisamoto, M.; Hirose, K.; Akiyama, K.; Taniguchi, H. Antioxidant properties of ferulic acid and its related compounds. *J. Agric. Food Chem.* **2002**, *50*, 2161–2168.
19. Cheng, Z.; Moore, J.; Yu, L. A high-throughput relative DPPH radical scavenging capacity (RDSC) assay. *J. Agric. Food Chem.* **2006**, 54, 7429–7436.
20. Moore, J.; Yin, J.; Yu, L. Novel fluorometri assay for hydroxyl radical scavenging capacity (HOSC) estimation. *J. Agric. Food Chem.* **2006**, *54*, 617–626.
21. Graf, E. Antioxidant potential of ferulic acid. *Free Rad. Biol. Med.* **1992**, *13*, 435–448.
22. Trombino, S.; Serini, S.; Nicuolo, F. D.; Celleno, L.; Andò, S.; Picci, N.; Calviello, G.; Palozza, P. Antioxidant effect of ferulic acid in isolated membranes and intact cells: synergistic interactions with α-tocopherol, β-carotene, and ascorbic acid. *J. Agric. Food Chem.* **2004**, *52*, 2411–2420.
23. Natella, F.; Nardini, M.; Felice, M. D.; Scaccini, C. Benzoic and cinnamic acid derivatives as antioxidants: structure-activity relation. *J. Agric. Food Chem.* **1999**, *47*, 1453–1459.

CHAPTER 6

EFFECTS OF POSTHARVEST TREATMENTS, FOOD FORMULATION, AND PROCESSING CONDITIONS ON WHEAT ANTIOXIDANT PROPERTIES

Chunjian (C.J.) Lin
Gang Guo
Don L. Mennel

6.1 INTRODUCTION

Wheat as a potential source of dietary antioxidants and the significance of antioxidants to human health have been described in detail in previous chapters. Wheat is one of the most important cereal grains, contributing to energy intake in humans (1). When harvested and leaving the production system, wheat is in the form of grains or kernels. A wheat kernel is made of endosperm, germ, and bran (Fig. 6.1). Each part consists of many other components, but what is worth mentioning is aleurone layer because of its nutrient composition and effect on novel flour quality. The aleurone layer is a part of endosperm, but often separated along with bran in modern flour milling. Endosperm is the inner part serving as storage of energy (starch) and functioning proteins (gluten components). Bran is the outer layer protecting the grain. Germ is the kernel's reproduction system.

Wheat grains when entering into human food supply chain are treated as raw materials and must be processed. The processed grains are often used as food ingredients in formulations before further processed into foods for consumption. Antioxidants in wheat grains must withstand various postharvest treatments, food formulations, and processing conditions throughout the food supply chain before exerting their potential health benefits. Therefore, this chapter will focus on how the common postharvest processing techniques of wheat grains, food formulations, and processing conditions affect antioxidant activity of wheat.

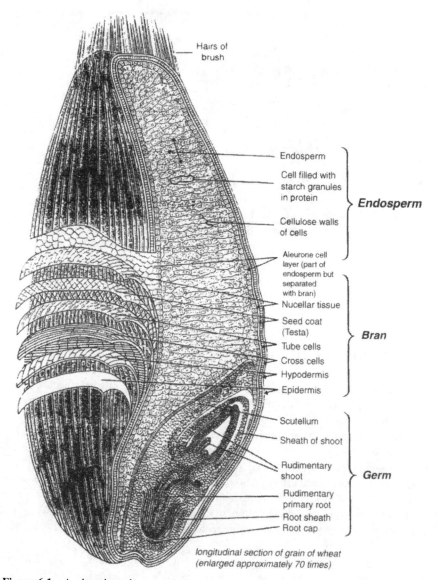

Figure 6.1   A wheat kernel.

6.2   WHEAT POSTHARVEST TREATMENTS

6.2.1   Wheat Flour Milling

Milling is the most common processing technique for wheat grains for human consumption. Roller mill is the primary technology used for milling wheat into flour.

The technology is a gradual reduction system so that damage to the components of wheat is minimized during flour milling. The objective of milling is to separate the bran and germ from the starchy endosperm so that the endosperm can be ground into flour. A modern flour milling system is divided into three parts based on their functions—breaking, reduction, and sifting. The break system consists of a number of pairs of corrugated rolls that break wheat kernels and subsequently separate the endosperm from bran (kernel outer layer) and germ. The reduction system is made of a number of pairs of smooth rolls that reduce the endosperm into fine particles. The sifting system contains sieves with different screen and cloth meshes separating flours based on particle size. In an industrial mill a wheat kernel is separated into more than thirty different parts called flour streams, each having its own unique characteristics. At the end of the process, the various streams are blended and mixed to make various grades of flour. Flour extraction defined as the proportion of flour by weight milled from a known quantity of wheat is commonly used in the milling industry to define different types of wheat flour. When the extraction rate is 75% or less, typical white flour is generated. If the extraction rate exceeds 80%, the flour will contain significant amount of nonendosperm particles, and if the flour extraction approaches 100%, wholemeal flour is produced. A commercial flour mill with flour extraction rate less than 100% yields at least two products — straight grade flour (patent and clear flour) and a milling by-product. The milling by-product includes germ, bran (coarse), shorts (fine bran), and red dog (finer bran and unreduced endosperm).

Flour milling process causes redistribution or concentration of phytochemicals in the milling products owing to the fact that phytochemicals in the wheat kernels are not evenly distributed (1). Prakash (2) reported that the refined wheat flour was about 7.7 times less than the bran in antioxidant activity and whole-wheat meal was in between the milling fractions. Using an experimental roller mill, Adom et al. (3) milled three different soft wheat cultivars into two fractions, flour and bran/germ, and analyzed each of the fractions for phytochemicals. They concluded that in whole-wheat meal the bran/germ fraction contributed 83% of total phenolic content, 79% of flavonoid, 51% of lutein, 78% of zeaxanthin, 42% of the β–cryptoxanthin, 85% of hydrophilic antioxidant activity, and 94% of lipophilic antioxidant activity (Table 6.1). Similarly, Beta et al. (4) reported that the bran fraction produced by roller milling of wheat contained much higher level of phenolics than the largest flour stream representing endosperm. Out of the

TABLE 6.1 Ranges of Phytochemicals and Antioxidant Activity of Milling Fractions of Three Soft Wheat Cultivars (3)

Phytochemicals[a]	Flour	Bran/germ
Total phenolics, μmol/100 g	176–195	2867–3120
Flavonoids, μmol/100 g	60–80	740–940
Total carotenoids, μmol/100 g	42–78	192–228
Hydrophilic antioxidants, μmol/g	7–16	261–545
Lipophilic antioxidants, nmol/100 g	45–65	1785–4669

[a]Total phenolics measured in gallic acid equivalent; flavonoids in catechin equivalent; hydrophilic antioxidant activity in vitamin C equivalent; lipophilic antioxidant activity in vitamin E equivalent.

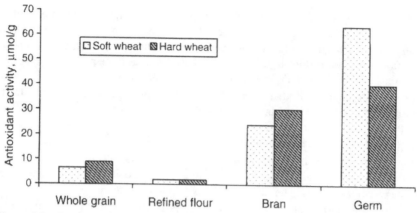

Figure 6.2 Comparison of antioxidant activity measured as DPPH scavenging capacity (μmol/g defatted material) among whole grain and the corresponding milling fractions of soft and hard wheat (5).

three bran fractions, shorts had the highest phenolics concentration, followed by bran and bran flour. Antioxidant activity of whole grain after milling is concentrated in germ and bran with refined flour showing the lowest activity (Fig. 6.2) (5). Liyana-Pathirana and Shahidi (6) compared the total phenolics content and total antioxidant capacity between whole grains of Canadian Western Amber Durum (CWAD) and Canadian Western Red Spring (CWRS) wheat and their respective milling fractions — bran, shorts, feed flour, and refined white flour. Bran was 2.7–3.0 times more in total phenolic compounds and 2.4–2.6 times higher in total antioxidant capacity than the original grains, followed by shorts (2.4–2.5 and 2.1–2.3 times, respectively), and feed flour (1.6–1.8 times each), with the refined white flour being the lowest accounting for 20–30% of total phenolics and 50% of total antioxidant capacity.

6.2.2 Pearling or Debranning

Pearling or debranning is a pretreatment process prior to roller milling of wheat and has recently gained interest of flour millers. Wheat grains are physiologically composed of 84% endosperm, 13.5% bran, and 2.5% germ. However, commercial roller mills are rarely able to extract all the endosperm into flour without bran contamination in part because of their limited ability to separate the endosperm from the bran. The actual flour extraction rates is normally less than 78% by weight depending on wheat class, cultivar, and milling equipment. Thus, the debranning or pearling process was developed to remove bran prior to milling to improve flour extraction and prevent bran contamination of refined flour products. A debranner or pearler is a machine with both friction and abrasion action to remove wheat bran layers sequentially. Wheat is preconditioned with water for a short period of time before debranning so that water penetrates into the outer regions of the seed coat. The friction action created through rotation of a vaned hollow shaft causes kernels to rub against each other as the wheat kernels pass through the machine while the abrasion action makes kernels rub against an abrasive stone. Air is introduced in both actions to assist the removal of the by-products through screens. Depending on how the friction and abrasion actions are sequenced in a machine, two types of commercial wheat debranning systems are available. In both processes the wheat bran layers are removed by successive passages. The by-products generated by each passage are characterized with unique functional and nutritional properties (7). The debranning process has been practiced in some of the modern flour mills. Both milling and debranning affect antioxidant content and activity of wheat milling products because of the unevenly distributed phytochemicals in wheat grain (2,6).

Like milling, the debranning or pearling process causes redistribution of antioxidants of wheat grains among the products. Fellers et al. (8) reported that the phenolic compounds measured as phytate phosphorus in the debranned wheat decreased at each passage of the debranner. Similar results were observed by Beta et al. (4) who successively pearled wheat grain of each of the six cultivars at 5% increment using a commercial debranner and examined total phenolic content and antioxidant activity assayed with free radical 2,2-dipheny-1-picryhydrazyl (DPPH) method of the pearled fractions. They observed that total phenolic contents were highly correlated to antioxidant activity ($R^2 = 0.94$) and differed significantly among the pearled wheat fractions with the first 10% pearled portion being the highest (4160–5300 mg/kg), decreasing as more bran layers being pearled off, and with the remaining fraction representing endosperm being the lowest (1350–1620 mg/kg). Liyana-Pathirana et al. (9) also reported that the degree of pearling adversely affected antioxidant properties of the pearled wheat and the resultant by-product was always higher in antioxidant activity than the respective pearled wheat regardless of wheat class.

6.2.3   Storage of Wheat and Wheat Products

Wheat is harvested once a year, but it is consumed year-round. Normally wheat is stored in bins until being used. Likewise the processed grains or ingredients are also put in storage before being further processed into foods. Under normal storage conditions wheat can retain its grain quality after years of storage, while the processed wheat ingredients can maintain their baking quality for months. Factors such as humidity, temperature and initial moisture content of stored products directly affect shelf life of wheat grains and processed wheat ingredients. Poor storage conditions or high moisture content could lead to heat damaged grain and/or moldy wheat, which render them unsuitable for human consumption.

Effect of storage on antioxidants is not well studied although bread-baking quality improves as wheat grain or flour ages during proper storage. The aging effect is in part attributed to oxidation that facilitates disulfide bond formation. Oxidation can expend antioxidants resulting in reduced antioxidant activity. Cheng et al. (10) stored wheat grain under different temperatures (25, 60, and 100°C) for 9 days in an accelerated storage test model and found that the total phenolic compound and antioxidant activity did not change during the entire period regardless of the assay methods used (oxygen radical absorbing capacity (ORAC), DPPH scavenging capacity, 2,2-azinobis (3-ethylbenothiazoline-6-sulfonic acid) (ABTS) cation scavenging capacity). But the wheat bran exhibited a downward trend losing more antioxidant activity with increased storage time and elevated temperature. In contrary, Martinez-Tome et al. (11) reported no changes in antioxidant activity of wheat bran stored at 40°C for 28 days when measuring linoleic acid oxidation as an indicator. Effect of temperature under storage on total phenolic compounds and antioxidant activity varied significantly depending on the material stored, particle size, wheat variety, or assay methods used for antioxidant activity (Table 6.2) (10).

6.2.4   Other Postharvest Treatments

Grinding of wheat bran to smaller particles is sometimes needed in food processing to improve uniformity, quality, mouthfeel of food products, and so on. Reduction of wheat bran particle size facilitates phytochemical release, thus enhancing the available and maybe bioavailable antioxidant activity (11). But grinding exposes phytochemicals to oxidation if not protected resulting in shorter shelf life and loss of antioxidant activity (10).

Preheat treatment of wheat bran or germ sometimes is necessary because of its vulnerability to rancidity. Some of the indigenous enzymes such as lipase and lipoxygenase in these materials have been identified as the culprit. Heat treatment under low moisture will likely inactivate the responsible enzymes without significantly changing the baking property of these products. Additionally, wheat flour used for ice cream dough, baby foods, etc. requires low microbial counts. Thermal treatment of wheat flours is also practiced in the industry for such applications. Normally, the temperature used in the preheat treatment is relatively low not exceeding 100°C, and heat exposure time is short. However, severe preheat treatment conditions (higher temperature and longer processing time) are also attempted as an alternative to chlorine bleaching to alter flour properties for cake application in countries where chlorine gas is banned. Thermal treatment causes phytochemical degradation, oxidation, and Maillard reactions resulting in changes in antioxidant property (10). Steaming wheat milling by-products for 15 min induces 25–37% drop in ferulic and p-coumaric acid content of the reconstituted whole-wheat flour, but total phenolic compounds and antioxidant activity measured by ORAC, DPPH, and ABTS scavenging capacity are not significantly changed by the preheat treatment (Table 6.3) (12). Further, effect of thermal treatment on antioxidant property is discussed later in the chapter.

TABLE 6.2   Percent Changes of Antioxidant Activities Between Two Temperature Regimens, 25 and 100°C for 9 days, Affected by Wheat Cultivars and Samples[a] (10)

Wheat cultivar	Samples	ORAC, %	DPPC quenched, %	ABTS, %	TPC, %
Trego	Grain	+32.50	+14.40	−52.65	−29.27
	Bran	−87.65	−47.02	−95.85	−51.28
	40-M bran	−96.95	−59.37	−97.85	−62.50
Ankor	Grain	−6.71	+17.50	−25.59	−13.16
	Bran	−47.92	−37.96	−80.15	−50.72
	40-M bran	−70.87	−54.67	−54.95	−48.00

[a]Percent changes were calculated as $100 \times (\text{value}_{25\,C} - \text{value}_{100\,C})/\text{value}_{25\,C}$.

TABLE 6.3 Phenolic acids, Total Phenolics, and Antioxidant Activities of Whole-wheat Flours Containing Milling By-products Treated With or Without Steam for a Soft Wheat Blend and Madison Cultivar (12)

	Blend		Madison	
Soft red winter wheat	Control	Steamed[a]	Control	Steamed[a]
Ferulic acid, µg/g	841.57	632.01	718.67	451.87
p-Coumaric acid, µg/g	27.04	20.29	21.54	14.02
Syringic acid, µg/g	13.59	16.36	18.44	16.24
Vanillic acid, µg/g	15.36	13.86	12.76	12.08
Total phenolics, GA µg/g[b]	874.29	843.56	941.17	979.23
ORAC, TE µmol/g[c]	58.80	55.17	57.97	58.06
ABTS, TE µmol/g[d]	5.74	5.85	5.55	5.78
DPPH, % remaining[e]	29.46	27.23	36.17	30.05
Iron chelating, EDTA Eq µg/g[f]	134.11	176.23	161.58	181.54

[a]Milling by-products were heat treated with steam for 15 min and reconstituted with the respective refined
[b]flour to make the whole-wheat flour.
[c]Total phenolics were measured as gallic acid equivalent in µg/g.
[d]ORAC values were expressed as trolox equivalent in µmol/g.
[e]DPPH radical scavenging capacity was expressed as trolox equivalent in µmol/g.
[f]Radical cation ABTS was measured as trolox equivalent in µmol/g.
Iron chelating activity was expressed as EDTA equivalent in µg/g.

Bioavailability of phytochemicals is a concern because only free and soluble conjugated phenolic acids are thought to be available for absorption to optimize potential health benefits of antioxidants. Pretreatment of wheat or wheat products with enzymes in a liquid environment is reported to free phenolic acids from fiber esters (13,14), thus improving the bioavailability of wheat antioxidants. Moore et al. (15) treated wheat bran with different commercially available enzyme preparations including β-glucanase, carboxylic esterase, polygalacturonase, aminopepridase, and cellulase in a solid-state fermentation model. They found that there were significant variations among the enzyme preparations in their ability to increase free phenolic acid concentration with the β-glucanase from *Humicola insolens* exhibiting the highest activity, converting as much as 50% of insoluble bound ferulic acid in wheat bran to the soluble free form. Significant dose-dependent increase in extractable antioxidant properties and interaction between enzyme concentration and moisture level were conspicuous. It was concluded that solid-state enzyme treatment of wheat bran may be a commercially viable postharvest procedure for improving the bioavailability of wheat antioxidants (15).

6.3 FOOD FORMULATIONS

With a few exceptions, wheat-based foods consumed in the United States are formulated with other ingredients prior to further food processing to improve appearance, product quality, flavor, taste, mouthfeel, and/or nutrition. Inclusion of other food ingredients is likely to influence antioxidant property of wheat through dilution, addition, and synergistic and/or antagonistic reaction. Martinez-Tome et al. (11) formulated a bran-based breakfast cereal containing 87% wheat bran, 2% malt flour, 4% brown sugar, and 1.4% salt. The formulated cereal had 4%–8% less antioxidant activities than 100% wheat bran, using various assays including lipid peroxidation, deoxyribose, peroxidase, Rancimat, and ABTS$^{\cdot+}$ scavenging capacity. After extrusion at 2 atm, and 150°C, the finished cereal product when compared to the unprocessed bran was 42% better in protection of butter from oxidation using Rancimat method, similar in cation ABTS radical scavenging capacity, 1% and 2% lower in inhibition of lipid peroxidation and deoxyribose damage, respectively, and lost 92% of inhibition using peroxidase assay. A different formulation using higher percentage of wheat bran (94.8%) had the same antioxidant properties as 100% bran, but the extruded products were different from original bran and the previous cereal product in protection of lipid oxidation, deoxyribose damage, scavenging of hydrogen peroxide, and butter oxidation as well as cation ABTS radical scavenging capacity. Yu et al. (12) used a standard cookie formulation (16) of which whole-wheat flours accounted for 50% of the total dry weight. They found that the levels of individual phenolic acids were about 50% less in cookie doughs than the corresponding flours. But total phenolic compounds and antioxidant activities in the formulated cookie doughs were not just the percentage of respective whole-wheat flours. Total phenolic compounds were only 3–5% less than the flours, while antioxidant activities varied dramatically depending on the assay methods. DPPH radical scavenging capacity of the cookie doughs was three to four times less, ORAC values and radical cation ABTS radical scavenging activity ranged from 48% to 74% lower, and iron chelating activity was either 22% less or two times higher when compared to the respective flours. The differences in total phenolic concentration and antioxidant activity between the formulated doughs and the flours were attributed to the addition of phytochemicals from other ingredients and possible interactions between these antioxidants.

Food formulation determines available reducing sugars, water activity, and pH value prior to thermal treatments. Reducing sugars are the substrates for Maillard reaction, whereas pH and water activity may significantly alter the degree of Maillard reaction that takes place under high temperature. Chemicals produced by Maillard reaction contribute to significantly increased antioxidant activity of final food products. High level of available reducing sugars, optimal water activity (a_w) (0.3–0.7), and alkaline condition will favor the Maillard reaction and thus increase the antioxidant activity of the foods (17–19). Furthermore, Maillard reaction products may protect phytochemicals from oxidation. Billaud et al. (20) reported that Maillard reaction products derived from mixtures of glucose or fructose and cysteine or glutathione greatly inhibited activities of polyphenoloxidases and oxidoreductases. Sugar type affected the inhibitory effect with pentoses being the best, followed by sucrose, hexoses, and maltose. In the presence of glucose, simultaneous presence of an amino group, a carboxyl group, and a free thiol group on the same molecule seemed essential for the production of highly inhibitory compounds (20).

Inclusion of other ingredients containing antioxidants in food formulations helps preserve original wheat phytochemicals from loss resulted from oxidation. Oil or fat imparts volume, texture, and eating property to wheat-based bakery products, and thus is a common ingredient in food formulations. Many shortening products used in food formulations are fortified with pure natural or synthetic antioxidants. These extraneous antioxidants vary in their capability of extending oil and ultimately the food shelf life. Ottaway and Coppock (21) using a biscuit formulation found that dodecyl gallate alone and butylated hydroxyanisole added with synergists had approximate similar antioxidant properties when incorporated into the oil component at the maximum permissible level. Butylated hydroxytoluene, alone or with synergists, was somewhat inferior to dodecyl gallate and butylated hydroxyanisole under comparable conditions. Butylated hydroxyanisole was slightly superior to dodecyl gallate in animal fats and vice versa in vegetable fats. Replacing 2% of the wheat flour in the formula with full-fat debittered soy flour increased the resistance of biscuits to oxidative deterioration. Hsu et al. (22) included up to 20% of yam flour to replace wheat flour in a bread formulation. They discovered that the bread containing yum flour had higher antioxidant capacity as tested with both DPPH free radical scavenging capacity and total antioxidant concentration although the bread volume decreased with increased yum flour. The increased level of antioxidant activity was attributed to the yum flour.

Some minerals, being added in food formulation or contaminated from wear or tear of processing equipment, are not antioxidants. However, they may be directly or indirectly involved in oxidation-reducing reactions, and thus alter the antioxidant properties. For example, iron is a catalyst for oxidations (23), and selenium is a component of antioxidant enzymes (24,25). Wheat is one of the major human dietary sources of selenium.

Balance of acid and alkaline in food formulations to achieve certain pH levels is one of important factors imparting foods' unique characteristics. Without exception, pH conditions influence wheat antioxidant activity and bioavailability. Platt and Clydesdale (26) simulated gastrointestinal pH conditions *in vitro* and studied their effect on the antioxidant activity of ready-to-eat breakfast cereals. They found that gastrointestinal pH conditions significantly increased antioxidant activity of the high bran and whole-grain cereal extracts. Acid conditions changed the antioxidant property in activity, composition, and concentration of water-soluble low molecular weight phytochemicals (27,28). It is suggested that acid hydrolysis influences soluble phenolic concentrations by causing the release of free phenolic acid from fiber in a similar manner to the increase in phenolic solubility from corn bran by alkaline hydrolysis as reported by Ohta and coworkers (27,29).

Wheat antioxidant activity is highly correlated to the total phenolic contents (4,30). These phenolic compounds exist in three forms — bound, free, and conjugated. The bound phenolics account for 99% of the total wheat phenolics, but contribute only to about 82% of total antioxidant activity (25,31). Bulk of wheat phenolics including ferulic acid and flavonoids are more likely to exert their health benefits when they are released. For example, rutin and naringin are not absorbed into the blood until they are hydrolyzed from their sugar counterparts by enzymes originating from intestinal microbes (27,32). Enzymes as part of dough conditioner have been widely used in the baking industry to improve dough property, baking quality, and shelf life (33–35). Xylanase, a common additive in food formulations, promotes hydrolysis of arabinoxylans to produce feruloylated oligosaccharides that in turn are hydrolyzed by ferulic acid esterase to produce free ferulic acid (36). Feruloylated oligosaccharides and free ferulic acid have higher antioxidant activity than feruloylated polysaccharides. However, some oxidative enzymes such as glucose oxidase promote linkage between two ferulic acid molecules, causing possible reduction of total phenolics content. Several wheat indigenous oxidases including lipoxygenase, polyphenol oxidase, and peroxidase that are highly concentrated in wheat bran (37) are also known to be involved in the destruction of carotenoid pigments during dough mixing, resulting in decreased antioxidant activity (38).

Wheat antioxidants can be divided into soluble (water soluble and fat soluble) and insoluble compounds. Both hydrophilic and lipophilic phytochemicals make significant contributions to total wheat antioxidant activity (3). Food formulations depending on water, oil, or combinations of both could potentially influence the bioavailability and activity of antioxidants. Food ingredient forms and the ways to be added into formulations may also alter antioxidant properties (20,39).

6.4   FOOD HEAT PROCESSING

Wheat grain and the milling products for human consumption in the United States are subjected to some type of heat processing. This may include baking, frying, extrusion, flaking, and puffing. Baking is defined as cooking by dry heat in an oven, for example, bread and/or cookie baking. Frying is a cooking method in fat or oil in a pan or griddle. Extrusion cooking is an industrial process used widely in production of foods. During the processing, simultaneous actions of temperature, pressure, and shear occur and their intensities are varied enormously depending on the types of ingredients and characteristics the end products are to have. Flaking is used to make wheat flakes for breakfast cereal. Wheat grains are cooked with water under pressure followed by delumping, partial drying, heating, and tempering. The grits are then rolled into flakes and subjected to an additional cooking process. Wheat can also be puffed for foods. A portion of the bran may be removed in puffing whole grains. The wetted wheat is loaded into puffing guns where the grains are heated at very high pressure before being rapidly released into ambient pressure. The internal pressurized steam in the grain is released suddenly causing puffing to occur. There are many cooking methods for food processing, but they all involve heat or temperature.

Heat processing is known to affect nutrients such as carbohydrates, protein in foods, but its effect on phytochemical content and antioxidant activity is not well investigated. Results from other food systems showed that thermal treatment significantly reduced concentration of natural antioxidants, but the overall antioxidant properties of food products were maintained or even enhanced by the development of Maillard products (1).

Maillard reaction, also called sugar-amino browning reaction, which is a form of nonenzymatic browning, is a chemical reaction between an amino acid and reducing sugar under heating conditions. The reactive carbonyl group of the sugar interacts with the nucleophilic amino group of the amino acid to create hundreds of different compounds. Maillard reaction products are effective antioxidants. The extent of Maillard reaction ultimately determining the amount of Maillard products depends on the available substrates, water activity, pH condition, and temperature. Therefore, any formulations and/or process conditions to accelerate Maillard reaction may exert antioxidant activity of the final foods. According to Lee and Kwon (23), the optimum reaction conditions predicted to produce the greatest levels of active antioxidative and antimutagenic melanoidins were 155.0°C for 6.3 h at a pH of 12.1 and 155.5–160.4°C for 4.6–9.7 h at pH 10.4–12.5, respectively.

Exposure of wheat bran to high temperature (100°C) resulted in partial loss of total phenolic compounds and significant decline in antioxidant activity (10). The change in antioxidant properties was attributed to chemical reactions such as thermal degradation, oxidation, and Maillard reactions. Petrofsky et al. (40) studied the processing effect on antioxidant activity of white wheat flour using an extrusion model system simulating a puffed whole-grain cereal or snack piece. They found that antioxidant activity measured as DPPH scavenging capacity in extruded samples increased with increasing extruder barrel temperature from 180 to 200°C. Water feeding rates during the extrusion process also influenced antioxidant activity of extruded wheat through its effect on the rate of Maillard browning, for example, antioxidant activity increased from 2092 trolox equivalent (TE) per 100 g at 6 kg water per hour to 5762 TE/100 g at 1.5 kg/h under a constant barrel temperature (180°C). They concluded that extrusion processing of whole-wheat flour increased the antioxidant activity, and as extrusion process conditions increased in severity, the antioxidant activity of wheat was enhanced.

Baking as one of the primary cooking methods for wheat-based food products also affects phytochemical concentration and antioxidant activity. Yu et al. (12) studied the cookie-baking effect on phytochemical concentration and antioxidant activity of whole-wheat cookies. Two whole-wheat flours were selected: one being milled from a commercial wheat blend and the other from a pure Madison cultivar. A standard procedure (16) was used to prepare the dough that was made of flour, 47.6%, sucrose, 27.5%, shortening, 13.5%, water, 10%, sodium bicarbonate, 0.5%, sodium chloride, 0.4%, and dextrose, 0.4%. Part of the dough was freeze-dried and the rest was baked into cookies at 205°C for 10 min. The freeze-dried dough and cookies were analyzed for phenolic acids, total phenolic compounds, and antioxidant activities using ORAC, DPPH, ABTS, and iron chelating. Baking markedly reduced the concentration of ferulic and p-coumaric acid and total phenolic compounds, while the minor phenolic acids—syringic and vanillic acids, were not affected (Table 6.4). However, antioxidant activities increased by 13–118% and iron chelating activity went up by 91–287% during the baking process. Partial loss of major phenolic acids and total phenolic compounds for antioxidant activity may have been compensated by the formation of new antioxidants such as Maillard reaction products during baking. Krings (41) indicated that Amadori compounds, amino reductones, and Maillard-type polymers accumulated during Maillard reaction were responsible for the increase in antioxidant activity of heat-treated wheat germ obtained at elevated roasting temperature from 160 to 200°C for 20 min although the original stock of antioxidative compounds were found to have greatly reduced by roasting. However, Gelinas and McKinnon (42) reported that bread baking at 177°C for 20 min slightly increased the concentration of total phenolic compounds. Crust of white bread was higher than the respective crumb in total phenolic compound, but not of the whole-wheat bread. Extending the baking time did not change the total phenolic contents.

TABLE 6.4   Effect of Cookie Baking on Phenolic Acids, Total Phenolics, and Antioxidant Activities of Dough and Cookie Made of Whole-wheat Flours from a Soft Wheat Blend and Madison Cultivar[a] (12)

Soft red winter wheat	Blend		Madison	
	Dough	Cookie	Dough	Cookie
Ferulic acid, µg/g	397.65	359.22	419.91	364.08
p-Coumaric acid, µg/g	13.02	11.53	11.73	10.09
Syringic acid, µg/g	6.88	6.43	7.91	7.31
Vanillic acid, µg/g	7.56	7.54	7.62	6.69
Total phenolics, GA µg/g[b]	843.89	696.97	894.20	708.35
ORAC, TE µmol/g[c]	30.48	34.53	15.16	33.08
ABTS, TE µmol/g[d]	2.05	3.35	2.24	3.16
DPPH, % remaining[e]	124.90	113.91	122.48	115.80
Iron chelating, EDTA Eq µg/g[f]	95.91	371.58	85.14	162.91

[a]Cookies were baked according to AACC 10–50D. All values were on a dry basis.
[b]Total phenolics were measured as gallic acid equivalent in µg/g.
[c]ORAC values were expressed as trolox equivalent in µmol/g.
[d]DPPH radical scavenging capacity was expressed as trolox equivalent in µmol/g.
[e]Radical cation ABTS was measured as trolox equivalent in µmol/g.
[f]Iron chelating activity was expressed as EDTA equivalent in µg/g.

Bread-making process causes significant loss of carotenoids of wheat. Leenhardt et al. (38) reported that dough mixing resulted in the loss of 49% of total carotenoids present in the whole durum wheat meal and 15% in semolina (the endosperm of durum wheat grain). The disappearance of the carotenoids was attributed to the presence of the wheat indigenous oxidases and the difference in the level of the enzymes present. Baking caused further reduction in carotenoid. Overall loss of carotenoid content during the entire bread-baking process was about 70% in whole-wheat bread and 33–50% in refined wheat flour bread compared to the original raw ingredients.

Baking conditions, primarily temperature and time, affect the antioxidant property of wheat-based foods. Moore and Yu (43) studied the changes in total phenolic contents and antioxidant property of whole-wheat pizza crust baked at different temperature (204 and 288°C) and time (7 and 14 min) for two wheat cultivars. Elevating temperature from 204 to 288°C while keeping baking time the same increased the total phenolic contents by 41–72%, whereas extending baking time by onefold under the same temperature raised the total phenolic concentrations by 14–25%. Same trends were observed in antioxidant activities regardless of the assay methods used when pizza crust was baked under increased temperatures for a constant time. Doubling the baking time at 204°C improved the radical DPPH scavenging capacity (RDSC) by 50–60% and ABTS scavenging capacity by 42–43%. However, ORAC and hydroxyl scavenging capacity (HOSC) either remained the same or declined when baking time was extended from 7 to 14 min at 204°C, depending on the source of whole-wheat flours (Figs. 6.3 and 6.4).

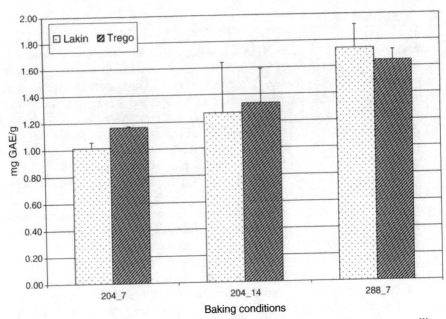

Figure 6.3   Effect of baking conditions on total phenolic contents expressed as mg gallic acid equivalent (GAE) per g of pizza crust. 204_7 = baked at 204°C for 7 min; 204_14 = baked at 204°C for 14 min; 288_7 = baked at 288°C for 7 min (43).

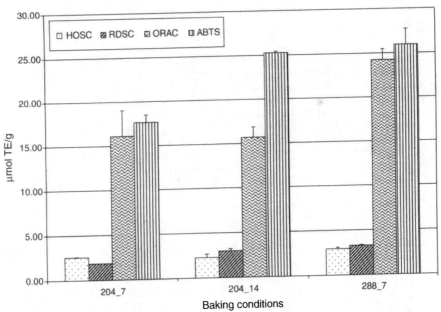

Figure 6.4   Effect of baking conditions on antioxidant activities expressed as μmol trolox equivalent (TE) per g of pizza crust. 204_7 = baked at 204°C for 7 min; 204_14 = baked at 204°C for 14 min; 288_7 = baked at 288°C for 7 min (43).

6.5 SUMMARY

Wheat grain is generally subject to various postharvest treatments before being used as an ingredient in food formulations and further processed into foods for human consumption using various cooking methods. As wheat goes through each step of the food supply chain, its antioxidant property changes and the resultant foods will have different antioxidant activity. Milling is the most common posttreatment technique in the United States, whereas debranning is a new technology that can augment flour milling. Both milling and debranning produces products with various concentrations of antioxidants. Proper selection of wheat bran or other milling by-products rich in phytochemicals will take advantages of wheat natural antioxidant properties. Other posttreatments, for example, grinding, storage conditions, preheat treatment, enzymatic hydrolysis affect wheat-indigenous phytochemical concentrations and bioavailability. Food formulations influence antioxidant property of wheat and the resultant foods either through dilution, addition, synergism, and/or antagonism. Thermal processing of various cooking methods generally reduces wheat's natural phytochemical contents, but markedly increases the antioxidant activity primarily through Maillard reaction. Many factors such as cooking temperature, time, water activity, and pH value of food preparations impact the extent of Maillard reactions and consequently the antioxidant activities. Therefore, postharvest treatments, food formulation, and processing must be considered when wheat antioxidant property is studied.

REFERENCES

1. Slavin, J. L.; Jacobs, D.; Marquart, L. *Crit. Rev. Food Sci. Nutr.* **2000**, *40* (4), 309–326.
2. Prakash, A. *Medallion Lab Anal. Prog.* **2001**, *19*, 2.
3. Adom, K. K.; Sorrells, M.; Liu, R. H. *J. Agric. Food Chem.* **2005**, *53*, 2297–2306.
4. Beta, T.; Nam, S.; Dexter, J. E.; Sapirsteim, H. D. *Cereal Chem.* **2005**, *82* (4), 390–393.
5. Liyana-Pathirana, C. M.; Shahidi, F. *J. Agric. Food Chem.* **2005**, *53*, 2433–2440.
6. Liyana-Pathirana, C. M.; Shahidi, F. *Food Chem.* **2007**, *101*, 1151–1157.
7. Dexter, J. E.; Wood, P. J. *Trends Food Sci. Technol.* **1996**, *7*, 35–41.
8. Fellers, D. A.; Mossman, A. P.; Johnston, P. H.; Wheeler, E. L. *Cereal Chem.* **1976**, *53* (3), 308–317.
9. Liyana-Pathirana, C. L.; Dexter, J.; Shahidi, F. *J. Agric. Food Chem.* **2006**, *54*, 6177–6184.
10. Cheng, Z.; Su, L.; Moore, J.; Zhou, K.; Luther, M.; Yin, J. J.; Yu, L. *J. Agric. Food Chem.* **2006**, *53*, 2433–2440.
11. Martinez-Tome, M.; Murcia, A.; Frega, N.; Ruggieri, S.; Jimenez, A. M.; Roses, F.; Parras, P. *J. Agric. Food Chem.* **2004**, *52*, 4690–4699.
12. Yu, L. L.; Guo, G.; Lin, C. J. **2005**. Unpublished data.
13. Kroon, P. A.; Garcicia-Conesa, M. T.; Hazlewood, G. P.; Williamson, G. *J. Sci. Food Agr.* **1999**, *79*, 428–434.
14. Sorensen, H. R.; Meyer, A. S.; Pedersen, S. *Biotechnol. Bioeng.* **2003**, *81*, 726–731.
15. Moore, J.; Cheng, Z. H.; Su, L.; Yu, L. L. *J. Agric. Food Chem.* **2006**, *54*, 9032–9045.
16. American Association of Cereal Chemists. AACC Approved Methods, 10th edn. Minneapolis, MN, **2000**.
17. Camire, M. E.; Dougherty, M. P.; Briggs, J. L. *Cereal Chem.* **2005**, *82*, 666–670.
18. Yokota, A.; Miyata, K.; Muraguchi, H.; Takahashi, A. *Nippon Nogeikagaku Kaishi.* **1987**, *61*, 1273–1278.

19. Eichner, K.; Karel, M. *J. Agric. Food Chem.* **1972**, *20*, 218–223.
20. Billaud, C.; Maraschin, C.; Peyrat-Maillard, M. N.; Nicolas, J. *Ann. N. Y. Acad. Sci.* **2005**, *1043*, 876–885.
21. Ottaway, F. J. H.; Coppock, J. B. M. *J. Sci. Food Agric.* **1958**, *9*, 294–299.
22. Hsu, C. L.; Hurang, S. L.; Chen, W.; Weng, Y. M.; Tseng, C. Y. *Int. J. Food Sci. Tech.* **2004**, *39*, 231–238.
23. Lee, G. D.; Kwon, J. H. *Int. J. Food Sci. Tech.* **1998**, *33*, 375–383.
24. The world healthiest foods *www.whfoods.com.*
25. Liu, R. H. *J. Nutr.* **2004**, *134*, 3479–3485.
26. Platt, S. R.; Clydesdale, F. M. *J. Food Sci.* **1984**, *49*, 531–535.
27. Baublis, A. J.; Lu, C.; Clydesdale, F. M.; Decker, E. A. *J. Am. Coll. Nutr.* **2000**, *19*, 308S–311S.
28. Baublis, A.; Decker, E. A.; Clydesdale, F. M. *Food Chem* .**2000**, *68*, 1–6.
29. Ohta, T.; Yamasaki, S.; Egashira, Y.; Sanada, H. *J. Agric. Food Chem.* **1994**, *42*, 653–656.
30. Velioglu, Y. S.; Mazza, G.; Gao, L.; Oomah, B. D. *J. Agric. Food Chem.* **1998**, *46*, 4113–4117.
31. Manach, C.; Morand, C.; Texier, O.; Favier, M. L.; Agullo, G.; Demigne, C.; Regerat, F.; Remesy, C. *J. Nutr.* **1995**, *125*, 1911–1922.
32. Adom, K. K.; Sorrells, M. E.; Liu, R. H. *J. Sci. Food Agric.* **2003**, *51*, 7825–7834.
33. Van Benschop, C.; Fatula, B. Proceedings of World Grains Summit: Foods and Beverages, San Francisco, 18–20 September, **2006**. p. 141.
34. Hille, J. D. R. *Innovation Food Tech* .**2005**, *26*, 56–58.
35. Yu, L.; Perret, J. *J. Agric. Food Chem.* **2003**, *51*, 492–495.
36. Pietta, P.; Simonetti, P.; Mauri, P. *J. Agric. Food Chem.* **1998**, *46*, 4487–4490.
37. Rani, K. U.; Prasada Rao, U. J. S.; Leelavathi, K.; Haridas Rao, P. *J. Cereal Sci.* **2001**, *34*, 233–242.
38. Leenhardt, F.; Mijalovsky, A.; Lyan, B.; Chanliaud, E.; Remesy, C. Proceedings International Workshop on Modeling Quality Traits and Their Genetic Variability for Wheat, Session 3, Clermont-Ferrant, France, 18–21 July, **2004**.
39. Park, H.; Seib, P. A.; Chung, O. K. *Cereal Chem* .**1997**, *74*, 207–211.
40. Petrofsky, K. E.; Chevrel, M. E.; Everson, K. A.; Rigelhof, F.; Fulcher, R.; Marquart, L. Proceedings of World Grains Summit: Foods and Beverages, San Francisco, 18–20 September, **2006**, p.162.
41. Krings, U.; El-Saharty, Y. S.; El-Zeany, B.A.; Pabel, B.; Berger, R. G. *Food Chem.* **2000**, *71*, 91–95.
42. Gélinas, P.; McKinnon, C. M. *Int. J. Food Sci. Tech.* **2006**, *41*, 329–332.
43. Moore, J.; Yu, L. L. Personal communications.

CHAPTER 7

ANTIOXIDANT PROPERTIES OF WHEAT-BASED BREAKFAST FOODS

Elizabeth A. Arndt
Marla W. Luther

7.1 INTRODUCTION

Of the cereal grains grown around the world, the most common is wheat, which accounts for one-third of the total grain crop world wide (1). In addition to providing energy, whole grains contain several key nutrients, including fibers, lipids, B vitamins, and minerals as well as other beneficial components such as phenolic acids, flavonoids, lignans, and phytosterols. Several of these components function as antioxidants. While multiple studies have been conducted to analyze antioxidants in different grains, the focus of this chapter is to review the antioxidants in wheat-based breakfast foods, including the effects of formulation, processing, and storage on the quantity and bioavailability of antioxidants in the final product.

7.2 WHOLE GRAINS FOR HEALTH & WELLNESS

On the basis of scientific evidence, the Dietary Guidelines for Americans 2005 report recommends increased consumption of whole grains as part of a health-promoting diet. Research indicates that a diet rich in whole-grain foods may reduce the risk of coronary heart disease, type 2 diabetes, and certain cancers and help with weight management. At least half of daily grain intake should be from whole grains, or at least three ounce-equivalent servings for a 2000-calorie diet (2). Increased consumption of whole-grain foods is also recommended by the American Heart Association (3), the American Diabetes Association (4), and the American Cancer Society (5). The FDA has approved the health claim for foods containing at least 51% whole grain by product weight and meeting other criteria: "Diets rich in whole-grain foods and other plant foods and low in total fat, saturated fat, and cholesterol may help reduce the risk of heart disease and some cancers" (6,7).

Wheat Antioxidants, Edited by Liangli Yu
Copyright © 2008 John Wiley & Sons, Inc.

7.3 GRAINS CLASSIFICATION AND CONSUMPTION

Cereal grains are the seeds (fruits) of grasses from the Poaceae family. Wheat, corn, rice, oats, barley, rye, millet, sorghum, teff, wild rice, canary seed, fonio, and Job's tears are cereal grains. Amaranth, buckwheat, and quinoa are pseudocereal grains, which are not cereal grasses but are included in the grain category because of similarities in composition and in use. Legumes and oilseeds are not considered as grains (8,9).

Based on 2004 USDA ERS data (10), total grain per capita consumption adjusted for losses is 167 g/day. More than 70% of the total grain intake is wheat-based grain at 117.5 g/day. Consumption of corn, rice, and oats followed at 27, 17.9, and 3.05 g/day, respectively. Barley and rye are each consumed at less than 0.5 g/day.

Whole grains are the entire seed, also known as a caryopsis or kernel, and must contain all three parts of the seed—the endosperm, bran, and germ—along with the naturally occurring nutrients of the entire seed. In addition to whole kernel forms, whole grains can be processed by cutting, cracking, crushing, flaking, or grinding into flour or cooked by methods including steaming and extrusion (11,12). Food manufacturers are encouraged to use the term "whole" or "whole grain" in the ingredient name to help consumers identify foods that are formulated with whole-grain ingredients.

According to Mintel Global New Products Database (12,13), launches of new food products with a "whole-grain" claim have grown considerably since 2000. Nearly 1400 products made with whole grains were launched through mid-November 2006 compared to 154 products in 2000. More than 90% of the product introductions in 2006 were bakery products, breakfast cereals, snacks, and side dishes (Table 7.1). Despite

TABLE 7.1 New Product Introductions with a "Whole Grain" Claim[a]

Category	2000	2001	2002	2003	2004	2005	2006
Baby food	3	3	6	1	8	0	20
Bakery	80	109	149	205	309	291	570
Beverages	0	0	0	1	2	2	3
Breakfast cereals	31	80	68	98	138	192	347
Confectionery	3	2	1	1	2	1	0
Dairy	3	4	4	5	3	3	10
Dessert and ice cream	0	0	0	0	0	2	2
Fruit and vegetables	0	0	0	0	2	0	1
Meals and meal centers	7	3	12	9	25	20	65
Pet food	1	1	0	0	6	7	14
Processed fish, meat, egg	7	1	0	2	8	2	2
Sauces and seasonings	0	1	1	1	1	6	9
Side dishes	16	18	45	33	44	93	117
Snacks	3	22	15	30	58	83	227
Soup	0	0	1	0	3	1	0
Spreads	0	0	0	1	0	1	0
Total	154	244	302	386	609	704	1387

Adapted from References (12) and (13).
[a]Through November 10, 2006.

increased availability of foods made with whole grains and strong recommendations to increase whole-grain consumption, Americans currently consume an average of one daily serving of whole grains, which is one third of the recommended intake. Less than 10% of adult Americans meet the recommended whole-grain consumption level (14).

7.4   WHEAT TYPES, MORPHOLOGY, AND COMPOSITION

Wheat has a long history of cultivation, dating back to at least 6750 B.C., and is the most important grain food in the temperate zones of the world.

Wheat is classified into at least 10 Triticum species, of which six are cultivated. *Triticum aestivum L.* includes five subspecies, most notably common wheat (*Triticum aestivum L.* subspecies *aestivum*). Common wheat is the source of most of the cultivated wheat varieties in the United States and may be spring or winter type, red or white, as well as hard or soft. *Club* is another subspecies of common wheat that is cultivated in the United States. Other Triticum species include durum, spelt, emmer, einkorn, and Kamut®. Einkorn, emmer, and spelt are hulled wheats (15).

The wheat endosperm is approximately 82% of the kernel weight and primarily consists of starch as well as protein. The endosperm functions as a source of energy used in the germination and the early growth of a new plant. The aleurone is the outermost cell layer of the endosperm; however, the composition is more similar to the outer bran layers. The aleurone cell walls are known to contain phenolic acids, in particular ferulic acid. The bran is about 15% of the kernel weight and is comprised of several cell layers. The bran contains approximately 16% protein, 65% carbohydrates (40% fiber), 6% ash, and 4% lipid, as well as several vitamins, minerals, and other phytonutrients, including ferulic acid. A 15 g serving of wheat bran provides 25% of the daily value of fiber. The germ is 2–3% of the kernel weight and is the embryonic plant. The germ consists of the embryonic axis (primary root and shoot) and scutellum (storage and transport organ). The germ contains approximately 25% protein, 52% carbohydrates (13% fiber), 5% ash, and 10% lipid. The germ is also an important source of several vitamins, minerals, and other phytonutrients, including apigenin and luteolin-based flavonoids. A 15 g serving of wheat germ provides 13% of the daily value of vitamin E (16).

Whole-grain wheat contributes several important nutrients to the diet including dietary fiber, several B vitamins (thiamin, riboflavin, niacin, and folic acid), vitamin E, iron, magnesium, selenium, manganese, and chromium. Three whole-grain ounce-equivalents based on whole wheat (16 g per ounce-equivalent) provides needed nutrients including 23% of the daily value of dietary fiber, 15% of magnesium, 6% of potassium, 50% of selenium, and 90% of manganese (16).

7.5   ROLE OF ANTIOXIDANTS IN WHEAT AND OTHER CEREAL GRAINS

In wheat and other grains, naturally occurring phytonutrients with antioxidant activity include phenolic acids, flavonoids, coumarins, polyphenols, phytates, terpenes, carotenoids, tocopherols, and tocotrienols (17). Potential roles of the naturally occurring antioxidants in grains include free radical scavenger, reducing agent, complexation of prooxidant metals, quenchers of singlet-oxygen molecules and antioxidant defense enzyme system activators for biological systems (18,19).

As discussed by Baublis and coworkers (20), grains, such as wheat, contain several components that have potential antioxidant activity and thus may have a protective role in human health. These components include phytates, polysaccharides, proteins, and phenolics. Compared to phytate and phenolics, much less is known about the role and mechanisms of grain-based polysaccharides and proteins as antioxidants in food systems or in human health.

7.6   WHEAT MILLING AND DISTRIBUTION OF ANTIOXIDANTS

In the traditional wheat milling process, the goal is to produce refined wheat flour by maximizing the separation of the starchy endosperm from the endosperm aleurone layer, bran, and germ.

Research has shown that antioxidants are concentrated in the outer parts of the kernel, particularly the endosperm aleurone layer, pericarp, nucellar envelope, and germ. The combined bran and germ fractions from three wheat varieties, two of which were each grown in different locations, were found to contain 15–18 times more total phenolics, 52–70 times more ferulic acid, and on average 4 times more lutein, 12 times more zeaxanthin, and 2 times more cryptoxanthin as compared to the endosperm fractions (21,22).

Leenhardt and coworkers (23) found that 86% of the carotenoids were recovered in the refined flour portion that represented 74.3% of the kernel weight. In this work, the germ was not recovered as a separate fraction. Lutein is reported to be distributed throughout the kernel, whereas zeaxanthin is localized in the germ. The carotenoid content of wheat appears to vary based on genetics with diploid and tetraploid wheat varieties, such as einkorn and durum, respectively, containing higher levels of lutein compared to common bread wheat varieties.

TABLE 7.2   Total Antioxidant Capacity of Wheat Milling Fractions[a]

Wheat Milling Fraction	µM TE/100 g
Refined flour	1450 ($n = 2$)
Coarse wheat bran	3500 ($n = 3$)
Red dog flour	4300 ($n = 1$)
Shorts	4200 ($n = 1$)
Wheat bran and germ flour	7400 ($n = 1$)
Wheat germ	8400 ($n = 6$)

[a]Total antioxidant capacity (TAC) is the sum of the hydrophilic and lipophilic portions expressed as (mole Trolox equivalents/100 g of roller mill separated wheat fractions.

Wheat milling fractions were analyzed for both hydrophilic and lipophilic antioxidant fractions and were reported as a collective total antioxidant capacity (μmole Trolox equivalents/100 g) (Table 7.2). Wheat germ had the highest average total antioxidant capacity (8400; 7300–9400), which was 2.4 times higher compared to coarse wheat bran (3500; 3200–3600) and more than 5.7 times higher than the refined wheat flour fraction (1450; 1000–1900). The bran samples analyzed were from hard red winter, hard red spring, and soft white winter wheat classes, while the germ analyzed also included a sample from hard white winter wheat. The total antioxidant capacities of red dog flour, shorts, and wheat bran and germ flour were intermediate between the germ and coarse bran fractions, which is reflective of the mixed composition of each of these mill fractions.

7.7   WHEAT-BASED BREAKFAST FOODS

Traditional wheat-based breakfast foods include hot and cold ready-to-eat (RTE) cereals, as well as a variety of yeast-leavened and chemically leavened bakery products. Popular yeast-leavened products that may be eaten for the breakfast meal include pan and artisan-style breads, buns, rolls, pastries, danishes, and croissants. Likewise, popular chemically leavened baked goods include quick breads, muffins, pancakes, waffles, and tortillas. Wheat-based foods used for the breakfast meal include refined, partial whole-wheat, and 100% whole-wheat and multigrain forms. High bran content cereals are readily available. Wheat bran and germ are available as individually packaged foods, which are generally used by consumers as breakfast cereals or as ingredients for home baking. Since 2000, there has been a considerable increase in the number of bakery and cereal product introductions that are making a "whole grain" claim. This includes both partial and 100% wholegrain products (Table 7.1).

7.8   BREAKFAST MEAL CONSUMPTION AND DEMOGRAPHICS

Breakfast consumption and its effects on health have been widely studied and accepted as a dietary pattern indicative of better food choices that may result in improved cognitive function and overall health. Despite recommendations, breakfast consumption has declined for all age groups in the United States, particularly for adolescents (24–26).

Multiple studies have analyzed the effects of commensalism, gender, age, race, education, economic disposition, and migration on breakfast consumption and food choices. Commensal eating patterns demonstrate that breakfast is the primary meal eaten alone which may suggest why it is the most frequently skipped meal, reported at a frequency of 10–30% in the United States and Europe. Some societal standards identify a true meal as one consumed in the presence of others (25,27). Gender is not considered to be a significant factor influencing breakfast consumption.

Reports on the effects of education and economic status on breakfast consumption and food choices have been inconsistent, possibly due to the availability of school breakfast programs and cost versus perceived healthfulness of RTE cereal products (28,29). Adolescents are more likely to skip breakfast than other age groups (26,30,31). A recent study by Niemeier and coworkers (32) have found that across demographic groups, the reduced incidence of breakfast consumption from adolescence to early adulthood is often coupled by increased fast food consumption. Fast food consumption was highest in African Americans among all ethnic groups. This aligns with the study by Affenito and coworkers, which suggests that dietetic professionals should promote the importance of breakfast to adolescents, particularly to African American females who are more likely to skip breakfast and consume a higher fat diet (31).

It is unclear whether breakfast consumption, particularly of whole-grain cereals and other foods affects health status, or whether health status affects food choice. A study of Irish adults found that subjects who consumed more whole-meal breads, rolls, and, breakfast cereals had a positive attitude about their efforts to eat a healthy diet, while subjects who ate white breads and had a lower cereal intake had a negative attitude about their efforts to eat a healthy diet (33).

7.9 ANTIOXIDANTS IN RTE BREAKFAST CEREALS

While the frequency of breakfast consumption has declined overall, consumption of RTE breakfast cereal products increased 60% between 1977–78 and 1994 (34). RTE breakfast cereals may increase macronutrient and micronutrient intake in individuals, as they are generally high in carbohydrates, and many are high in fiber and often contain or are fortified with vitamins and minerals (35).

Studies conducted on adolescent females found that breakfast consumption, particularly RTE cereals, increased micronutrient intake and reduced total fat and cholesterol intakes (26,30). A study of US adults concluded that fortified RTE breakfast cereals are in the top 10 food selections for 15 of 18 micronutrients. Populations that consume breakfast cereals may also have a decreased serum cholesterol and a reduced risk of calcium (largely associated with increased milk consumption), iron, riboflavin, and folate dietary inadequacy (35). RTE cereals are also sources of fiber and many other phytonutrients; however, the phytonutrient or antioxidant load of these foods has rarely been taken into account for their ability to improve overall health.

Miller and coworkers (36) determined the antioxidant content in a variety of fruits, vegetables, breakfast cereals, and breads. The average antioxidant content of the RTE breakfast cereals was higher than that of the fruits and vegetables tested, except for berries. The antioxidant content of whole-grain cereal with raisins was only slightly lower than that of berries. Among the grain-based foods tested, whole-grain cereals and breads were higher in antioxidant content than their refined grain counterparts. Also, bran-based RTE cereals were higher in antioxidant content than 100% whole-wheat cereals.

On average, a 41 g serving of RTE breakfast cereal is comparable in antioxidant activity to the average daily intake of fruits or vegetables. Antioxidant activity measured in μmoles Trolox equivalents/100 g ranged from 1300 to 3900 among 17

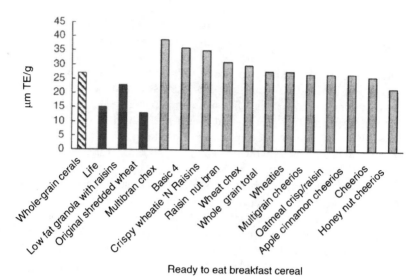

Figure 7.1 Trolox equivalent values for RTE breakfast cereals in the United States (striped bar 49, solid black bars 39, shaded bars 36).

different refined, whole-grain, and bran-based RTE cereals (36) (Fig. 7.1). Products produced with wheat bran have higher levels of antioxidant activity, hydroxyl and hydrogen peroxide scavenging activity, and lipid peroxidation inhibition than those produced from oat bran (37). Additionally, RTE breakfast cereals appear to be a stable source of antioxidants that do not substantially diminish during normal storage conditions, with limited losses even under temperature abuse (36).

Researchers have compared the antioxidant activities of a variety of foods, including vegetables, fruits, chocolate, nuts and seeds, spices, along with breakfast cereals, breads, and other grain-based products (36,38,39). When compared on an equal weight basis, spices are generally very high in antioxidant activity, followed by cocoa products and berries. There is a wide range of antioxidant activity among fruits and vegetables, although berries are usually among the highest of fruits and vegetables. The antioxidant activity range of whole grain and bran-based RTE cereals is generally higher than many common vegetables and comparable to many fruits.

Baublis and coworkers (40) studied the antioxidant effects of RTE cereals based on wheat bran, whole-grain wheat, and refined wheat using aqueous extracts prepared from crushed cereal samples. Breakfast cereals are commonly fortified with iron, which may function as a prooxidant. Using a phosphatidylcholine liposome model, water-soluble extracts of both whole wheat and wheat bran-based RTE breakfast cereals were shown to inhibit lipid oxidation promoted by iron and peroxyl radicals. This system was used to emulate an artificial cellular membrane system, that was considered more predictive of in vivo activity compared to other bulk or emulsified lipid systems. The wheat bran-based aqueous extract had the highest antioxidant activity, although not significantly different from the whole wheat-based extract. Under the conditions tested, the refined wheat RTE cereal aqueous extract showed little antioxidant capacity. The wheat bran-based RTE cereal extract was separated into low and high molecular weight fractions using ultrafiltration. Both fractions inhibited lipid oxidation, although higher concentrations of low molecular weight extract were required. The antioxidant capacities were greatly increased by subjecting the whole wheat and wheat bran-based extracts to pH changes similar to the human gastrointestinal system, suggesting that antioxidant capacity in grain-based foods may be enhanced by digestion. The components found in whole wheat and other whole grains that may have antioxidant activity include phytate, polysaccharides, proteins, and phenolics (in decreasing order of activity: ferulic acid, vanillic acid, and *p*-coumaric acid). The free and polysaccharide-esterified phenolic acids appear to have the greatest potential of these components. Importantly, the extract levels tested were representative of concentrations that would be consumed in a normal serving of wheat-based RTE cereal.

It is difficult to make direct comparisons between products analyzed in different studies since there is variability among product formulas and resulting compositions, as well as in the analytical methods used. A descriptor such as whole-grain cereal or high fiber bran cereal does not reflect potential formulation differences among these products that may affect antioxidant activity. Therefore, it is important to look for overall trends when comparing antioxidant contents of similar products reported in the literature.

7.10 ANTIOXIDANTS AND BREAD MAKING

Bread making results in losses of vitamins and antioxidants (41). A study on French wheat bread making found that there were significant losses in tocopherols and tocotrienols during the dough formation and baking stages (41). Leenhardt and coworkers (23) measured carotenoids and vitamin E during kneading, fermentation, and baking using bread, einkorn, and durum wheat. Vitamin E content decreased 30% during bread preparation with losses attributed to oxygenation during dough production and heat destruction during baking. There were minimal differences between wheat species. On the contrary, carotenoid losses in the kneading stage varied by species and were directly correlated with lipoxygenase activity. In bread wheat, the carotenoid content decreased 66% in whole and 40% in refined grain dough after kneading. By comparison, there was less than 10% loss in carotenoids in whole-grain einkorn dough after kneading. Einkorn has a much higher carotenoid content and a lower lipoxygenase activity compared to bread wheat. Carotenoid losses during fermentation and baking were about 10% and 36–45%, respectively, and did not differ appreciably between wheat species. Choice of wheat species, as well as breeding common (bread) wheat varieties with higher carotenoid levels may serve as approaches to increase the overall phytonutrient content of breads. The losses reported in the study by Leenhardt and coworkers (23) are higher than other baking studies reporting carotenoid losses of approximately 20% for bagels and cakes, 30% for cookies, and total carotene losses of 4–15% in bread products and 18–23% in crackers (42,43). Vitamin E and tocopherol levels were also measured in various Finnish bread products, which comprise a substantial contribution of vitamin E intake of Finns (Table 7.3). Bread-making formulations and practices may have a significant effect on the retention or loss of many phytonutrients.

TABLE 7.3 Tocopherols in Wheat Bread[*]

Product	α–T	α–T3	–T	–T3	–T	–T3	–T	Vitamin E mg/α–T eq.
Wheat flour (1.2–1.4% ash)[a]	1.6	0.3	0.8	1.7				2.1
Wheat bran[a]	1.6	1.5	0.8	5.6				2.7
Wheat germ[a]	22.1	0.3	8.6	1.0			<0.1	25.7
Wheat bread[b]	0.4	0.1	0.2	0.9	0.3		0.1	0.6
Wheat bread, dark[b]	0.6	0.1	0.3	1.0	0.2		0.1	0.8
Sweet wheat bread (7% fat)[b]	0.4	0.1	0.1	0.8	0.4	<0.1	0.1	0.5
Sweet wheat bread (10% fat)[b]	0.8	0.1	0.2	0.8	1.0	<0.1	0.2	1.1

[*]Adapted from Reference (48). Reported as Tocopherols (T) and Tocotrienols (T3), in mg/100 g fresh product.
[a]Data averaged from 1981 and 1982.
[b]Data averaged across 1981–1983.

Fermentation processes, such as those used in sourdough bread production, have variable effects on specific bioactive compounds. While limited data are available, flour that has undergone sourdough fermentation has higher folate and total phenolic content after extraction compared with unfermented flour (44). In contrast, the amount of alkylresorcinols (ARs) was diminished when wheat flour was made into sourdough wheat bread; however, it is important to note that complete extraction of ARs may be difficult due to complexation with starch in the food matrix (45). In a whole-wheat bread system, sourdough fermentation resulted in greater reduction in phytate (62%) compared to yeast fermentation (38%), while increasing the solubility of magnesium and phosphorus (46). The dough fermentation stage of bread making may maintain more phytonutrients than other bread processing conditions such as kneading, which introduces oxygen.

Baking conditions may also alter antioxidant content. Increasing baking time and temperature increases antioxidant activity in the bread crust and lowers antioxidant activity in the bread crumb portion (47). Maillard reaction products have received significant attention as they may produce potential carcinogens; however, study of bakery product fractions also indicates that areas subject to the browning reaction have higher antioxidative potential in contrast to their corresponding inner counterparts (47).

Halvorsen and coworkers (38) found that the antioxidant content of whole-wheat bread is more than 1.9 times higher compared to refined wheat bread. Additionally, the process of toasting bread was found to increase the antioxidant content 134–367% higher in bagels, French bread, white bread, and whole-wheat bread compared to their untoasted counterparts. The nature of the increase in antioxidants after toasting was not determined.

7.11 CONCLUSION

Wheat and wheat-based products, particularly partial and 100% whole-grain products, have been shown to provide a variety of beneficial components including fiber, vitamins, minerals, and other phytonutrients with antioxidant activity. Further work is needed to understand the effects of formulation, processing, and storage on the quantity and bioavailability of the naturally occurring antioxidants in finished products. An additional critical factor is that the final product must have sensory properties that are acceptable to consumers. Recently, whole-wheat and other whole-grain products have become more commonplace as research suggests that consumption of these products may have health beneficial effects. As reviewed in the Dietary Guidelines for Americans 2005 report, scientific evidence indicates that consumption of three or more ounce-equivalents of whole-grain foods per day may reduce the risk of coronary heart disease, type 2 diabetes, and certain cancers and help with weight management.

REFERENCES

1. Slavin, J. L.; Jacobs, D.; Marquart, L. Grain processing and nutrition. *Crit. Rev. in Biotech.* **2001**, *21*, 49–66.
2. U.S. Departments of Health and Human Services and U.S. Department of Agriculture. Dietary Guidelines for Americans, 2005, 6th. edn. U.S. Government Printing Office, Washington, DC, January **2005**.
3. American Heart Association. Our 2006 Diet and Lifestyle Recommendations, **2006**. http://www.americanheart.org/presenter.jhtml?identifier=851.
4. American Diabetes Association. Making Healthy Food Choices, **2007**. http://www.diabetes.org/nutrition-and-recipes/nutrition/healthyfoodchoices.jsp.
5. American Cancer Society. Diet and Physical Activity: What's the Cancer Connection? **2006**. http://www.cancer.org/docroot/PED/content/PED_3_1x_Link_Between_Lifestyle_and_CancerMarch03.asp.
6. FDA and CFSAN. Health claim notification for whole grain foods. Docket No. 99P-2209.www.cfsan.fda.gov/dms/flgrains.html.U.S. Government Printing Office,Washington, DC, **1999**.
7. FDA and CFSAN. Health claim notification for whole grain foods with moderate fat content. Docket No. 03Q-0547.www.cfsan.fda.gov/dms/flgrain2.html.U.S. Government Printing Office,Washington, DC, **2003**.
8. Food Labeling and Standards Staff; Office of Nutritional Products, Labeling, and Dietary Supplements; Center for Food Safety and Applied Nutrition; Food and Drug Administration. Whole Grain Label Statements – Draft Guidance. **2006**. http://www.cfsan.fda.gov/~dms/flgragui.html.
9. US Food and Drug Administration. 2006D-0066: Guidance for Industry and FDA Staff: Whole Grains Label Statements. C15 AACC International Task Force, **2006**. http://www.fda.gov/ohrms/dockets/dockets/06d0066/06d-0066-c000015-01-vol1.pdf.
10. U.S. Department of Agriculture, Economic Research Services. Food Availability (Per Capita) Data System: Food Guide Pyramid, **2007**. http://www.ers.usda.gov/Data/FoodConsumption/FoodGuideIndex.htm#grain.
11. American Association of Cereal Chemists. 'Letter to the U.S. Food and Drug Administration', **2004**. http://www.aaccnet.org/definitions/wholegrain.asp.
12. Whole Grains Council, A Consumer Definition of Whole Grains, **2006**. http://www.wholegrainscouncil.org/.
13. Mintel Global New Products Database. Global New Product Database, Monitoring New Trends and Innovation, **2007**. http://www.gnpd.com/sinatra/gnpd/frontpage/.
14. Cleveland, L. E.; Moshfegh, A. J.; Albertson, A. M.; Goldman, J. D. Dietary intake of whole grains. *J. Am. Coll. Nutr.* **2000**, *19*, 331–338.
15. Purdue University, Center for New Crops and Plants Products. Wheat, **1997**. http://www.hort.purdue.edu/newcrop/crops/wheat.html.
16. USDA National Nutrient Database for Standard Reference, Release 19, **2007**. http://www.ars.usda.gov/main/site_main.htm?modecode=12-35-45-00.

17. Shahidi, F. Functional foods: Their role in health promotion and disease prevention. *J. Food Sci.* **2004**, *69*, 146–149.

18. Andlauer, W.; Furst, P. Antioxidant power of phytochemicals with special reference to cereals. *Cereal Foods World.* **1998**, *43*, 356–360.

19. Zielinski, H.; Kozlowska, H. Antioxidant activity and total phenolics in selected cereal grains and their different morphological fractions. *J. Agric. Food Chem.* **2000**, *48*, 2008–2016.

20. Baublis, A. J.; Clydesdale, F. M.; Decker, E. A. Antioxidants in wheat-based breakfast cereals. *Cereal Foods World.* **2000**, *45*, 71.

21. Adom, K. K.; Sorrells, M. E.; Liu, R. H. Phytochemicals and antioxidant activity of milled fractions of different wheat varieties. *J. Agric. Food Chem.* **2005**, *53*, 2297–2306.

22. Moore, J.; Hao, Z.; Zhou, K.; Luther, M.; Costa, J.; Yu, L. Carotenoid, tocopherol, phenolic acids, and antioxidant properties of Maryland-grown soft wheat. *J. Agric. Food Chem.* **2005**, *53*, 6649–6657.

23. Leenhardt, F.; Lyan, B.; Rock, E.; Boussard, A.; Potus, J.; Chanliaud, E.; Remesy, C. Wheat lipoxygenase activity induces greater loss of carotenoids than vitamin E during breadmaking. *J. Agric. Food Chem.* **2006**, *54*, 1710–1715.

24. Nicklas, T. A.; Myers, L.; Reger, C.; Beech, B.; Berenson, G. S. Impact of breakfast consumption on nutritional adequacy of the diets of young adults in Bogalusa, Louisiana: Ethnic and gender contracts. *J. Am. Diet. Assoc.* **1998**, *98*, 1432–1438.

25. Rampersaud, G. C.; Pereira, M. A.; Girard, B. L.; Adams, J.; Metzl, J. D. Breakfast habits, nutritional status, body weight, and academic performance in children and adolescents. *J. Amer. Diet. Assoc.* **2005**, *105*, 742–760.

26. Song, W. O.; Chun, O. K.; Obayashi, S.; Cho, S.; Chung, C. E. Is consumption of breakfast associated with body mass index in US adults? *J. Amer. Diet. Assoc.* **2005**, *105*, 1373–1382.

27. Sobal, J.; Nelson, M. K. Commensal eating patterns: A community study. *Appetite* **2003**, *41*, 181–190.

28. Golub, A.; Binkley, J. Determinants of household choice of breakfast cereals: Healthy or unhealthy? *American Agricultural Economics Association Annual Meeting*, Providence Rhode Island, July 24–27. **2005**.

29. Crepinsek, M. K.; Singh, A.; Bernstein, L. S.; McLaughlin, J. E. Dietary effects of universal-free school breakfast: Findings from the evaluation of the school breakfast program pilot project. *J. Amer. Diet. Assoc.* **2006**, *106*, 1796–1803.

30. Haines, P. S.; Guilkey, D. K.; Popkin, B. M. Trends in breakfast consumption of US adults between 1965 and 1991. *J. Amer. Diet. Assoc.* **1996**, *96*, 464–470.

31. Affenito, S. G.; Thompson, D. R.; Barton, B. A.; Franko, D. L.; Daniels, S. R.; Obarzanek, E.; Schreiber, G. B.; Striegel-Moore, R. H. Breakfast consumption by African-American and White adolescent girls correlates positively with calcium and fiber intake and negatively with body mass index. *J. Amer. Diet. Assoc.* **2005**, *105*, 938–945.

32. Niemeier, H. M.; Raynor, H. A.; Lloyd-Richardson, E. E.; Rogers, M. L.; Wing, R. R. Fast food consumption and breakfast skipping: Predictors of weight gain from adolescence to adulthood in a nationally representative sample. *J. Adol. Health.* **2006**, *39*, 842–849.

33. Hearty, A. P.; McCarthy, S. N.; Kearney, J. M.; Gibney, M. J. Relationship between healthy eating and dietary behavior, lifestyle, and demographic factors in a representative sample of Irish adults. *Appetite* **2007**, *48*, 1–11.

34. U.S. Department of Agriculture. Agricultural Fact Book 1998, **1998**. http://www.usda.gov/news/pubs/fbook98/content.htm.

35. Galvin, M. A.; Kiely, M.; Flynn, A. Impact of ready-to-eat breakfast cereal (RTEBC) consumption on adequacy of micronutrient intakes and compliances with dietary recommendations in Irish adults. *Public Health Nutr.* **2003**, *6*, 351–363.

36. Miller, H. E.; Rigelhof, F.; Marquart, L.; Prakash, A.; Kanter, M. Antioxidant content of whole grain breakfast cereals, fruits and vegetables. *J. Am. Col. Nutr.* **2000**, *19*, 312S–319S.

37. Martinez-Tome, M.; Murcia, M. A.; Frega, N.; Ruggieri, S.; Jiménez, A. M.; Roses, F.; Parras, P. Evaluation of antioxidant capacity of cereal brans. *J. Agric. Food Chem.* **2004**, *52*, 4690–4699.

38. Halvorsen, B. L.; Carlsen, M. H.; Phillips, K. M.; Bohn, S. K.; Holte, K.; Jacobs, D. R., Jr.; Blomhoff, R. Content of redox-active compounds (ie, antioxidants) in foods consumed in the United States. *Am. J. Clin. Nutr.* **2006**, *84*, 95–135.

39. Wu, X.; Beecher, G. R.; Holden, J. M.; Haytowitz, D. B.; Gebhardt, S. E.; Prior, R. L. Lipophilic and hydrophilic antioxidant capacities of common foods in the United States. *J. Agric. Food Chem.* **2004**, *52*, 4026–4037.

40. Baublis, A.; Decker, E. A.; Clydesdale, F. M. Antioxidant effect of aqueous extracts from wheat based ready-to-eat breakfast cereals. *Food Chem.* **2000**, *68*, 1–6.

41. Wennermark, B.; Jagerstad, M. Breadmaking and storage of various wheat fractions affect vitamin E. *J. Food Sci.* **1992**, *57*, 1205–1209.

42. Park, H.; Seib, P. A.; Chung, O. K.; Seitz, L. M. Fortifying bread with each of three antioxidants. *Cereal Chem* .**1997**, *74*, 202–206.

43. Ranhotra, G. S.; Gelroth, J. A.; Langemeier, J.; Rogers, D. E. Stability and contribution of beta carotene added to whole wheat bread and crackers. *Cereal Chem.* **1995**, *72*, 139–141.

44. Katina, K.; Arendt, E.; Liukkonen, K.-H.; Autio, K.; Flander, L.; Poutanen, K. Potential of sourdough for healthier cereal products. *Trend Food Sci. Technol.* **2005**, *16*, 104–112.

45. Ross, A. B.; Kamal-Eldin, A.; Aman, P. Dietary Alkylresorcinols: Absorption, bioactives, and possible use as biomarkers of whole-grain wheat- and rye-rich foods. *Nutr. Rev.* **2004**, *62*, 81–95.

46. Lopez, H. W.; Krespine, V.; Guy, C.; Messager, A.; Demigne, C.; Remesy, C. Prolonged fermentation of whole wheat sourdough reduces phytate level and increases soluble magnesium. *J. Agric. Food Chem.* **2001**, *49*, 2657–2662.

47. Lindenmeier, M.; Hofmann, T. Influence of baking conditions and precursor supplementation on the amounts of the antioxidant pronyl-L-lysine in bakery products. *J. Agric. Food Chem.* **2004**, *52*, 350–354.

48. Piironen, V.; Syvaoja, E.-L.; Varo, P.; Salminen, K.; Koivistoinen, P. Tocopherols and tocotrienols in cereal products from Finland. *Cereal Chem.* **1986**, *63*, 78–81.

49. Slavin, J. Why whole grains are protective: biological mechanisms. *Proc Nut. Soc.* **2003**, *62*, 129–134.

CHAPTER 8

EFFECTS OF EXTRACTION METHOD AND CONDITIONS ON WHEAT ANTIOXIDANT ACTIVITY ESTIMATION

Jeffrey Moore
Liangli (Lucy) Yu

8.1 INTRODUCTION

Generally, any analytical work involves two steps, sample preparation and measurement. It is well recognized that mistakes in the sample preparation step cannot be corrected during measurement or determination step. Extraction is a critical step involved in antioxidant property estimation of wheat-based food ingredients and food products. The extraction of antioxidants from wheat and wheat-based agricultural materials may also be a critical step for possible commercial production of antioxidant ingredients. Several factors are commonly considered when selecting an appropriate extraction procedure. These factors include, but are not limited to, the purpose of extraction such as for analysis or production, nature of the components to be extracted, physicochemical natural of the matrix, duration, equipment(s) needed, required personal training, repeatability, waste disposal, safety, and cost. An extraction procedure generally includes the extraction methods such as Soxhlet extraction and the extracting conditions including solvent system, time, pH, and temperature during the extraction. While an ideal extraction usually transfers all desired compounds from a sample matrix into the extract, the diverse antioxidant compositions of wheat and wheat-based food products makes the situation more complicated. Wheat naturally contains numerous classes of antioxidant compounds briefly including phenolic acids, steryl ferulates, lignans, carotenoids, tocopherols, and tocotrienols (1–13). Many of these compounds such as phenolic acids may exist in multiple forms as free, esterified, or glycosylated, and additionally as complexes with proteins and other plant materials. In wheat-based food products, food matrix and processing effects are also likely to affect the physicochemical nature and extractability of wheat antioxidants. Additionally, the concentration of each antioxidant compound in wheat grain and fractions also depends on the genotype, growing conditions, and interactions between genotype and

Wheat Antioxidants, Edited by Liangli Yu
Copyright © 2008 John Wiley & Sons, Inc.

environment. Because of these factors, a number of extraction conditions have been reported and used by different wheat antioxidant research groups for several purposes and for a large number of wheat samples. This has made interlaboratory comparisons of wheat antioxidant properties difficult, and created a need for standardized and optimized extraction procedure(s) specific for different wheat fractions or wheat matrix and antioxidant assay combinations. In addition to optimized extraction procedures for wheat antioxidant property estimations, it is also of interest for commercial purposes to develop optimized extraction procedures that can be safely and economically utilized to produce wheat-based antioxidant ingredients to be used as food ingredients or dietary supplements.

This chapter will review and discuss the different extraction methods and extraction solvents along with other extraction conditions reported for wheat antioxidant extraction, their influences on antioxidant property estimation, and potential implications for commercial extraction processes. In addition, general considerations for pre-extraction sample preparation and their potential effects on wheat antioxidant extraction will be discussed. Recommendations for extraction methods and conditions will be made for specific wheat matrix and antioxidant assay combinations, as well as for commercial extraction purposes. The aim of this chapter is to promote the standardization of wheat antioxidant extraction procedures allowing for inter-laboratory comparisons of research results, to better understand and explain the research results on wheat antioxidants from individual groups, and to provide recommendations for those interested in the commercial extraction of wheat antioxidants.

8.2   EXTRACTION METHODS AND CONDITIONS

Reports to date evaluating wheat antioxidant properties have utilized several different extraction methods and conditions as summarized in Table 8.1. The multiple extraction procedures used in these studies can partially be attributed to the diversity of the targeted antioxidant compounds. For example, some studies are specifically interested in lipophilic carotenoids or tocopherols for quantitative analysis versus studies while others are designed to extract nonspecific antioxidant compounds and evaluate the general antioxidant properties of a wheat-based material. No systematic validation of extraction procedure for wheat antioxidants was conducted until Zhou and Yu reported the effects of extraction solvent on wheat antioxidant property estimation in 2004 (14). It is therefore not surprising that even for a specific wheat material being analyzed using a specific antioxidant property assay, multiple extraction procedures might have been used. Zhou and coworkers (5), for example, reported the use of a Soxhlet extraction with ethanol used for total phenolic content (TPC) measurements of wheat bran, while Zielinski and coworkers (15) reported a simple solvent extraction with water for the same wheat material, a different extraction method and an extraction solvent with significantly different properties. This has made it difficult to compare results for similar wheat materials using similar assays between laboratories.

To help understand the effects of these factors on wheat antioxidant extraction, studies evaluating the effects of individual factors such as extraction method or solvent

TABLE 8.1 Reported Wheat Antioxidant Extraction Methods and Conditions

Wheat material	Antioxidant estimation assay(s)	Extraction method	Extraction solvent composition	Extraction time, temp, pH			Particle size, mesh	Solid: solvent ratio	References
Red wheat and fractions	ORAC, DPPH·, ABTS·+, TPC, $O_2^{·-}$, Fe^{2+} chelating, phenolic acid composition	Solvent extraction	Acetone/water (50:50 v/v)	15 h	Ambient	NA	FP	1:10 w/v	5
Hard wheat grain, wheat cereal products	Rancimat OSI, DPPH, Fe^{2+} Chelating, ABTS·+, TPC, DPPH·, ESR DPPH·	Soxhlet extraction	100% Ethanol	3 h	NA	NA	FP	NA	16,24,25
Wheat bran, flour	LDL, ORAC, DPPH·, Fe^{2+} Chelating, ABTS·+, $O_2^{·-}$, TPC, phenolic acid composition	Solvent extraction	100% Ethanol	15 h	Ambient	NA	FP	1:10 w/v	3,14,16,25–27
Wheat bran, soft wheat grain	ORAC, DPPH·, ESR DPPH·, ESR ·OH, ABTS·+, TPC, $O_2^{·-}$, Fe^{2+} chelating, Fe^{2+} chelating with ESR, Cu^{2+} chelating with ESR, phenolic acid composition, ·OH.	Solvent extraction	50% acetone	15 h	Ambient	NA	100 mesh	1:10 w/v	1,2,4,20
Wheat bran	TPC, phenolic acids composition	Defatted bran used in simple solvent extraction	80% MeOH	1 h	Ambient	NA	32 mesh	1:5 w/v	12
Wheat grain	Phenolic acids content, TPC, flavonoid content, TOSC	Solvent extraction in blender	80% MeOH	10 min	Chilled ethanol	NA	NA	1:2 w/v, twice	6
Wheat grains	TPC, total flavonoid content, TOSC, phenolic acids composition	Solvent extraction in blender	80% MeOH	10 min	Chilled ethanol	NA	60 mesh	1:4 w/v, twice	7
Wheat grain and fractions	TPC, flavonoids content, Hydro-PSC	Solvent extraction under basic conditions to hydrolyze bound and conjugated antioxidants	2 M NaOH solution	1 h	NA	NA	60 mesh	NA	8

Sample	Assays	Extraction method	Solvent	Time	Temperature		Mesh	Ratio	Ref.
Wheat grain and fractions	Lipo-PSC	Solvent extraction of lipophilic wheat compounds	MeOH/THF (1:1, v/v)	5 min	75°C	NA	60 mesh	3:10 w/v, twice	8
Wheat and fractions	ABTS•+, PC liposome, TPC	solvent extraction with shaking	Water	40 min	0°C	NA	NA	1:10 w/v	15
Wheat and fractions	ABTS•+, PC liposome, TPC	solvent extraction with shaking	80% MeOH	40 min	20°C	NA	NA	1:10 w/v	15
Red and white wheat grains	TPC	Solvent extraction	HCl/MeOH/water (1:80:10, v/v/v)	2 h	Ambient	NA	20 mesh	1:20 w/v	28
Red and white wheat grains	DPPH•	Solvent extraction	100% MeOH	2 h	Ambient	NA	NA	NA	28
Chinese black-grained wheat	TPC, DPPH•	Solvent extraction with shaking	100% MeOH	16 h	Ambient	NA	NA	1:10 w/v	11
Soft and hard winter wheat and fractions	TPC, TEAC, DPPH•, β-C-L, Fe^{2+} chelating, LDL	Defatted wheat samples followed by solvent extraction	80% EtOH	16 h	4°C	NA	16 mesh	1:10 w/v	29
Soft wheat grain and fractions	TPC, TEAC, DPPH•, LDL, reducing power, ORAC, Fe^{2+} chelating, PCL, rancimat OSI, DNA,	Defatted wheat samples followed by solvent extraction using homogenizer	MeOH/Acetone/Water (7:7:6, v/v/v)	15 s	Ambient	NA	16 mesh	1:20 w/v, six times	22
Durum wheat and fractions	PV, phenolic acids composition	Solvent extraction of defatted wheat materials with shaking	95% EtOH	1 h	Ambient	NA	NA	1:5 w/v	23
Wheat bran and bran cereal products	Liposome oxidation, OH, rancimat OSI, peroxides, TEAC,	Solvent extraction while stirring	Water	30 min	NA	NA	NA	1:5 w/v	30
Wheat-based cereal products	Lipid oxidation using TBARS	Solvent extraction with stirring	Water	30 min	NA	NA	NA	~1:15 w/v	31
Wheat and fractions	β-C-L, TPC, DPPH•, reducing Power	Solvent extraction	80% MeOH	15 min	80°C	NA	NA	NA	32,33
Wheat flour and bran	FRAP, DPPH•, TPC	Solvent extraction with shaking	Acidic methanol/water (50:50)	1 h	Ambient	2	NA	1:10 w/v	34
Soft wheat grain	Phenolic acid composition	Solvent extraction	Acetone/MeOH/water (7:7:6, v/v/v)	15 h	Ambient	NA	NA	1:10 w/v	2

(continued)

TABLE 8.1 (*Continued*)

Wheat material	Antioxidant estimation assay(s)	Extraction method	Extraction solvent composition	Extraction time, temp. pH			Particle size, mesh	Solid: solvent ratio	References
Wheat grain	Tocols	Saponification followed by solvent extraction	n-hexane/ethyl acetate (9:1, v/v)	NA	NA	NA	NA	2:15 w/v, twice	9
Wheat grain	Carotenoids content	Solvent extraction with stirring	100% MeOH	2 h	NA	NA	50 mesh	1:20 w/v	35
Wheat bran, soft wheat grain	Carotenoids and tocopherols contents	Solvent extraction with sonication	MeOH/THF (50:50, v/v)	15 h	Ambient	NA	FP	1:50 w/v	1,2,4
Wheat bran	Steryl ferulate content	Solvent extraction with shaking	100% Acetone	1 h	NA	NA	NA	1:10 w/v	10
Wheat bran	Lignans	Solvent extraction	Methanol/water (70:30, v/v)	4 h	70 C	NA	64 grit	1:10 w/v	36
Blue-grained wheat grain	Anthocyanins	Solvent extraction	MeOH/1M HCl (85:15, v/v)	NA	NA	NA	NA	NA	37

OSI stands for oxidative stability index assay, LDL stands for low-density lipoprotein oxidation inhibition assay, β-C-L stands for β-carotene linoleate oxidation inhibition assay, ORAC stands for oxygen radical absorbing capacity, DPPH⁺ stands for DPPH radical scavenging capacity, ESR stands for electron spin resonance assay, ˙OH stands for hydroxyl radical scavenging assay, ABTS˙⁺ stands for cation ABTS radial scavenging capacity, TEAC stands for trolox equivalent antioxidant capacity, TPC stands for total phenolic contents using Folin–Ciocalteu reagent, β-CLAMS stands for β-carotene linoleic acid model system, PV stands for peroxide value, HC stands for conjugated diene hydroperoxides content, PC liposome stands for L-α-phosphatidylcholine liposome system, Lipo-PSC stands for lipophilic peroxyradical scavenging capacity, Hydro-PSC stands for hydrophilic peroxyradical scavenging capacity, PCL stands for photochemiluminescence antioxidant activity assay, DNA stands for inhibition of supercoiled DNA strand breakage, NA stands for information not available, and FP stands for fine powder.

will be discussed in the following sections. For the purpose of laboratory antioxidant property evaluations of wheat samples, an ideal extraction procedure should selectively and effectively extract the antioxidant compounds that are measured in the assay without solvent interference with the assay reaction system, while being simple enough to carry out efficiently on a large number of samples. When considering an extraction for commercial purposes, however, other factors must be taken into account including the availability and cost of equipment and solvents, along with solvent cost, safety, and recyclability that affect the viability of the commercial application (16).

8.2.1 Effects of Extraction Method

As seen in Table 8.1, simple continuous solvent extraction at ambient temperature and neutral pH is the most commonly used extraction method for wheat antioxidants, while a few studies utilizing Soxhlet extraction have been reported. The simple continuous solvent extraction generally includes (a) mixing the solid material with known weight and desired particle size with certain volume of a selected solvent system, (b) keeping the mixture at ambient temperature for a certain time period for extraction, generally 15 h, and (c) collecting the supernatant by filtration or centrifugation. The simple continuous extraction requires neither special apparatus nor operation, may be performed with single or mixtures of solvents, and is safe and ideal for extraction of a large number of samples. This procedure also involves no heat and can be performed under nitrogen if necessary, a good choice for antioxidant extraction because of their thermal and oxidative instability. The Soxhlet extraction uses the Soxhlet extractor and is a semicontinuous procedure involving heat. Soxhlet extraction has higher extracting efficiency and sometimes may use less solvent especially for a large sample volume. Two studies have compared these two extraction methods for different wheat fractions and antioxidant property assays. A study by Krings and others (17) compared the antioxidant activities of wheat germ extracts prepared by simple continuous solvent or Soxhlet extraction methods for 16 h with ethanol. Antioxidant activities were expressed as the capacity to suppress the lipid peroxidation of stripped corn oil by measuring the formation of peroxides, conjugated dienes, and the loss of α-tocopherols (17). This study showed no significant differences between the two extraction methods, indicating that Soxhlet extraction had no clear advantage for extracting wheat germ antioxidants capable of inhibiting lipid oxidation in corn oil and other possible edible oils and fats. This observation may be due to the solid–solvent ratio, physicochemical nature and concentration of the antioxidants in the wheat germ, and the conditions of the accelerated lipid peroxidation testing conditions. It needs to be pointed out that Soxhlet extraction with same solvent generally has better extraction efficiency.

A second study by Zhou and Yu (14) evaluated two varieties of hard wheat bran and compared Soxhlet extraction using ethanol to a variety of simple solvent extractions including pure ethanol for their resulting extract antioxidant properties using four popular antioxidant assays. Antioxidant properties measured included oxygen radical absorbing capacity (ORAC), TPC, DPPH˙ scavenging capacity, and

TABLE 8.2   Effect of Extraction Method on Hard Wheat Bran Antioxidant Activity Estimation from Zhou and Yu (14)

Method–variety	TPC, mg GE/g bran	DPPH·, % scavenged	ABTS·+, μmol TE/g bran	ORAC, μmol TE/g bran
Soxhlet–Akron	1.35	81.79	4.68	35.92
Ethanol–Arkon	0.65	58.57	3.09	23.63
Soxhlet–Trego	1.04	72.80	4.03	32.23
Ethanol–Trego	0.50	46.15	2.74	23.70

All results expressed per gram wheat bran on dry weight basis. TPC stands for total phenolic contents measured using Folin–Ciocalteu reagent with results expressed in mg gallic acid equivalent (GE) per g wheat bran. % DPPH· scavenged measured DPPH· scavenging capacity. ABTS·+ measured the scavenging capacity against this cation radical expressed as μmol trolox equivalent (TE) per g wheat bran. ORAC stands for oxygen radical absorbing capacity measuring peroxyl radical scavenging capacity.

ABTS·+ scavenging capacity. Their results, summarized in Table 8.2 showed that Soxhlet-ethanol extracts had higher antioxidant properties compared to the ethanol extracts prepared using the simple continuous extraction procedure, indicating that the Soxhlet method is more efficient in antioxidant extraction. For the purposes of commercial wheat bran antioxidant extraction, this study concluded that the Soxhlet method using ethanol was the best choice since ethanol is safe for food ingredient processing, and it allows for solvent recycling to reduce cost and solvent waste. For rapid laboratory antioxidant property estimation of many wheat bran samples, however, this study suggested simple solvent extractions using aqueous organic solvents to be a better approach since their extracts showed comparable or higher antioxidant properties, and are much easier for rapid laboratory analysis of large numbers of analytical samples.

8.2.2   Effects of Extraction Conditions

Extraction conditions including solvent system, time, temperature, pH, particle size, and solid-to-solvent ratio may significantly alter the extracting efficacy. The extraction conditions reported in wheat antioxidant extractions are summarized in Table 8.1. Of these, solvent selection is one of the most important factors, and has been most often studied for its effects on wheat antioxidant extraction and wheat antioxidant property estimation. Solvent system is selected according to the purpose of extraction, chemical nature of the interested components, and physicochemical nature of the matrix. For wheat and wheat-based edible materials, solvent systems are generally designed to extract antioxidants for antioxidant property estimations, or to quantify specific antioxidant compounds such as carotenoids, tocols, phenolic acids, lignans, or steryl ferulates. These later extraction solvent systems for specific antioxidant compound quantification will not be discussed in detail in this chapter, but will be discussed in the later chapters of this book dealing with individual antioxidant assays.

Seven recent studies to date have dealt with different extraction solvents and their effects on wheat antioxidant property estimation. In 1999, Mageed and Fadel (18) successively extracted antioxidants from wheat bran with hexane, chloroform, and ethanol using the simple continuous extraction method for 12 h at a solid-to-solvent ratio of 1 : 5 (w/v). These extracts were then compared for their capacities in suppressing lipid peroxidation in the cooked beef kept at 4°C for 7 days. The antioxidant activities were expressed as the reduced TBARS formation and loss of unsaturated fatty acids using the TBA reagent and gas chromatography, respectively. TBA and TBARS stand for thiobarbituric acid and TBA reactive substances, respectively. Their results showed ethanol to have the highest extraction yield at 1.7% (w/w) followed by hexane and chloroform at 1.5% and 1.06% (w/w), respectively. Results also showed ethanol extracts to be more effective than both hexane and chloroform extracts at inhibiting lipid oxidation during 7 days of storage. This indicates that ethanol may be more effective than less polar solvents for extracting lipophilic antioxidants capable of inhibiting the formation of aldehyde compounds generated from lipid oxidation that are detectable with the TBA test.

A second, similar study by Krings and others (17) compared ethanol, acetone, and diethyl ether extracts of roasted wheat germ for their inhibitory capacities against lipid oxidation in stripped corn oil for over 10 days. Lipid peroxidation was determined by measuring the formation of peroxides (peroxide value, PV) and diene hydroperoxides. Extraction conditions were a simple continuous solvent extraction for 16 h at a solid-to-solvent ratio of 1 : 8 (w/v). Results from this study showed ethanol extracts to be most effective in inhibiting lipid peroxidation measured using both PV and diene hydroperoxide assays, followed by diethyl ether and acetone. Similar to the study by Mageed and Fadel (18), this study found ethanol, a moderately polar solvent, to be a more effective extraction solvent than less polar solvents, although a different fraction of wheat was analyzed. This study also found ethanol to be a more effective extraction solvent than acetone for extracting lipid soluble antioxidants capable of inhibiting lipid oxidation, similar to the findings of Mageed and Fadel (18).

The next study by Zielinski and Kozlowska (15) compared water and 80% methanol (v/v) extracts of two whole wheat grain samples for their antioxidant properties including TPC measured using the Folin–Ciocalteu reagents, and ABTS[·+] scavenging capacity. Their results shown in Table 8.3, indicate that a moderately polar extraction solvent such as 80% methanol may be more effective for extracting phenolic antioxidants and ABTS[·+] scavengers from wheat grain than highly polar extraction solvents such as water.

TABLE 8.3 Effect of Extraction Solvent on Total Phenolic Contents and ABTS[·+] Scavenging Capacity Antioxidant for Whole-Wheat Grain from Zielinksi and Kozlowska (15)

Wheat variety	Extraction solvent	Total phenolic content, μg catechin/g grain	ABTS[·+] scavenging activity, μmol trolox/ g grain
Almari	Water	483.3	1.78
Henika	Water	460.0	1.33
Almari	80% methanol	471.5	2.81
Henika	80% methanol	546.6	8.07

Results calculated from the Zielinski and Kozlowska (15) report using data extraction yield data.

TABLE 8.4   Estimation of the Effect of Extraction Solvent on (a) Akron and (b) Trego Wheat Bran Antioxidant Property from Zhou and Yu (14)

Solvent/Test	ABTS$^{\cdot+}$, μmol TE/g	ORAC, μmol TE/g	TPC, mg GE/g	DPPH$^{\cdot}$, % scavenged
		Akron		
70% methanol	7.83c	34.94b	1.00c	83.13c
70% ethanol	9.23d	60.04d	0.84b	62.17b
Ethanol	3.09a	23.63a	0.65a	58.57a
50% acetone	15.26e	55.75c	2.63e	61.52b
		Trego		
70% methanol	8.83c	30.87b	1.10bc	75.4d
70% ethanol	9.76d	48.27c	1.20c	53.44c
Ethanol	2.74a	23.70a	0.50a	46.15a
50% acetone	12.04e	44.16c	2.09d	52.32b

ABTS$^{\cdot+}$, ORAC, and TPC results expressed per gram wheat bran on dry weight basis. TPC stands for total phenolic contents measured using Folin–Ciocalteu reagent with results expressed in mg gallic acid equivalent (GE) per g wheat bran. % DPPH$^{\cdot}$ scavenged measured DPPH$^{\cdot}$ scavenging capacity, calculated by subtracting % DPPH$^{\cdot}$ remaining value from 100%. ABTS$^{\cdot+}$ measured the scavenging capacity against this cation radical expressed as μmol trolox equivalent (TE) per g wheat bran. ORAC stands for oxygen radical absorbing capacity measuring peroxyl radical scavenging capacity. Values marked by the same letter in each table are not significantly different ($P < 0.05$).

Later in 2004, Zhou and coworkers (14) examined the effects of extraction solvent on wheat bran antioxidant estimation by comparing 50% acetone, 100% ethanol, 70% ethanol, and 70% methanol wheat bran extracts for their DPPH$^{\cdot}$ and ABTS$^{\cdot+}$ scavenging capacities, ORAC, and TPC values. Extraction conditions were a simple continuous solvent extraction using 80 mesh bran samples extracted for 15 h at ambient temperature with a solid-to-solvent ratio of 1 : 10 (w/v). Results (Table 8.4) for all four antioxidant assays and both wheat bran varieties tested showed that 100% ethanol had the lowest antioxidant properties. This indicated that ethanol is the least effective solvent for extracting natural phenolics, and radical scavenging agents against DPPH$^{\cdot}$, ABTS$^{\cdot+}$ and peroxyl radicals from wheat bran samples. Their results also indicated that the most effective extraction solvent was dependent on the antioxidant activity measured, with 50% acetone extracts showing the highest scavenging capacity against ABTS$^{\cdot+}$ and TPC values, while 70% ethanol extracts had slightly higher ORAC values than 50% acetone extracts. In addition, 70% methanol was the most effective solvent for extracting DPPH$^{\cdot}$ scavenging agents from both wheat bran samples. This report demonstrated the effect of extraction solvent on antioxidant capacity estimations and showed a clear need to optimize extraction solvents for each antioxidant property.

As a continuous effort, a similar study was performed in our group to examine the effects of extraction solvent on antioxidant property estimation for soft wheat grain. Eight varieties of soft wheat grain samples were ground and extracted with 50% acetone, 100% ethanol, and 80% methanol. The extracts were determined for their ABTS$^{\cdot+}$ and O$_2^{\cdot-}$ scavenging capacities, ORAC, and TPC values. The extraction

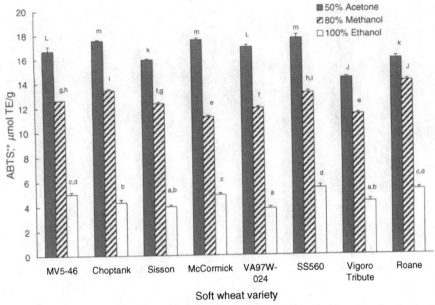

Figure 8.1 Effects of extraction solvent on the ABTS·+ scavenging capacity of soft wheat grain. Grains of eight Maryland-grown soft wheat varieties were extracted with 50% acetone, 100% ethanol, and 80% methanol at the extraction ratio of 1 : 10 (w/v) for 15 h. Grain samples and ABTS·+ scavenging capacity assay conditions were same as previously reported by Moore and coworkers (2). TE stands for trolox equivalents. All tests were conducted in triplicate, and mean values are reported. The vertical bars represent the standard deviation of each data point. Values marked by the same letter are not significantly different ($P < 0.05$), determined using ANOVA with Tukey's HSD *post hoc* testing in SPSS version 10.0.5.

conditions were similar to that used by Zhou and coworkers (14) involving a simple continuous solvent extraction for 15 h at ambient temperature with a solid-to-solvent ratio of 1 : 10 (w/v). Results, shown in Figures 8.1–8.4, indicate significant differences between extraction solvents for all antioxidant properties evaluated. 50% acetone was shown to be the most effective extraction solvent for estimating ABTS·+ scavenging capacity and ORAC values, and 80% methanol the most effective for the O_2^- scavenging components for all samples tested. Results for the TPC values showed fewer significant differences between extraction solvents, but showed a trend of 80% methanol being a more effective extraction solvent for the majority of the samples tested. Agreeing with the study from Zhou and coworkers (14), this study showed that ethanol was the least effective extraction solvent for ORAC and ABTS·+ scavenging components, and found that extraction solvents should be optimized for each antioxidant property. This notion can be understood in that each solvent system extracts different antioxidant components depending on their polarities, and each type of antioxidant compound is likely to exhibit different free radical scavenging or chelating properties depending on the nature and mechanism of the free radicals used and their reactivities with different antioxidant compounds. The different optimal extraction solvents described in this study and Zhou's report (14) also emphasize the

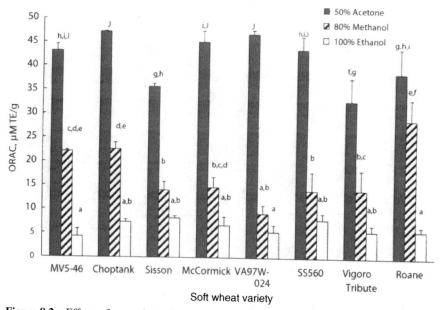

Figure 8.2 Effects of extraction solvent on the oxygen radical absorbing capacity (ORAC) of soft wheat grain. Grains of eight Maryland-grown soft wheat varieties were extracted with 50% acetone, 100% ethanol, and 80% methanol at extraction ratio of 1 : 10 (w/v) for 15 h. Grain samples and ORAC assay conditions were same as previously reported by Moore and others (2), using fluorescein as the fluorescent probe and a Victor[3] microplate reader with fluorescence detector (Perkin-Elmer, Turku, Finland). TE stands for trolox equivalents. All tests were conducted in triplicate, and mean values are reported. The vertical bars represent the standard deviation of each data point. Values marked by the same letter are not significantly different ($P < 0.05$), determined using ANOVA with Tukey's HSD *post hoc* testing in SPSS version 10.0.5.

point that extraction solvents need to be optimized for each type of wheat material analyzed. This can be understood in that each type of wheat material is likely to have different antioxidant compositions and matrices that will affect extraction.

Recently, Liyana-Pathirana and Shahidi (19) used response surface methodology (RSM) to examine the effects of solvent composition, extraction temperature, and extraction time on the ABTS[·+] scavenging capacity (trolox equivalent antioxidant capacity, TEAC) values for defatted soft and hard wheat whole-grain and bran samples. This study first involved a set of three preliminary experiments to determine appropriate extraction conditions to be tested in the RSM model. The first experiment determined an appropriate extraction solvent by comparing a range of organic solvents in combination with different ratios of water (0–100% v/v; water/ethanol, methanol, or acetone). These extraction conditions included simple solvent extractions for 20 min at 80°C using a solid-to-solvent ratio of 3 : 50 (w/v) for previously defatted samples. The proportion of organic solvent to water was found to significantly influence TEAC antioxidant properties of wheat extracts as a quadratic function with increasing organic solvent leading to increased TEAC values to a certain point, followed by decreasing TEAC values with further increased organic solvent. For all tested solvents

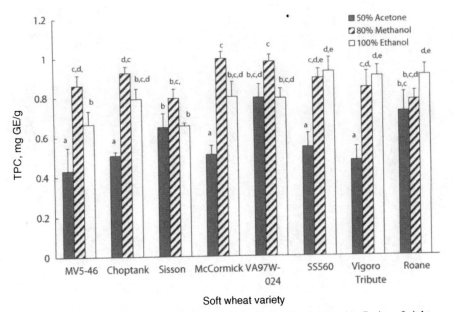

Figure 8.3 Effects of extraction solvent on the TPC of soft wheat grain. Grains of eight Maryland-grown soft wheat varieties were extracted with 50% acetone, 100% ethanol, and 80% methanol at extraction ratio of 1 : 10 (w/v) for 15 h. Grain samples and TPC assay conditions were same as previously reported by Moore and coworkers 2 using the Folin–Ciocalteu reagent. GE stands for gallic acid equivalents. All tests were conducted in triplicate, and mean values are reported. The vertical bars represent the standard deviation of each data point. Values marked by the same letter are not significantly different ($P < 0.05$), determined using ANOVA with Tukey's HSD *post hoc* testing in SPSS version 10.0.5.

and wheat materials, an approximately 1 : 1 (v/v) ratio of organic to water found to be the optimal ratio for all three organic solvents. Of the three organic solvents, ethanol was reported to be the most effective and was used in the remaining preliminary experiments and in the surface response method study. The second preliminary experiment for the Liyana-Pathirana and Shahidi (19) study examined extraction temperatures from 15 to 95°C using 50% ethanol extraction as the solvent for 20 min, and showed increasing TEAC values with increasing temperatures up to 60°C followed by a decline in TEAC values with increase in temperature for all wheat materials. This was consistent with the notion that increasing extraction temperatures up to 60°C might increase the mobilization of TEAC active compounds, followed by thermal degradation of these compounds at higher temperatures. The third preliminary experiment of this study evaluated the effects of extraction times from 15 to 105 min using a 50 % acetone extraction at 60°C. It was found that increasing extraction time from 15 to 60 min increased the TEAC values, followed by decreasing TEAC values after 70 min.

On the basis of the preliminary experiments, Liyana-Pathirana and Shahidi (19) utilized three coded levels for each extraction factor in the RSM model, including 30%, 50%, and 70% for solvent ethanol concentration, 40, 60, and 80°C for extraction temperatures, and 45, 60, and 75 min for extraction times. The results of their RSM model predicted similar optimized extraction conditions for all four TEAC-analyzed

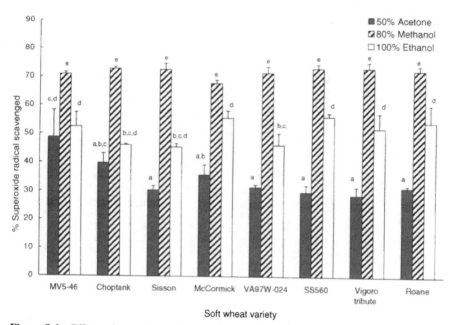

Figure 8.4   Effects of extraction solvent on the superoxide anion radical scavenging capacity of soft wheat grain. Grains of eight Maryland-grown soft wheat varieties were extracted with 50% acetone, 100% ethanol, and 80% methanol at extraction ratio of 1 : 10 (w/v) for 15 h. Grain samples were same as previously reported by Moore and coworkers 2, while $O_2^{\cdot-}$ scavenging capacity assays were conducted following the same conditions reported by Zhou and coworkers 1. This ESR method used the xanthine–xanthine oxidase system to generate $O_2^{\cdot-}$ and directly measure $O_2^{\cdot-}$ using 5-*tert*-butoxycarbonyl 5-methyl-1-pyrroline *N*-oxide (BMPO) as a spin-trapping agent. Results were calculated using amplitude of ESR spectra relative to a control, and expressed as percentage of inhibition. All tests were conducted in triplicate, and mean values are reported. The vertical bars represent the standard deviation of each data point. Values marked by the same letter are not significantly different ($P < 0.05$), determined using ANOVA with Tukey's HSD *post hoc* testing in SPSS version 10.0.5.

wheat materials, ranging from 49% to 54% for ethanol concentration, 61–64°C for optimized extraction temperatures, and 60 to 65 min for optimized extraction time. Overall, this study agrees with finding from the other discussed studies that solvent composition can significantly alter antioxidant property estimation, and that optimized extraction conditions may differ for each type of wheat material analyzed. In addition, this study demonstrated that extraction time and temperatures can have an effect on wheat antioxidant property estimation, and that response surface methodology can successfully be utilized to optimize wheat antioxidant extraction conditions. Such modeling, while not practical for optimizing extractions for laboratory antioxidant estimations of large numbers of wheat materials, may prove beneficial for commercial applications where such modeling could significantly improve manufacturing efficiency for extraction of antioxidants from a single wheat material.

Lastly, a previous study suggested the potential effects of extraction solvents on antioxidant activity estimation (20). Moore and coworkers (20) described the development of a new hydroxyl radical scavenging capacity (HOSC) assay and reported interferences that common extraction solvents could have on hydroxyl radical scavenging capacity measurements. The study described interferences encountered when attempting to utilize acetone extracts in the existing 2-deoxyribose hydroxyl radical scavenging assay. The authors found that the addition of acetone to this assay changed the absorption spectrum of the chromagen used in the assay, and thereby significantly interfered with the assay ability to quantify hydroxyl radical scavenging effects for samples extracted and analyzed with acetone. Their study also examined potential solvent interferences with hydroxyl radical scavenging estimations using an ESR (electron spin resonance) spin-trapping assay, the most direct method to examine free radical reactions. The results demonstrated that common extraction solvents such as ethanol, DMSO, or methanol are not compatible with any hydroxyl radical scavenging assay, as hydroxyl radicals can easily abstract protons from these solvent molecules generating carbon-centered radicals that interfere with hydroxyl radical scavenging measurements (20).

In summary, extraction conditions including solvent polarity, extraction temperature, and extraction time have been shown to affect antioxidant property estimation, with optimized conditions dependent on the individual antioxidant assay as well as the wheat material being analyzed. In addition, interference of extraction solvents with free radical reaction systems in radial scavenging assays has been demonstrated by hydroxyl radical scavenging assays, and is likely to occur with other types of free radicals. Therefore, it is very important for any research publications on wheat antioxidant to provide details of extraction conditions including solvent composition, extraction procedure, extraction temperature and time, ratio of wheat sample to solvent, and any pretreatments of wheat samples.

8.3 GENERAL CONSIDERATIONS FOR SAMPLE PREPARATION AND EXTRACTION

It is well recognized that numerous factors during sample extract preparation can affect the efficiency of antioxidant compound extraction from food matrices and, therefore, have a significant effect on antioxidant property evaluation (21). The other factors during wheat antioxidant extract preparation (besides the extraction solvent, time, and temperature effects already discussed) that have the potential to influence antioxidant property estimations include sampling, sample homogenization, pretreatment, ratio of wheat sample to solvent, filtration, and extract storage.

The objective of sample homogenization is to decrease particle size for optimal extraction efficiency, while minimizing thermal and oxidation sample degradation during grinding. While wheat and its milled fraction typically have moisture contents low enough to allow for efficient grinding, higher moisture wheat-based food products may require lyophilization before grinding. Grinding may be conducted using comercial coffee grinders, laboratory grinders such as the Micro-mill (Bel-Art Products, Pequannock, NJ) and the cyclone grinding systems such as the Udy Cyclone (Udy Corp., Fort Collins, CO) with built-in sieves. Particle size is important for wheat antioxidant extraction, as decreased particle size increases extraction surface area and is likely to affect extraction efficiency. Recent wheat antioxidant investigations have used particle sizes ranging from 16 to 60 mesh (1190–177 µm), as seen in Table 8.1. Achieving this small mesh size for though fractions like wheat bran is often difficult with coffee or Micro-mill type grinders and can require extensive grinding. Our lab has recently found the use of liquid nitrogen to aid in efficiently grinding wheat bran samples with small particle size while minimizing thermal and oxidative damage. Regardless of what grinding method is chosen, it is recommended that particle size be consistent and reduced to at least 20 mesh.

Several previous studies involved a defatting step prior to wheat antioxidant extraction, seen in Table 8.1. The purpose of this step is to remove fat using a nonpolar solvent. The most commonly reported defatting procedures involve blending wheat samples three times with hexane (1 : 5, w/v) for 5 min in a Waring blender (Model 33BL73, Waring, Newhartford, CT) (8,22) or petroleum ether for 1 h with shaking (23) followed by filtration and air drying of defatted wheat samples to eliminate any remaining organic solvent. It needs to be pointed out that some lipophilic compounds including lipophilic antioxidants may be extracted by hexane or petroleum ether, and lead to under estimation of overall antioxidant activities of the original samples. In addition, the defatting procedure may be utilized to analyze the hydrophilic antioxidant properties of wheat or wheat-based materials.

Extraction solid-to-solvent ratios and the number of repeated extractions performed are well recognized to influence extraction efficiency for any analysis including antioxidant property estimations. It is generally accepted that the ratio of raw material to solvent and the number of repeated extractions are chosen according to the purpose of extraction, sample number and size, and cost. While no studies have evaluated or optimized this for wheat antioxidant extractions, a range has been reported as seen in Table 8.1. It is recommended to use similar extraction ratio of solid-to-solvent and the number of repeated extractions as that used in the previous studies for the same or similar purpose in future research, so that the data can be compared across laboratories and time.

Once extractions are complete, samples are typically centrifuged and or filtered to remove any solid residue remaining in the liquid extract that may interfere with antioxidant property measurements. Filtration can be done with qualitative cellulose filter papers such as Whatman #1 filter paper (Whatman Int., UK) while centrifugation is typically done at 2000–5000 × g for 10–15 min. Centrifugation is preferred when the solvent system contains two or more solvents such as 50% acetone or 80% methanol, since the solvent composition may vary due to evaporation during filtration. Extracts should be dried with nitrogen or stored under nitrogen in the absence of light until analysis, and can also be stored at low temperatures such as −20 or −80°C.

8.4 EXTRACTION CONDITION RECOMMENDATIONS FOR WHEAT ANTIOXIDANT PROPERTY ESTIMATION

On the basis of the results and conclusions from the above discussed extraction condition optimization studies, recommendations are presented for individual wheat materials and antioxidant estimation assays in Table 8.5. Wheat materials or methods

TABLE 8.5 Recommended Extraction Conditions for General Wheat Antioxidant Estimation

Wheat material	Antioxidant estimation assay(s)	Solvent composition	Solid-to-solvent ratio, w:v	Extraction time	Extraction temperature	References
Hard wheat bran	ORAC, DPPH$^\cdot$, ABTS$^{\cdot+}$	Acetone/water (50 : 50)	1 : 10	15 h	Ambient	14
Soft or hard wheat bran or whole grain	TEAC	Ethanol/water (1 : 1 v/v)	3 : 50	60 min	60°C	19
Soft wheat grain	ORAC, ABTS$^{\cdot+}$	Acetone/water (50 : 50)	1 : 10	15 h	Ambient	Figs. 8.1 and 8.2
Soft wheat grain	TPC, O$_2^{\cdot-}$	Methanol/water (80 : 20)	1 : 10	15 h	Ambient	Figs. 8.3 and 8.4
Hard and soft wheat bran	TPC, β-CLAMS	Methanol/water (80 : 20)	1 : 5	1 h	Ambient	12
Roasted wheat germ	Inhibition of lipid oxidation measured by PV and HC	100% ethanol	1 : 8	16 h	Ambient	17
Wheat bran	Lipid oxidation measured by TBA	100% ethanol	1 : 5	12 h	NA	18

ORAC stands for oxygen radical absorbing capacity, DPPH$^\cdot$ stands for DPPH radical scavenging capacity, TEAC stands for trolox equivalent antioxidant capacity, ABTS$^{\cdot+}$ stands for cation ABTS radial scavenging capacity, TEAC stands for trolox equivalent antioxidant capacity, TPC stands for total phenolic contents using Folin–Ciocalteu reagent, O$_2^{\cdot-}$ stands for superoxide anion radical scavenging, β-CLAMS stands for β-carotene linoleic acid model system, PV stands for peroxide value, HC stands for conjugated diene hydroperoxides content, TBA stands for thiobarbituric acid assay, NA stands for information not available.

not listed in this table have not been reported in extraction optimization studies, but may be reported in other studies in Table 8.1. In addition, while recommendations are not listed for quantitative antioxidant methods such as phenolic, tocopherol, or carotenoid compositions, details for these methods will be discussed in the individual methods chapters in this book.

REFERENCES

1. Zhou, K.; Yin, J.; Yu, L. Phenolic acid, tocopherol and carotenoid compositions, and antioxidant functions of hard red winter wheat bran. *J. Agric. Food Chem.* **2005**, *53*, 3916–3922.
2. Moore, J.; Hao, Z.; Zhou, K.; Luther, M.; Costa, J.; Yu, L. Carotenoid, tocopherol, phenolic acid, and antioxidant properties of Maryland-grown soft wheat. *J. Agric. Food Chem.* **2005**, *53*, 6649–6657.
3. Zhou, K.; Yu, L. Antioxidant properties of bran extracts from Trego wheat grown at different locations. *J. Agric. Food Chem.* **2004**, *52*, 1112–1117.
4. Zhou, K.; Su, L.; Yu, L. Phytochemicals and antioxidant properties in wheat bran. *J. Agric. Food Chem.* **2004**, *52*, 6108–6114.
5. Zhou, K.; Laux, J. J.; Yu, L. Comparison of Swiss red wheat grain, and fractions for their antioxidant properties. *J. Agric. Food Chem.* **2004**, *52*, 1118–1123.
6. Adom, K. K.; Liu R. H. Antioxidant activity of grains. *J. Agric. Food Chem.* **2002**, *50*, 6182–6187.
7. Adom, K. K.; Sorrells, M. E.; Liu, R. H. Phytochemical profiles and antioxidant activity of wheat varieties. *J. Agric. Food Chem.* **2003**, *51*, 7825–7834.
8. Adom, K. K.; Sorrells, M. E.; Liu, R. H. Phytochemicals and antioxidant activity of milled fractions of different wheat varieties. *J. Agric. Food Chem.* **2005**, *53*, 2297–2306.
9. Panfili, G.; Fratianni, A.; Irano, M. Normal phase high-performance liquid chromatography method for the determination of tocopherols and tocotrienols in cereals. *J. Agric. Food Chem.* **2003**, *51*, 3940–3944.
10. Nyström, L.; Mäkinen, M.; Lampi, A.; Piironen, V. Antioxidant activity of steryl ferulate extracts from rye and wheat bran. *J. Agric. Food Chem.* **2005**, *53*, 2503–2510.
11. Li, W.; Shan, F.; Sun, S.; Corke, H.; Beta, T. Free radical scavenging properties and phenolic content of Chinese black-grained wheat. *J. Agric. Food Chem.* **2005**, *53*, 8533–8536.
12. Kim, K.; Tsao, R.; Yang, R.; Cui, S. W. Phenolic acid profiles and antioxidant activities of wheat bran extracts and the effects of hydrolysis conditions. *Food Chem* .**2006**, *95*, 466–473.
13. Crosby, G. Lignans in food and nutrition. *Food Tech.* **2005**, *59*, 32–36.
14. Zhou, K.; Liangli, Y. Effects of extraction solvent on wheat bran antioxidant activity estimation. *Lebensm.-Wiss. Technol.* **2004**, *37*, 717–721.
15. Zielinski, H.; Kozlowska, H. Antioxidant activity and total phenolics in selected cereal grains and their different morphological fractions. *J. Agric. Food Chem.* **2000**, *48*, 2008–2016.
16. Yu, L.; Haley, S.; Perret, J.; Harris, M.; Wilson, J.; Qian, M. Free radical scavenging properties of wheat extracts. *J. Agric. Food Chem.* **2002**, *50*, 1619–1624.
17. Krings, U.; El-Saharty, Y. S.; El-Zeany, B. A.; Pable, B.; Berger, R. G. Antioxidant activity of extracts from roasted wheat germ. *Food Chem.* **2000**, *71*, 91–95.
18. Mageed, M. A. A. E.; Fadel, H. M. F. Evaluation of the antioxidant activity of wheat bran. *Indian J. Chem. Techn.* **1999**, *6*, 117–120.
19. Liyana-Pathirana, C. M.; Shahidi, F. Optimization of extraction of phenolic compounds from wheat using response surface methodology. *Food Chem.* **2005**, *93*, 47–56.
20. Moore, J.; Yin, J.; Yu, L. Novel fluorometric assay for hydroxyl radical scavenging capacity (HOSC) estimation. *J. Agric. Food Chem.* **2006**, *54*, 617–626.
21. Escribano-Bailon, M.; Santos-Buelga, C. In: Santos-Buelga, C.; Williamson, G. (Eds.)., *Methods in Polyphenol Analysis: Polyphenol Extraction from Foods.*, Cambridge, 1st edn. **2003**. pp.1–47.
22. Liyana-Pathirana, C. M.; Shahidi, F. Importance of insoluble-bound phenolics to antioxidant properties of wheat. *J. Agric. Food Chem.* **2006**, *54*, 1256–1264.
23. Onyeneho, S. N.; Hettiarachchy, N. S. Antioxidant activity of durum wheat bran. *J. Agric. Food Chem.* **1992**, *40*, 1496–1500.

24. Yu, L.; Haley, S.; Perret, J.; Harris, M. Antioxidant properties of hard winter wheat extracts. *Food Chem.* **2002**, *78*, 457–461.

25. Yu, L.; Perret, J.; Davy, D.; Wilson, J.; Melby, C. L. Antioxidant properties of cereal products. *J. Food Sci.* **2002**, *67*, 2600–2603.

26. Yu, L.; Perret, J.; Harris, M.; Wilson, J.; Haley, S. Antioxidant properties of bran extracts from "Akon" wheat grown at different locations. *J. Agric. Food Chem.* **2003**, *51*, 1566–1570.

27. Yu, L.; Zhou, K.; Parry, J. W. Inhibitory effects of wheat bran extracts on human LDL oxidation and free radicals. *Lebensm.-Wiss. Technol.* **2005**, *38*, 463–470.

28. Mpofu, A.; Sapirstein, H. D.; Beta, T. Genotype and environmental variation in phenolic content, phenolic acid composition, and antioxidant activity of hard spring wheat. *J. Agric. Food Chem.* **2006**, *54*, 1265–1270.

29. Liyana-Pathirana, C. M.; Shahidi, F. Antioxidant properties of commercial soft and hard wheats (Triticum aestivum L.) and their milling fractions. *J. Sci. Food Agr.* **2006**, *86*, 477–485.

30. Martínez-Tomé, M.; Murcia, M. A.; Frega, N.; Ruggieri, S.; Jimenez, A. M.; Roses, F.; Parras, P. Evaluation of antioxidant capacity of cereal brans. *J. Agric. Food Chem.* **2004**, *52*, 4690–4699.

31. Baublis, A.; Decker, E. A.; Clydesdale, F. M. Antioxidant effect of aqueous extracts from wheat based ready-to-eat breakfast cereals. *Food Chem.* **2000**, *68*, 1–6.

32. Marama, M.; Amarowicz, R.; Weidner, S.; Abe, S.; Shahidi, F. Antioxidant activity of Triticale caryopses and embryos extracts. *Food Sci. Biotechnol.* **2004**, *13*, 421–424.

33. Amarowicz, R.; Marama, M. Antioxidant activity of wheat caryopses and embryo extracts. *J. Food Lipids* **2002**, *9*, 201–210.

34. Pérez-Jiménez, J.; Saura-Calixto, S. Literature data may underestimate the actual antioxidant capacity of cereals. *J. Agric. Food Chem.* **2005**, *53*, 5036–5040.

35. Pinzino, C. Aging, free radicals, and antioxidants in wheat seeds. *J. Agric. Food Chem.* **1999**, *47*, 1333–1339.

36. Qu, H.; Madl, R. L.; Takemoto, D. J.; Baybutt, R. C.; Wang, W. Lignans are involved in the antitumor activity of wheat bran in colon cancer SW480 cells. *J. Nutr.* **2005**, *135*, 598–602.

37. Abdel-Aal, E. M.; Hucl, P. Composition and stability of anthocyanins in blue-grained wheat. *J. Agric. Food Chem.* **2003**, *51*, 2174–2180.

CHAPTER 9

METHODS FOR ANTIOXIDANT CAPACITY ESTIMATION OF WHEAT AND WHEAT-BASED FOOD PRODUCTS

Jeffrey Moore
Liangli (Lucy) Yu

9.1 INTRODUCTION

Recent interest in the health promoting and food preserving properties of antioxidants has resulted in the development of numerous antioxidant capacity assays. A simple and broadly accepted definition of an antioxidant described by Halliwell is "a molecule which, when present in small concentrations compared to that of an oxidizable substrate, significantly delays or prevents the oxidation of that substrate" (1). While a variety of terms have been utilized to describe the effectiveness of an antioxidant including terms such as antioxidant activity, efficiency, power, or potential, the term antioxidant capacity (AOC) has recently been suggested as the most appropriate term and will be used throughout this chapter (2).

Antioxidants from a food science perspective function to inhibit rancidity development in food systems by preventing the initiation of lipid oxidation through metal ion chelation or reduction of existing peroxides, and/or stopping free radical chain reactions through free radical scavenging. This includes mostly small molecular antioxidants scavenging peroxyl radicals or chelating iron or copper ions. From a biology and nutrition perspective, antioxidants can function *in vivo* to prevent the oxidation of a variety of biological targets and can include substances which (1) chelate metal ions to prevent the formation of reactive oxygen/nitrogen species, (2) directly react with and scavenge reactive oxygen or nitrogen species, (3) inhibit oxidative enzymes (i.e., cyclooxygenases), or (4) induce antioxidant enzyme activities. This expands the scope of antioxidants to include more complex molecules such as enzymes, and numerous reactive oxygen species (ROS) such as peroxyl ($ROO^{\bullet}$), hydroxyl ($^{\bullet}OH$), and superoxide anion ($O_2^{\bullet-}$) radicals, singlet oxygen ($^1\Delta gO_2$), and reactive nitrogen species such as nitric oxide ($^{\bullet}NO$), nitrogen dioxide ($^{\bullet}NO_2$), and peroxynitrite ($OONO^-$).

Wheat Antioxidants, Edited by Liangli Yu
Copyright © 2008 John Wiley & Sons, Inc.

It is important to note that the antioxidant activity estimation is highly affected by the reactive oxygen or nitrogen species employed in the assay, although the chemical structure of the selected antioxidant molecule determines its antioxidant capacity. For example, while an antioxidant may be extremely effective in scavenging one type of free radical, it may relatively ineffective towards another type of radical or chelating a metal ion.

For both food and biological systems, it has been recognized that in addition to the diverse ways in which antioxidants can function, the properties of the assay system can also greatly influence the effectiveness of an antioxidant (3). Physicochemical properties of the testing system including viscosity, polarity, pH, and presence of organic solvent and other molecules including enzymes and solubilizing agents can alter not only the physical location and concentration of antioxidants in the system but also the chemical mechanisms through which they interact with free radicals or other reactive oxygen and nitrogen species. It has therefore become necessary to employ not one but two or more antioxidant assays to characterize the diverse and multifaceted nature of a selected antioxidant preparation.

Potential interferences for antioxidant capacity assays are often overlooked when the methods are initially developed and can lead to errors in the reported results. While each assay described in this chapter will be discussed for potential interferences in detail, some general background should be noted. Antioxidants can scavenge free radicals through several mechanism including hydrogen atom transfer (HAT), single electron transfer (ET or SET), and the recently described sequential proton loss electron transfer (SPLET) mechanism (4). While details of these mechanisms are presented elsewhere (5), it is important to note that each mechanism involves different kinetics. While in most cases free radical scavenging reactions occur through a combination of mechanisms, factors that influence which mechanism predominates can greatly affect the kinetics and therefore results of antioxidant capacity assays. Such factors can include solvent properties such as polarity, pH, and hydrogen bond accepting ability (6,7). Another common potential interference for antioxidant assays is the ability of solvent to donate hydrogen atoms to free radicals and themselves become radical species that can alter overall results. Evidence of this has been demonstrated recently by Moore and others (8) using electron spin resonance spin trapping assays. Another recent report by Perez-Jimenez and Saura-Calixto (9) examined the effect of sample solvent on results for several antioxidant capacity assays, and found significant interferences. These and other specific interferences described for individual assays below emphasize the importance of avoiding potential interferences with antioxidant capacity assays to ensure the accuracy of results.

The objective for this chapter is to summarize common antioxidant capacity assays that have been used to evaluate the antioxidant properties of wheat and wheat-based food products. Eleven antioxidant activity estimation methods are presented, each including three sections: (1) the principles, chemistry, and history of the development for the method; (2) the detailed protocol and calculations; and (3) a discussion including reported values for wheat and wheat-based food products and the advantages and limitations for the method.

9.2 DPPH RADICAL SCAVENGING CAPACITY ASSAY

9.2.1 Principles and Background

The DPPH radical (DDPH$^\bullet$) scavenging capacity assay is a decolorization assay that measures the capacity of antioxidants to directly react with (scavenge) DPPH radicals by monitoring its absorbance at 517 nm with a spectrophotometer. The DPPH radical is a stable organic nitrogen centered free radical (Fig. 9.1) with a dark purple color that when reduced to its nonradical form by antioxidants becomes colorless.

Use of DPPH$^\bullet$ to measure the antioxidant properties of compounds dates back to the 1950s (10), while Brand-Williams and coworkers in 1995 (11) published the first method to quantify scavenging capacities against this radical using a spectrophotometer. This original method quantified scavenging capacity by measuring a test compound at different concentrations and calculating the compound concentration required to reduce the initial DPPH$^\bullet$ concentration by 50% at steady-state (EC$_{50}$). A lower EC$_{50}$ value is associated with a stronger DPPH radical scavenging capacity under the same testing conditions. These results for clarity were also expressed in terms of antiradical power (ARP) calculated as 1/EC$_{50}$, in which larger ARP values represented a larger scavenging capacity. Reporting results as EC$_{50}$ value or ARP, however, has been criticized because results are highly dependent on initial DPPH$^\bullet$ concentration and definition of steady-state time point, both of which have differed between groups (12,13). This has made comparison of EC$_{50}$ values of different antioxidant samples across individual research laboratories impossible. Given the time required to examine the multiple concentrations (generally 8 concentrations) of a selected antioxidant sample necessary to determine EC$_{50}$, many researchers have also published results simply as % DPPH$^\bullet$ remaining at an arbitrary time. This approach makes it impossible

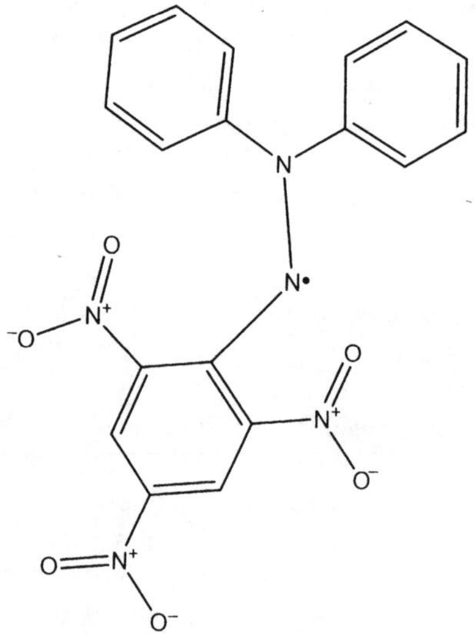

Figure 9.1 Structure of DPPH radical.

to compare reactions performed with different concentrations of antioxidants and the results are still dependent on initial DPPH$^\bullet$ concentration, meaning that it is not possible to compare results across laboratories or obtained at different points.

Sanchez-Moreno and coworkers in 1998 (14) proposed to use a calculation termed "antiradical efficiency" (AE), where $AE = 1/EC_{50} \times T_{EC50}$, where T_{EC50} was defined as the time to reach steady state. This calculation method was shown to be a more discriminatory calculation of antioxidant scavenging capacity than EC_{50} since it takes into account kinetic properties of the antioxidant reactions (reaction time) (14). A calculation of similar concept termed radical scavenging efficiency (RSE) was introduced in 2003 by De Beer and coworkers (15), which took into account the initial reaction rate and EC_{50}. Both RSE and AE calculations have been criticized because the actual chemical or scientific meaning of these terms is unclear. Also, since both rely on EC_{50} calculations, their results are still highly dependent on reagent concentrations and how to choose the steady-state time, and therefore not reproducible or comparable between laboratories (12,13). A recently published relative DPPH$^\bullet$ scavenging capacity (RDSC) estimation method by Cheng and coworkers (13) reports the DPPH radical scavenging capacity of a selected antioxidant preparation relative to that of an antioxidant standard such as trolox. This method addresses the many criticisms of previous methods and allows for interlaboratory comparison of results. This RDSC method measures the area under the antioxidant-DPPH$^\bullet$ reaction curve (AUC) for antioxidant activity calculation, taking into account both kinetic and thermodynamic properties of antioxidant-DPPH$^\bullet$ reactions. It also expresses results relative to a standard antioxidant, trolox. Cheng's RDSC method (13) was also adapted to microplate format allowing for high-throughput analysis with a microplate reader and validated for linearity, accuracy, precision, and reproducibility.

The RDSC protocol described below is for use with either a single read or microplate spectrophotometers. This protocol has been successfully used for wheat-based food ingredients and food products as well as other botanical materials. It is recommended as it allows for comparison of results of antioxidant samples tested at different concentrations and in different laboratories. This protocol can also be used to obtain the EC_{50} value under the experimental condition. Details of the required experimental approaches and calculation methods are provided.

9.2.2   Materials and Solutions Preparation

- 0.2 mM 2,2-diphenyl-1-picrylhydrazyl (DPPH$^\bullet$) working solution prepared in the selected solvent, which may be the same organic solvent as sample solutions.
- Testing sample solution preparation: sample solutions will be diluted with the same solvent so that the absorbances of the sample-DPPH$^\bullet$ reactions are within the linear range of the trolox standard curve. This is important for RDSC value calculation. Seven different dilutions of a selected antioxidant sample are required to obtain its EC_{50} value. The dilution factor (DF) is recorded and used for calculation.
- 50 mM primary stock solution of 6-hydroxy-2,5,7,8-tetramethylchoman-2-carboxylic acid (trolox) is prepared in the same solvent.

- 0.5 mM trolox secondary stock solution diluted with the same solvent.
- Trolox working standards (7–37 μM) are diluted with the same solvent.
- Clear 96-well plates compatible designed for absorbance readings.
- Microplate sealing tapes such as product number 232701 from Nunc (Rochester, NY).

9.2.2.1 Procedure for High Throughput 96-Well Microplate Method

- For example of plate layout recommended for triplicate samples, see Fig. 9.2;
- add 200 μL of the solvent to Blank wells;
- add 100 μL of the solvent to Control wells;
- add 100 μL of standard and sample solutions to appropriate wells;
- add 100 μL of 0.2 mM DPPH˙ solution to control, standard, and sample wells;
- gently shake plate for 5 s;
- read absorbance at 515 nm every minute for 1.5 h with lid on plate.

9.2.2.2 Procedure for Single Read Machine Method

- Set absorbance wavelength on spectrophotometer to 515 nm;
- measure absorbance of solvent (Blank);
- for measuring both RSDC and EC_{50} values, 500 μL of standard, sample, or solvent (for control) are combined with 500 μL of 0.2 mM DPPH˙ solution in test tube and vortexed for 5 s;

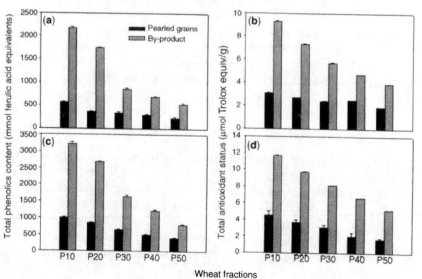

Figure 9.2 Recommended plate layout for RDSC assay. Layout for 26 sample, 5 trolox standards, control, and blank, all run in triplicate.

- for EC_{50} value estimation, after vortexing, each antioxidant-DPPH radical reaction mixture is kept in the test tubes in dark for 40 min (time point by which all reactions should have reached steady state) and transferred to a cuvette for absorbance measurement at 515 nm;
- for RSDC value estimation, after vortexing test tubes, each antioxidant-DPPH radical reaction mixture is transferred to a cuvette, then absorbance at 515 nm is recorded every minute for 40 min.

9.2.2.3 Calculations for RSDC Value

- Calculate % DPPH˙ quenched for all standards and samples at all time points using the following equation:

$$\% \ \ \text{DPPH˙ quenched at time } t \text{ min} = \left(1 - \frac{A_{\text{sample}} - A_{\text{blank}}}{A_{\text{control}} - A_{\text{blank}}}\right) \times 100$$

where A_{sample}, A_{blank}, and A_{control} represent the absorbance of sample, blank, and control wells at 515 nm at reaction time t.

- Calculate area under the curve of % DPPH˙ quenched versus antioxidant-DPPH˙ reaction time for each concentration of the standard and the testing sample using the following equation:

$$\text{AUC} = 0.5f_0 + (f_1 + f_2 + f_3 + \cdots + f_{i-1}) + 0.5f_i \tag{9.1}$$

where f_0 is the initial % DPPH˙ quenched reading at 0 min, and f_i is the % DPPH˙ quenched at time i min. Generally, $i = 40$ min, and trolox is used as the antioxidant standard.

- Generate the standard curve by plotting AUC values of trolox solutions against the trolox concentrations in μM as shown in Fig. 9.3.

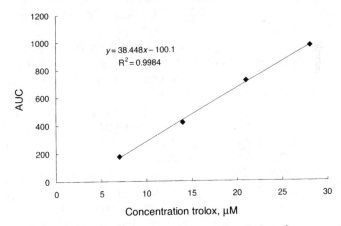

Figure 9.3 Standard curve for RDSC method using trolox. At least four concentrations of trolox are required to obtain a linear regression equation, which is used to calculate RDSC values

- Obtain the linear regression equation from standard curve. The equation is $y = 38.448x - 100.1$ from Fig. 9.3.
- Micromoles/L trolox equivalents can be calculated for each sample solution using linear regression equation from standard curve.
- Micromoles trolox equivalents/g sample calculated as

$$\mu\text{moles trolox equivalent}/g = (\mu\text{moles}/L) \times DF \times (L_{solvent}/g_{sample})$$

where DF is the dilution factor for sample extract and $L_{solvent}$ and g_{sample} are the liters of solvent and gram samples used for the sample extraction.

9.2.2.4 Calculations for EC$_{50}$ Value

- Calculate % DPPH˙ quenched for all samples at time = 40 min. (or later if steady state not reached by 40 min) using the following equation.

$$\% \quad \text{DPPH}^{\bullet} \text{ quenched} = \left(1 - \frac{A_{sample} - A_{blank}}{A_{control} - A_{blank}}\right) \times 100$$

- For each sample, plot the calculated % DPPH˙ quenched values against the sample concentrations tested.
- Determine the EC$_{50}$ value from the plot. The EC$_{50}$ value is the sample concentration at which 50% of the DPPH in the testing system can be scavenged.

9.2.3 Discussion

The DPPH˙ scavenging capacities of wheat and wheat-based food products have been reported using RSDC and EC$_{50}$ values in addition to % DPPH˙ remaining or % DPPH˙ quenched values. The available RSDC and EC$_{50}$ values for wheat related samples are summarized in Table 9.1.

The most critical step for the DPPH radical scavenging capacity estimation is consistent timing after DPPH˙ solution is added to start reactions. For high-throughput analysis, if automated liquid handling is not available, the DPPH˙ solution should be added quickly to wells using a multichannel pipetter. As organic solvents will evaporate, which changes the determining path length in microplates, suitable microplate sealing tapes should be utilized during readings to minimize this effect. For single read spectrophotometers, consistent vortexing and measurement times should be utilized. Proper storage of reagents is also important for reproducible results for DPPH radical scavenging capacity estimation. DPPH˙ working solutions should be prepared fresh, filtered before use to eliminate any precipitate and used within 48 h. Trolox working solution should be freshly prepared and used within 48 h.

The DPPH radical scavenging capacity estimation has found widespread use in screening the antioxidant properties of pure compounds and botanical extracts. Performing DPPH˙ scavenging capacity assays is technically simple and can be accomplished using a single read spectrophotometer, which explains its popularity. A

TABLE 9.1   Reported DPPH$^{\cdot}$ Scavenging Capacity Values for Wheat Grain, Its Fractions, and Wheat-based Food Products

Wheat material	DPPH$^{\cdot}$ Scavenging capacity value	Reporting method	Extraction solvent	References
Soft wheat grains	23–27 mg/mL	EC$_{50}$	MeOH	17
Hard wheat grains	2–5.2 mg/mL	EC$_{50}$	EtOH	18,19
Hard wheat brans	6–12 mg/mL	EC$_{50}$	EtOH	18,20
Hard wheat aleurone	6–8 mg/mL	EC$_{50}$	EtOH	18
Hard wheat bran	5–6 mg/mL	EC$_{50}$	50% Ace	13
Hard wheat grain	14–16 mg/mL	EC$_{50}$	50% Ace	13
Hard wheat bran	1.8–3.2 µmol TE/g	RDSC	50% Ace	13
Hard wheat grain	0.9–1.1 µmol TE/g	RDSC	50% Ace	13

EC$_{50}$ is the required concentration of a sample to scavenge 50% of DPPH radicals present in the reaction mixture. EC$_{50}$ values in this table are expressed as mg wheat equivalents per mL of final reaction mixture. RDSC stands for relative DPPH$^{\cdot}$ scavenging capacity with results expressed as trolox equivalents (TE) per gram material (13).

definite advantage of this assay over most other AOC assays has been its broad solvent compatibility with aqueous and polar and nonpolar organic solvents (13). This allows for evaluation of both hydrophilic and lipophilic antioxidant compounds for their DPPH$^{\cdot}$ scavenging capacities under the same experimental conditions without use of solubilizing agents such as the β-cyclodextrin that are necessary in other assays. Reaction mechanisms involved in antioxidant-DPPH$^{\cdot}$ reactions, however, have been recently studied and shown to be influenced by solvent properties and basic or acid impurities in solvents (4).

Major advantages of the RSDC method for determining DPPH$^{\cdot}$ scavenging capacity include that the RSDC determination is independent of either sample or initial DPPH$^{\cdot}$ concentrations in the assay system and that RSDC values make it possible to compare AOC results between samples and laboratories. The major disadvantage of the DPPH$^{\cdot}$ capacity scavenging assay, similar to the ABTS$^{\cdot+}$ assay is the use of a nonphysiologically relevant free radical, which has little resemblance to free radicals involved in oxidative processes in biological systems. Lastly, interferences from carotenoids that absorb light at 515 nm have been reported (16).

9.3   ABTS CATION RADICAL (ABTS$^{\cdot+}$) SCAVENGING CAPACITY ASSAY

9.3.1   Principles and Background

The ABTS cation radical (ABTS$^{\cdot+}$) scavenging capacity assay is a decolorization assay that measures the capacity of antioxidants to directly react with (scavenge) ABTS cation radicals generated by a chemical method. ABTS$^{\cdot+}$ is a nitrogen centered radical with a characteristic blue-green color, which when reduced by antioxidants to its nonradical (ABTS) form becomes colorless. The method quantifies scavenging capacity by measuring the absorbance of the antioxidant-radical reaction mixture at 734 nm at a selected time point with a spectrophotometer. Results are generally expressed relative to a standard, commonly trolox.

This method is a variation of the trolox equivalent antioxidant capacity (TEAC) assay first reported by Miller and coworkers in 1993 (21) and sold as a "total antioxidant status" kit by Randox Laboratories (Lakewood, CA). In this original TEAC method reported by Miller and coworkers (21), the ABTS$^{\cdot+}$ is generated enzymatically using the peroxidase activity of metmyoglobin. In this reaction, metmyoglobin (Fe^{3+}) is oxidized by H$_2$O$_2$ to the ferrylmyoglobin (Fe^{4+}) radical species, to which ABTS donates an electron and forms ABTS$^{\cdot+}$. During the assay, the test antioxidant sample was added before the ABTS$^{\cdot+}$ generation was accomplished and scavenging capacity was quantified based on absorption at 734 nm at a preselected time point. Arnao and coworkers in 1996 (22) soon after developed an alternative method using horseradish peroxidase instead of metmyoglobin and quantified scavenging capacity based on lag phase time using absorbance at 414 nm. Arnao and coworkers (22) described the advantages of their method to include better availability, higher reactivity towards ABTS, and easier concentration determinations for the horseradish peroxidase enzyme versus metmyoglobin (22). Both the original Miller (21) and Arnao (22) methods have been criticized because test compounds could both interfere with the radical formation by enzyme inhibition or reaction with H$_2$O$_2$ and/or scavenge ABTS$^{\cdot+}$. This means that these methods may not measure just radical scavenging properties and could lead to overestimations of radical scavenging capacity (23,25). Modified versions of both methods were later published by both groups that pregenerated the ABTS$^{\cdot+}$ before adding the test antioxidant (also called "endpoint" or "post-addition" methods) to eliminate this interference (23,24). Reported alternative approaches for generating ABTS$^{\cdot+}$ have included oxidation with manganese dioxide or potassium persulfate (24), or use of an azo radical initiator such as ABAP (25). More recent adaptations have included assays for lipophilic antioxidants (26,27) and use of a microplate reader (28,29).

The recommended protocol described below is based on the published improved method by Miller and coworkers (24) using MnO$_2$ to chemically pregenerate ABTS$^{\cdot+}$ and a preselected time-point for absorption measurement with a spectrophotometer. This chemical method of radical generation is simpler than enzymatic methods and eliminates potential interference during radical generation.

9.3.2 Materials and Solutions Preparation

- 0.5 M phosphate buffer (PSB), pH 7.4
- 0.5 mM trolox stock solution in the same solvent for sample solution(s)
- Trolox working standards of 1–120 μM, diluted with the same solvent

- Sample testing solutions/extracts (dilution may be required to bring the absorbance to the linear range of the standard curve)
- Blank consisting of 1 mL PBS and 80 μL of the solvent
- ABTS˙⁺ working solution
  - Prepare 5 mM ABTS (2,2′-azinobis (3-ethylbenzothiazoline-6-sulfonic acid) diammonium salt) in water;
  - add one to two spatulas of MnO_2 to oxidize ABTS to ABTS˙⁺;
  - filter solution through Whatman #1 filter paper;
  - dilute with PBS until absorbance in 1-cm cell at 734 nm is exactly 0.7.

9.3.3   Measuring Procedure

- Adjust absorbance wavelength of the spectrophotometer to 734 nm;
- blank spectrophotometer with blank solution;
- add 1 mL working ABTS˙⁺ solution and 80 μL standard or diluted sample extract to test tube;
- let tube stand for 30 s followed by vortexing for 1 min;
- transfer to cuvette and immediately take absorbance reading.

9.3.4   Calculations

- Generate standard curve using absorbance values of 4–5 concentrations of trolox standards as shown in Fig. 9.4.
- Obtain the linear regression equation from standard curve. The equation is $y = 0.0016x + 0.6306$ from Fig. 9.4.
- Micromoles/L trolox equivalents is calculated for each sample using the linear regression equation.

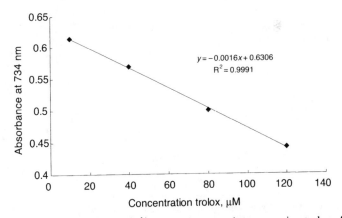

Figure 9.4   Standard curve for ABTS˙⁺ scavenging capacity assay using trolox. At least four concentrations of trolox are required to obtain a linear regression equation, which is used to calculate scavenging capacity values for potential antioxidant samples.

- If the experiment began with solid wheat sample, μmoles trolox equivalents/g sample is calculated as

$$\text{μmoles trolox equivalent}/g = (\text{μmoles}/L) \times DF \times (L_{solvent}/g_{sample})$$

where DF is the dilution factor for preparing the testing sample solution from the original wheat extract and $L_{solvent}$ and g_{sample} are the liters of solvent and grams of solid wheat sample used for the sample extraction, respectively.

9.3.5 Discussion

The structure of ABTS and its radical forms are shown in Fig. 9.5. It should be noted that the radical form of ABTS has been referred to as the cation radical (ABTS$^{\cdot+}$) and as the ABTS anion radical (ABTS$^{\cdot-}$) in some literature (25). The discrepancy between these notations is whether the negative charges on the sulfonate groups are considered.

Reported ABTS$^{\cdot+}$ scavenging capacities for wheat and wheat-based food products are shown in Table 9.2 using the radicals generated by either the metmyoglobin/H$_2$O$_2$ or the MnO$_2$ methods.

The most critical step for the ABTS$^{\cdot+}$ assay is the timing of the reactions since steady state is not necessarily achieved after the 1.5 min incubation time. Consistent timing for all samples and standards is necessary for reproducible results. Since temperature influences reaction rates, temperature fluctuations should be avoided within and between assays. Wheat extracted in 100% ethanol or other lipophilic

TABLE 9.2 Reported ABTS$^{\cdot+}$ Scavenging Capacity Values for Wheat Grain, Its Fractions, and Wheat-based Food Products

Wheat material	ABTS$^{\cdot+}$ Scavenging capacities, μmoles TE/g	ABTS$^{\cdot+}$ Generation method	References
Hard wheat grains	1.1–36	metMb/H$_2$O$_2$	31,32
Wheat-based breakfast cereals	2.3–2.4	metMb/H$_2$O$_2$	33
Hard wheat brans	24–35	metMb/H$_2$O$_2$	34,35
Hard wheat brans	5–22	MnO$_2$	18,20,36–38
Hard wheat aleurone	23–24	MnO$_2$	18
Soft wheat grains	14–18	MnO$_2$	17
Wheat grains	1.3–8[a]	metMb/H$_2$O$_2$	39
Soft and hard wheat grains	8–9[a]	ABAP	40
Soft and hard wheat germ	16–18[a]	ABAP	40
Soft and hard wheat bran and shorts	10–14[a]	ABAP	40

TE stands for trolox equivalents. MnO$_2$, ABAP, and metMb/H$_2$O$_2$ indicate methods using manganese dioxide, (2,2'-azobis-(2-amidinopropane) HCl, or the metmyoglobin/H$_2$O$_2$ systems, respectively, to generate ABTS cation radicals.

[a]Values calculated on per gram basis according to data in the literature(s).

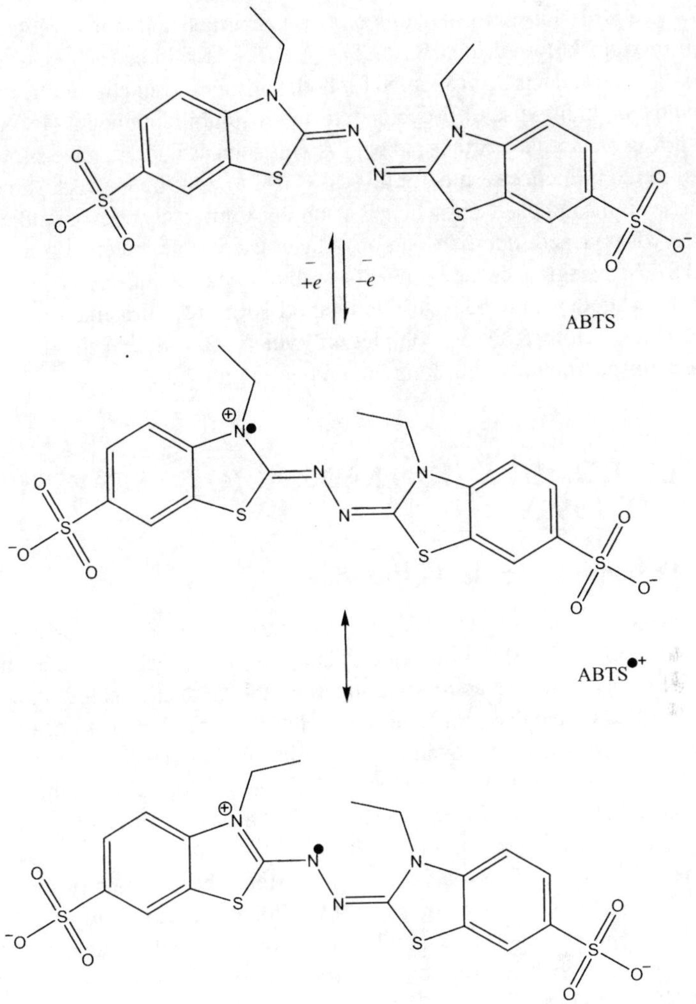

Figure 9.5   Structures of ABTS and cation ABTS radical.

extracts can form a precipitate that interferes with this assay. This issue can be solved by diluting sample extracts and standards in an ethanol solution containing 7% β-cyclodextrin (18). It needs to be pointed out that the possible interference of β-cyclodextrin has not been investigated for any ABTS⋅+ scavenging capacity assays. For reproducible results, proper storage of prepared solutions is also important. The ABTS⋅+ working solution is stable for several hours and must be prepared fresh daily.

The ABTS⋅+ scavenging capacity assay has been widely accepted and used in both food bioactive and clinical research. Due to its operational simplicity, analysis speed, and economical cost per sample. The procedure requires minimal operator training and experience and can be used with readily available and inexpensive single read spectrophotometers to analyze many samples per day. Compared to other AOC assays where pH is not controlled, this assay is performed under physiological pH in a reaction mixture buffered to pH 7.4. The $ABTS^{\bullet+}$ is soluble in both aqueous and organic solvents and has been adapted for both hydrophilic and lipophilic antioxidant compounds, including use of β-cyclodextrin as a solubilizing agent (18,30). Recent reports however, similar to other radical systems such as $DPPH^{\bullet}$, have shown solvent composition to influence results for this assay (9,25). One major disadvantage of the ABTS assay is the radical itself, which is a nonphysiologically relevant radical. Unlike peroxyl, hydroxyl, and superoxide anion radicals used in other methods, the relevance of $ABTS^{\bullet+}$ scavenging capacity results to biological situations is more difficult to interpret. Another drawback to this method is the use of a fixed time point for measurements before reaction completion, which may underestimate antioxidant properties for compounds with long lag phase reaction kinetics.

9.4   SUPEROXIDE ANION RADICAL ($O_2^{\bullet-}$) SCAVENGING CAPACITY ASSAY

9.4.1   Principles and Background

The superoxide anion radical ($O_2^{\bullet-}$) scavenging capacity assay was developed to evaluate the ability of hydrophilic antioxidants to directly react with this physiologically relevant radical. The assay measures the ability of a selected antioxidant to compete with a molecular probe, nitroblue tetrazolium (NBT), to scavenge $O_2^{\bullet-}$ generated by an enzymatic hypoxanthine-xanthine oxidase (HPX-XOD) system. NBT has a yellow color that upon reduction by $O_2^{\bullet-}$ forms formazan with a blue color measurable at 560 nm with a spectrophotometer. This assay reports $O_2^{\bullet-}$ scavenging capacity as percent $O_2^{\bullet-}$ remaining.

The anion superoxide radical is an important physiologically relevant reactive oxygen species that has been linked to cellular oxidative damage, may cause numerous chronic diseases, and may be mediators of cellular signaling (41). $O_2^{\bullet-}$ is generated *in vivo* primarily by electrons "leaking" from the mitochondrial electron transport chain that react with ground-state molecular oxygen to form this species (41). $O_2^{\bullet-}$ being charged cannot cross lipid membranes unlike its protonated counterpart, $HO_2^{\bullet}$, which exists in equilibrium *in vivo* with $O_2^{\bullet-}$. The superoxide anion radical is also involved in the formation of other reactive chemical species including H_2O_2, hydroxyl radical ($^{\bullet}OH$), and peroxnitrite (OONO$^-$). *In vivo* $O_2^{\bullet-}$ can also be dismutated enzymatically by superoxide dismutase (SOD).

The first method developed to evaluate the elimination of $O_2^{\bullet-}$ was reported by McCord and Fridovich in 1969 (42) and involved the generation of $O_2^{\bullet-}$ with the xanthine-xanthine oxidase system using ferricytochrome *c* as a reducible detector probe. This reaction system was originally developed to test samples for SOD activity, but was later utilized for measuring $O_2^{\bullet-}$ scavenging properties of samples (43). A number of issues with this original method were later reported including (a) antioxidant compounds directly reducing cytochrome c (44), (b) antioxidant compounds inhibiting enzymatic radical generating system (45), (c) limited sensitivity (46), and (d) solubility of xanthine in aqueous solutions (47). Because of these issues, numerous researchers have reported improvements including (a) enzymatic O$_2^{\cdot-}$ generating systems such as hypoxanthine-xanthine oxidase with better solubility (48) or horseradish peroxidase (49), (b) chemically generating O$_2^{\cdot-}$ such as the reaction of NADH with phenazine methosulfate (50), and (c) use of new detector probes such as NBT (51) or NBD-Cl (46). Besides the original method (43) on which a number of commercially available kits are based, two other methods have been popular for characterizing the O$_2^{\cdot-}$ scavenging properties of botanical extracts. The first one uses a xanthine-xanthine oxidase radical generator measuring nitrite formation from hydroxylammonium chloride during a selected reaction time period (52). The second one uses a hypoxanthine-xanthine oxidase radical generator with NBT as the detector probe and reaction rates to calculate the percent O$_2^{\cdot-}$ remaining (53). A method based on the latter has been used for investigating the O$_2^{\cdot-}$ scavenging properties of wheat-based food ingredients and food products and is described in details below.

9.4.2 Materials and Solutions Preparation

- 50 mM phosphate buffer (PBS), pH 7.4
- 2 mM hypoxanthine (HPX) solution prepared in PBS
- 0.56 U/mL xanthine oxidase (XOD) solution prepared in PBS
- 0.34 mM tretrazolium blue (NBT) solution prepared in PBS
- Sample extracts

9.4.3 Procedure

- Prepared blank solution by combining 300 μL PBS, 200 μL NBT solution, and 500 μL HPX solution;
- set to the absorbance at 560 nm to 0 with the blank solution;
- add 200 μL NBT solution, 500 μL HPX solution, and 100 μL sample solution (wheat extract) or the solvent for control;
- vortex for 5 s;
- add 200 μL XOD and immediately start timer;
- vortex for 30 s;
- measure absorbance every minute for 10 min.

9.4.4 Calculations

- Plot absorbance (y axis) of control or samples against the time (x axis) as shown in Fig. 9.6.
- Obtain the slope for control (Mc) and samples (Ms) using the linear regression equation. For instance, the slopes are 0.0716 and 0.0404 for control (Mc) and the sample (Ms), respectively (Fig. 9.6).

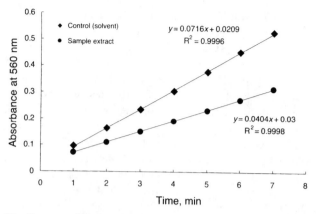

Figure 9.6 Kinetic curves for controls and samples used to calculate $O_2^{\bullet-}$ scavenging capacity. Absorbance measurements for controls and samples should be taken every minute for 7 min and plotted against time. Slopes for both control and samples are obtained from linear regression equations.

- Calculate the percent $O_2^{\bullet-}$ remaining using the following equation

$$\%O_2^{\bullet-} \text{ remaining} = 100 \times (Ms/Mc)$$

9.4.5 Discussion

$O_2^{\bullet-}$ scavenging capacity values have been reported for wheat and it fractions along with wheat-based food products. These results are summarized in Table 9.3.

The critical steps for this assay are the consistent timing for vortexing and taking absorption measurements. It is also important that the xanthine oxidase solution is prepared fresh daily, while HPX and NBT solutions can be stored under refrigerated conditions for at least 1 month.

The major advantage of this assay is the use of a physiologically relevant radical system and reaction pH. Other methods such as the DPPH and ABTS cation radical

TABLE 9.3 Reported $O_2^{\bullet-}$ Scavenging Capacity Values for Wheat Grain, and Its Fractions

Wheat material	$O_2^{\bullet-}$ Scavenging capacity	Units	$O_2^{\bullet-}$ Generating method	References
Hard wheat brans	21–44	% $O_2^{\bullet-}$ quenched[a]	HPX-XO/NBT	20,36,37
Swiss red wheat grain	22	% $O_2^{\bullet-}$ quenched[a]	HPX-XO/NBT	18
Swiss red wheat bran	41	% $O_2^{\bullet-}$ quenched[a]	HPX-XO/NBT	18
Swiss red wheat aleurone	41–57	% $O_2^{\bullet-}$ quenched[a]	HPX-XO/NBT	18
Sprouted wheat grains	7.8–9.9	μmols unreduced[a] NBT/g	HPX-XO/NBT	57

[a]% quenched values were calculated as 100 − % remaining.

scavenging capacity assays utilize free radical systems with no resemblance to those involved in oxidative processes in biological systems. This assay does, however, have several disadvantages compared to other available AOC assays. First, as mentioned above it is known that NBT can be directly reduced by certain antioxidants that lead to overestimation of AOC using this method. The use of NBT has also been criticized due to the limited water solubility of its oxidized product, formazan (54). Second, the enzymatic reaction used in this assay is known to generate not only $O_2^{\cdot-}$ but also another ROS, H_2O_2, which can cause interference with this assay (47). Third, common to any AOC method using an enzymatic radical generating system, AOC over-estimations can result from test compounds directly interfering with the enzyme reaction. Fourth, the calculation is solely based on the kinetics of the antioxidant-radical reaction and does not consider its thermodynamic properties. This calculation approach was designed for biochemical studies of enzyme inhibition and is not adequate for quantifying scavenging activities in chain processes (12). This makes it impossible to compare the results between samples such as wheat versus apple on a per sample weight basis or cross laboratories. Together these disadvantages suggest the need for an improved assay that not only addresses these issues, but is also validated with ESR for generation of only $O_2^{\cdot-}$, is compatible with both hydrophilic and lipophilic extracts, and evaluates the possible interferences of solvents. Lastly, although several high-throughput methods using a microplate reader have been reported (47,55,56), there is a need for the above discussed improvements to be able to efficiently screen the scavenging capacity of samples against this important free radical.

9.5 OXYGEN RADICAL ABSORBING CAPACITY (ORAC) ASSAY

9.5.1 Principles and Background

The oxygen radical absorbing capacity (ORAC) assay was originally developed to measure the hydrophilic chain-breaking capacity of antioxidants against the peroxyl radical using β-phycoerythrin as a molecular probe (58). This protocol was later adapted to semiautomated analysis using the Cobas Fara II system (59). The peroxyl radical is an important physiological radical that may be involved in the propagating steps of lipid peroxidation chian reactions. Peroxyl radicals can be formed *in vivo* through reactions of carbon centered radicals with other reactive oxygen species such as the hydroxyl radical. A study by Ou and coworkers (60) improved the method using a more stable and less expensive probe, fluorescein, and determined the reaction mechanism to follow a HAT mechanism by identifying oxidized products of FL with LC-MS. Given the later discontinuation of the Cobas Fara II system by the manufacturer and the time-consuming nature of this method, a study by Huang and coworkers (61) adapted the ORAC method for high-throughput completely automated analysis using a microplate reader with robotic liquid handling system. To measure the ORAC of lipophilic antioxidants and extracts, an alternative ORAC assay protocol was reported by Huang and coworkers (62) using β-cyclodextrin as a solubility enhancer.

The ORAC assay uses competitive kinetics to monitor the ability of antioxidants to compete with a molecular probe, fluorescein (FL), to scavenge peroxyl radicals generated by APPH, an azo compound. Reactions of peroxyl radicals with FL yield a nonfluorescent product and can be monitored by measuring the loss of fluorescence of FL with a fluorometer. The ORAC value for a selected antioxidant is determined based on the areas under the reaction kinetic curves for the antioxidant sample, at least 4–5 concentrations of an antioxidant standard, and a blank. Trolox is a commonly used standard with results expressed as micromoles trolox equivalents per unit of the sample.

The two methods presented below allow measurement of the hydrophilic ORAC of antioxidant samples such as wheat grain extracts. The first protocol is for high-throughput analysis using a microplate reader with fluorescence detector, while the second protocol can be carried out using single read fluorometers.

9.5.2 Materials and Solutions Preparation

- 75 mM phosphate buffer (PBS), pH 7.4
- 1 mM fluoroscein (FL) primary stock solution prepared in PBS
- 0.01 mM FL secondary stock diluted with PBS
- 81.6 nM FL working solution diluted with PBS
- 50 mM primary stock solution of 6-hydroxy-2,5,7,8-tetramethylchoman-2-carboxylic acid (trolox) prepared in the same solvent used for dissolving antioxidant sample
- 0.5 mM trolox secondary stock solution diluted with the same solvent
- Trolox working standard solutions from 10 to 100 µM prepared by diluting the stock solution with the same solvent
- 2,2′-Azobis(2-amidinopropane)dihydrochloride (AAPH) solution prepared in PBS, 0.36 M for high throughput method or 0.6 M for the single read protocol
- Sample solutions, diluted to be within the linear range of standard curve with the same solvent
- 96-well plates designed for fluorescence measurements for the high-throughput protocol

9.5.2.1 Procedure for High-Throughput Determination Using a Microplate Reader

- For example of plate layout recommended for triplicate samples, see Fig. 9.7;
- add 225 µL of 81.6 nM FL working solution to each well;
- add 30 µL of sample solution, standard, or blank (the solvent) to appropriate wells;
- incubate covered plate for 20 min in plate reader preheated to 37°C;
- add 25 µL of 0.36 M APPH solution to each well to start reaction;
- take fluorescence readings of uncovered plate every minute using excitation wavelength of 493 nm and emission wavelength of 535 nm at 37°C until all readings are less than 5% of the initial values.

	1	2	3	4	5	6	7	8	9	10	11	12
A												
B		Blank	Trolox 20 μM	Trolox 40 μM	Trolox 60 μM	Trolox 80 μM	Trolox 100 μM	Sample 1	Sample 2	Sample 3	Sample 4	
C		Blank	Trolox 20 μM	Trolox 40 μM	Trolox 60 μM	Trolox 80 μM	Trolox 100 μM	Sample 1	Sample 2	Sample 3	Sample 4	
D		Blank	Trolox 20 μM	Trolox 40 μM	Trolox 60 μM	Trolox 80 μM	Trolox 100 μM	Sample 1	Sample 2	Sample 3	Sample 4	
E		Sample 5	Sample 6	Sample 7	Sample 8	Sample 9	Sample 10	Sample 11	Sample 12	Sample 13	Sample 14	
F		Sample 5	Sample 6	Sample 7	Sample 8	Sample 9	Sample 10	Sample 11	Sample 12	Sample 13	Sample 14	
G		Sample 5	Sample 6	Sample 7	Sample 8	Sample 9	Sample 10	Sample 11	Sample 12	Sample 13	Sample 14	
H												

Figure 9.7 Recommended plate layout for ORAC assay. Layout for 14 sample, 5 trolox standards, and blank, all run in triplicate. Outer wells with shading filled with 300 μL water.

9.5.2.2 *Procedure for Using a Single Read Fluorometer*

- Add 2400 μL 81.6 nM FL working solution to test tube followed by
- 300 μL blank, standard, or sample extract;
- 300 μL 0.6 M APPH;
- vortex for 5 s;
- transfer the reaction mixture to cuvette and take fluorescene readings every minute using excitation wavelength of 493 nm and emission wavelength of 515 nm until all readings are less than 5% of the initial values;
- recommended processing order for standards and samples discussed below.

9.5.3 Calculations

- Relative area under the curve (RAUC) values for each well calculated as

$$\text{RAUC} = 0.5 + f_1/f_0 + f_2/f_0 + f_3/f_0 + \cdots + f_{i-1}/f_0 + 0.5(f_i/f_0)$$

where f_0 is the initial fluorescence reading at 0 min and f_i is the final fluorescence reading. f_i can be set as reading that is less than or equal to 5% of f_0.

- Net RAUC values for samples and standards are calculated as

$$\text{Net RAUC}_{sample} = \text{RAUC}_{sample} - \text{RAUC}_{blank} \text{ and}$$
$$\text{Net RAUC}_{standard} = \text{RAUC}_{standard} - \text{RAUC}_{blank}$$

- Generate a standard curve by plotting net $\text{RAUC}_{standards}$ values against trolox concentration in μM as shown in Fig. 9.8. Trolox may be replaced by other standard antioxidant compounds.

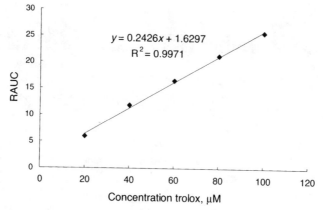

Figure 9.8 Standard curve for ORAC assay using trolox. At least four concentrations of trolox are required to obtain a linear regression equation, which is used to calculate ORAC values for potential antioxidant samples. RAUC stands for relative area under the curve and is calculated using the formula specified in the ORAC method section.

- Calculate trolox equivalents (TE) in μmoles/L for each sample solution using linear regression equation from standard curve ($y = 0.2426x + 1.6297$) as shown in Fig. 9.8. For example, if the net $RAUC_{sample}$ is 18.00, the ORAC value of the testing sample solution is 67.47 μmoles TE/L.
- ORAC value in μmoles trolox equivalents/g sample calculated as

$$\mu moles\ trolox\ equivalent/g = (\mu moles\ TE/L) \times DF \times (L_{solvent}/g_{sample})$$

where DF is the dilution factor for sample extract, and $L_{solvent}$ and g_{sample} are the liters of solvent to dissolve grams of sample used for preparing the sample solution (or the liters of solvent used to extract the grams of sample).

9.5.4 Discussion

ORAC values reported for wheat grain and its fractions along with wheat-based food products are summarized in Table 9.4.

The most critical step for the ORAC assay is the addition APPH to initiate the generation of peroxyl radicals and their attack on the fluorescent probe. Changes in the fluorescence of reaction mixtures in the initial few minutes of reaction are most rapid for blanks and samples with low antioxidant activity, and consistent timing between APPH addition and initial reading is critical to maintain reproducibility. Since peroxyl radical generation is a thermal decomposition reaction, temperature plays a critical role in maintaining assay reproducibility (30). For the single read method, it is important to maintain a constant room temperature throughout the entire analysis time frame. For high-throughput methods using microplates, use of temperature control at 37°C reduces the temperature fluctuations. Issues of variability for replicate reactions within the same plate for microplate methods have been reported (61) and observed in our laboratory. This variability has been attributed to the thermal conductivity

TABLE 9.4   Reported ORAC Values for Wheat Grain, Its Fractions, and Wheat-based Food Products

Wheat material	Reported ORAC value, μmoles TE/g	References
Maryland grown soft wheat grains	32–47	17
Swiss red wheat grain	51	18
Wheat bran	15–107	18,20,36,38,63
Swiss red wheat aleurone	125–136	18
Swiss red wheat micronized aleurone	126	18
White bread	20	63
Whole grain bread	12	63
Whole grain cereal	29	63

ORAC stands for oxygen radical absorbing capacity and TE stands for trolox equivalents.

properties of the polystyrene plates used, with reactions completing considerably faster for the outer wells (61). Preheating plates for 10–20 min before addition of APPH and filling the outer wells of the plate with blanks have been suggested to overcome this drawback.

Proper preparation and storage of solutions used for ORAC are important for accurate and reproducible results. Primary FL stock solution may be stored for up to 12 months, while secondary FL stock should be prepared weekly, both stored under nitrogen at 4°C in dark. Working FL solution should be prepared fresh. It is highly recommended that trolox solutions be prepared fresh, although the trolox standards prepared from 10 to 100 μM may be stable for about one week.

The ORAC assay has several advantages over other available antioxidant scavenging capacity assays that explain its popularity. Unlike ABTS$^{•+}$ and DPPH$^{•}$ scavenging capacity assays, ORAC measures scavenging activity against a physiologically relevant radical, the peroxyl radical, which is known to play a role in the oxidation of lipids in human bodies and food systems (64). Also, the ORAC assay is conducted under physiological pH and is adaptable for high throughput analysis. Lastly, the ORAC assay takes into account both kinetic and thermodynamic properties of antioxidant-radical reactions.

Similar to other antioxidant capacity estimation assays, there are several known issues and limitations with the ORAC assay. First, compared to methods such as the ABTS$^{•+}$ and DPPH$^{•}$, and TPC assays, ORAC is a more complicated assay and requires a more skilled operator. Second, given the time consuming nature of kinetic assays, expensive microplate readers are necessary for a large numbers of samples. Third, ORAC involves the complex reaction of an azo compound, APPH to generate peroxyl radicals. As already mentioned, this reaction is very temperature sensitive and can result in more variance than other methods even when performed under temperature controlled conditions.

It needs to be pointed out that the types of reactive radical species formed during ORAC assay conditions have never been verified using electron spin resonance (ESR) spin trapping techniques, although ORAC is used for measuring peroxyl radical scavenging capacity. It is known that the thermal decomposition of APPH yields carbon centered radicals which can react with ground-state molecular oxygen to yield peroxyl radicals. This means that at least two reactive radical species can exist and interact with antioxidant compounds. The addition of organic solvents commonly used to extract botanical materials also may generate other reactive species. It is known that reactive radical species can abstract protons from solvent molecules such as ethanol, forming carbon-centered radicals. Two recent studies in our laboratories (8,65) confirmed the formation of carbon-centered radicals as a result of hydroxyl radicals reacting with solubilizing agent such as β-cyclodextrin and organic solvents including ethanol, methanol, or DMSO. An alternative ORAC assay protocol has been reported using β-cyclodextrin as the solubilizing agent and following a similar protocol as described above (62). The purity of peroxyl radicals generated in the presence of β-cyclodextrin, however, has not been validated to date. It should be noted that the generation of other radicals such as the carbon-centered radicals may interfere the estimation of ORAC values for a selected antioxidant samples.

9.6   HYDROXYL RADICAL (˙OH) SCAVENGING CAPACITY (HOSC) ASSAY FOR HYDROPHILIC ANTIOXIDANTS

9.6.1   Principles and Background

The hydroxyl radical is one of the most reactive species known and is the most reactive species generated in biological systems (66). This species is thought to be generated *in vivo* through a number of reactions shown below including Fenton reactions (9.2), Fenton-like reaction (9.3), superoxide driven (or iron-catalyzed) Haber-Weiss reaction (9.4), and photodynamically (9.5), and is thought to be involved in cellular oxidative damage which may cause numerous chronic diseases (66).

$$Fe^{2+} + H_2O_2 \rightarrow {}^{\bullet}OH + OH^- + Fe^{3+} \tag{9.2}$$

$$LmLn^+ + H_2O_2 \rightarrow {}^{\bullet}OH + OH^- + LmMn^{2+} \tag{9.3}$$

(LmMn$^+$ includes numerous transition metals in oxidized states)

$$\text{Net reaction:} \ O_2^{\bullet-} + H_2O_2 + \text{metal catalyst} \rightarrow {}^{\bullet}OH + OH^- + O_2 \tag{9.4}$$

$$Fe^{3+} + O_2^{\bullet-} \rightarrow Fe^{2+} + O_2$$

$$Fe^{2+} + H_2O_2 \rightarrow {}^{\bullet}OH + OH^- + Fe^{3+}$$

$$H_2O_2 + \text{energy} \rightarrow 2\,{}^{\bullet}OH \tag{9.5a}$$

$$H_2O + \text{energy} \rightarrow {}^{\bullet}OH + H \tag{9.5b}$$

The hydroxyl radical scavenging capacity (HOSC) assay was developed to evaluate the ability of hydrophilic antioxidants to scavenge this highly reactive and physiologically relevant radical (8). This assay monitors the ability of the antioxidant sample to compete with a molecular probe, fluorescein (FL), and scavenge pure hydroxyl radicals generated by the Fenton-like Fe^{3+}/H_2O_2 system (8). Reactions of hydroxyl radicals with FL yield a nonfluorescent product that can be monitored by measuring the fluorescence reduction of FL with a fluorometer. HOSC values are calculated using the area under the reaction kinetic curve for samples, 4–5 concentrations of a standard, and a blank. Trolox may be used as the standard with results expressed as micromoles of trolox equivalents per unit of sample.

In this section, we provide details for two versions of the HOSC assay designed for hydrophilic antioxidant analysis (8). The first is performed using a microplate reader for high-throughput fluorescence measurements, while the second uses a single read fluorometer. Hydroxyl radical scavenging capacity for lipophilic antioxidants can be determined using an ESR method described later in this chapter (Section 9.7).

9.6.2   Materials and Solutions Preparation

- 75 mM phosphate buffer (PBS), pH 7.4
- 1 mM FL primary stock solution prepared in PBS
- 0.01 mM FL secondary stock solution prepared in PBS
- 92.8 nM FL working stock solution prepared in PBS
- 50 mM primary stock solution of 6-hydroxy-2,5,7,8-tetramethylchoman-2-carboxylic acid (trolox) prepared in the same solvent used to dissolve antioxidant sample
- 0.5 mM trolox secondary stock solution diluted with the same solvent
- Trolox standards from 10 to 100 μM prepared by diluting stock solution with the same solvent
- 3.43 mM $FeCl_3$ solution prepared in pure water
- 0.1990 M H_2O_2 solution prepared in pure water (prepared fresh for every run)
- Sample solutions, diluted to be within range of standard curve with the same solvent
- 96-well plates designed for fluorescence measurements

9.6.2.1   Procedure for High Throughput Assay Using 96-well Microplate

- For example of plate layout recommended for triplicate samples, see Fig. 9.9;
- add 170 μL of 92.8 nM FL solution to each well;
- add 30 μL of sample, standard, or blank (solvent) solution to appropriate wells;
- add 40 μL 0.1990M H_2O_2 solution to each well;
- add 60 μL 3.43 mM $FeCl_3$ solution to each well to start reaction;
- take fluorescence readings of the uncovered plate every minute using excitation wavelength of 485 nm and emission wavelength of 535 nm until all reading are less than 10% of initial values, typically 2–3 h.

	1	2	3	4	5	6	7	8	9	10	11	12
A	Blank	Blank	Blank	Trolox 20 μM	Trolox 20 μM	Trolox 20 μM	Trolox 40 μM	Trolox 40 μM	Trolox 40 μM	Trolox 60 μM	Trolox 60 μM	Trolox 60 μM
B	Trolox 80 μM	Trolox 80 μM	Trolox 80 μM	Trolox 100 μM	Trolox 100 μM	Trolox 100 μM	Sample 1	Sample 1	Sample 1	Sample 2	Sample 2	Sample 2
C	Sample 3	Sample 3	Sample 3	Sample 4	Sample 4	Sample 4	Sample 5	Sample 5	Sample 5	Sample 6	Sample 6	Sample 6
D	Sample 7	Sample 7	Sample 7	Sample 8	Sample 8	Sample 8	Sample 9	Sample 9	Sample 9	Sample 10	Sample 10	Sample 10
E	Sample 11	Sample 11	Sample 11	Sample 12	Sample 12	Sample 12	Sample 13	Sample 13	Sample 13	Sample 14	Sample 14	Sample 14
F	Sample 15	Sample 15	Sample 15	Sample 16	Sample 16	Sample 16	Sample 17	Sample 17	Sample 17	Sample 18	Sample 18	Sample 18
G	Sample 19	Sample 19	Sample 19	Sample 20	Sample 20	Sample 20	Sample 21	Sample 21	Sample 21	Sample 22	Sample 22	Sample 22
H	Sample 23	Sample 23	Sample 23	Sample 24	Sample 24	Sample 24	Sample 25	Sample 25	Sample 25	Sample 26	Sample 26	Sample 26

Figure 9.9   Recommended plate layout for HOSC assay. Layout for 26 sample, 5 trolox standards, and blank, all run in triplicate.

9.6.2.2   Procedure for Single Read Machine Method

- Add 1700 μL 92.8 nM FL working solution to test tube followed by
- 300 μL blank, standard, or sample solution;
- 400 μL 0.1990M H_2O_2 solution;
- 600 μL 3.43 mM H_2O_2 solution to start reaction;
- Vortex for 5 s;
- Transfer the reaction mixture to cuvette and take fluorescene readings every minute using excitation wavelength of 493 nm and emission wavelength of 515 nm until all reading are less than 10% of initial values, typically 2–3 h

9.6.3   Calculations

- Relative area under the curve (RAUC) values for each well calculated as

$$\text{RAUC} = 0.5 + f_1/f_0 + f_2/f_0 + f_3/f_0 + \cdots + f_{i-1}/f_0 + 0.5(f_i/f_0)$$

where f_0 is the initial fluorescence reading at 0 min and f_i is the final fluorescence reading, f_i can be set as reading that is less than or equal to 5% of f_0.
- Net area under the curve values for samples and standards calculated as

$$\text{Net RAUC}_{\text{sample}} = \text{RAUC}_{\text{sample}} - \text{RAUC}_{\text{blank}}$$

$$\text{Net RAUC}_{\text{standard}} = \text{RAUC}_{\text{standard}} - \text{RAUC}_{\text{blank}}$$

- Generate a standard curve by plotting net $\text{RAUC}_{\text{standard}}$ values against trolox concentrations in μM as shown in Fig. 9.10.

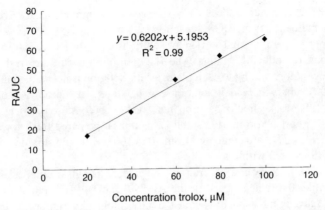

Figure 9.10   Standard curve for HOSC assay using trolox. At least four concentrations of trolox are required to obtain a linear regression equation, which is used to calculate HOSC values for potential antioxidant samples. RAUC stands for relative area under the curve and is calculated using the formula specified in the HOSC method section.

- Calculate trolox equivalents (TE) in μmoles/L for each sample solution using linear regression equation from standard curve ($y = 0.6202x + 5.1953$) as shown in Fig. 9.10. For example, if the net $RAUC_{sample}$ is 35.00, the HOSC value of the testing sample solution is 48.06 μmoles TE/L.
- HOSC value in μmoles trolox equivalents/g sample is calculated as

$$\mu\text{moles trolox equivalent}/g = (\mu\text{moles TE/L}) \times DF \times (L_{solvent}/g_{sample})$$

where DF is the dilution factor for sample extract, and $L_{solvent}$ and g_{sample} are the liters of solvent to dissolve grams of sample used for preparing the sample solution (or the liters of solvent used to extract the grams of sample).

9.6.4   Discussion

HOSC values reported for wheat and its fractions along with wheat-based food products are summarized in Table 9.5.

The most critical step for the HOSC assay is the addition of the H_2O_2 and $FeCl_3$ solutions to start the generation of hydroxyl radicals. The use of compatible solvents including either water or acetone or aqueous acetone mixtures is also important for this assay. Potential solvent interferences for this assay have been investigated and

TABLE 9.5   Reported Hydroxyl Radical Scavenging Capacity Values for Wheat Grain and Bran

Wheat material	Reported range	Units	Method used	References
Soft wheat grain	39	μmoles TE/g	HOSC	8
Hard wheat bran	75	μmoles TE/g	HOSC	8

HOSC stands for the hydroxyl radical scavenging capacity assay.

reported (8). These interferences were observed for commonly used solvents such as ethanol or methanol in the HOSC reaction system.

Proper preparation and storage of solutions used for HOSC are important for accurate and reproducible results. The H_2O_2 solution should be freshly prepared immediately before adding to each plate, as this solution is not stable under ambient conditions. Primary FL stock solution can be stored for up to 12 months, while secondary FL stock should be prepared weekly, both stored under refrigerated conditions in dark. Working FL solution should be prepared fresh for each set of samples. Trolox standards prepared from 10 to 100 μM should stored under refrigerated conditions used within a week.

The HOSC assay has several advantages compared to other available AOC assays. Unlike the radicals used in the $ABTS^{•+}$ and $DPPH^{•}$ methods, this method measures scavenging capacity against a physiologically important free radical. Second, this assay uses a radical generating system that has been validated with ESR spin trapping techniques to generate pure hydroxyl radicals. Other assays using physiological radicals including ORAC and the superoxide anion radical assay have not undergone this validation with ESR. Another advantage of HOSC over other assays is the solvent compatibility that has already been tested and verified with ESR spin trapping techniques. Lastly, similar to the ORAC and RDSC assays, HOSC uses the area under the curve to calculate the scavenging capacity that takes into account both kinetic and thermodynamic properties of the antioxidant-radical reaction.

Similar to other antioxidant capacity estimation assays, there are some disadvantages to the HOSC assay. First, compared to methods such as the $ABTS^{•+}$ and $DPPH^{•}$, and TPC assays, HOSC is more complicated assay and requires a more skilled operator. Second, given the time consuming nature of kinetic assays, expensive microplate readers are necessary for analysis of large numbers of samples. Third, as mentioned above, this assay is only compatible with two solvents, acetone and water. This limitation is common to all hydroxyl radical assays due to the reactive nature of hydroxyl radicals that are able to abstract hydrogen atoms from many organic solvent molecules forming carbon centered radicals that interfere with this assay (8). Lastly, due to the lack of compatibility with many organic solvents and β-cyclodextrin, this assay cannot be used to assess the scavenging properties of lipophilic compounds.

9.6.5   Other Reported Methods for Hydroxyl Radical Scavenging Capacity Estimation

In addition to the HOSC assay, several other protocols have been reported for estimation of hydroxyl radical scavenging capacity. The deoxyribose method developed by Aruoma (67) has been one of the most popular assays used to evaluate hydroxyl radical scavenging properties of compounds and botanical extracts (8). This method uses the classic Fenton reaction with Fe^{2+} (reduced from Fe^{3+} by ascorbic acid) and H_2O_2 in a buffered system to generate $^{•}OH$ and degrade 2-deoxyribose to malondialdehyde-like products that form a chromagen with thiobarbituric acid detectable at 532 nm. Criticisms of this assay have included the use of a fixed time point percent inhibition method to calculate scavenging activity, the possibility of interference from compounds chelating the Fe^{2+} in this system as opposed to scavenging, incompatibility of this assay with commonly used organic solvents, and reports showing minimal ˙OH production with this system under physiological pH (8,68). Other reported methods have included numerous hydroxyl radical generating systems such as the enzymatic superoxide-driven Fenton reaction using hypoxanthine-xanthine oxidase and Fe^{3+} (69), chemical superoxide-driven Fenton reaction (70), use of "Fenton-like" reagents including the Co(II) or Cu(II)/H_2O_2 systems (68), pulse radiolysis of water (66), and "Photo-Fenton" reagents that utilize photosensitizers (71). Reported detector probes have included DNA, aromatic compounds, fluorescein, coumarin derivatives, fluorescamine-derivatized nitroxide, β-phcoerythrin, and luminal (8,66). Numerous issues have been discovered with some of these methods including interferences when using enzymatic systems, generation of other ROS when using Co(II) or Cu(II)/H_2O_2 Fenton-like systems, and nonideal transient signals when using chemiluminescence (8).

9.7 HYDROXYL RADICAL SCAVENGING CAPACITY ASSAY FOR LIPOPHILIC ANTIOXIDANTS USING ESR

9.7.1 Principles and Background

This assay was developed to quantify the hydroxyl radical scavenging capacity of lipophilic antioxidants. The method utilizes a Fenton Fe^{2+}/H_2O_2 system to generate hydroxyl radicals under physiological pH and acetonitrile as the solvent to dissolve lipophilic antioxidants. Hydroxyl radical scavenging capacity is determined using electron spin resonance (ESR) spin trapping techniques and results are expressed relative to a standard such as trolox.

As described above in Section 6.1, the hydroxyl radical is one of the most reactive free radical species generated in biological systems. The HOSC method described above can be used to determine hydroxyl radical scavenging capacity of hydrophilic antioxidants that can be dissolved in aqueous acetone. This HOSC assay, however, is only compatible with hydrophilic but not lipophilic antioxidants due to their limited solubility in acetone or water or aqueous acetone (65). It is of interest, however, to characterize the ability of lipophilic compounds to scavenge free radical species. In biological systems, lipophilic antioxidants are localized primarily in hydrophobic domains of lipid membranes and lipoproteins, and may play an important role in protecting these biomolecules from oxidative damage.

Developing *in vitro* antioxidant capacity assays compatible with lipophilic antioxidants has been a challenge, given that most assays utilize aqueous phase reaction systems. One common strategy has been the use of solubilizing agents such as randomly methylated β-cylcodextrin (RMCD) or amphiphilic solvents such as DMSO (65). Both, however, have been shown to be incompatible with hydroxyl radical systems likely due to their abilities to directly react with hydroxyl radicals, resulting in interference (8,65).

The method described below is based on a recent report by Cheng and coworkers (65) that developed and validated an *in vitro* hydroxyl radical scavenging capacity assay for lipophilic antioxidants that operates under physiological pH. Cheng's study investigated different radical generating systems, detection techniques, and extraction solvents. Their report found most suitable the use of the classic Fenton (Fe^{2+}/H_2O_2) reaction system to generate a pure flux of hydroxyl radicals, ESR spin trapping as the detection technique, and acetonitrile as the sample solvent.

9.7.2 Materials and Solutions Preparation

- 100 mM phosphate buffer (PBS), pH 7.4
- 50 mM primary stock solution of 6-hydroxy-2,5,7,8-tetramethylchoman-2-carboxylic acid (trolox) prepared in acetonitrile
- 0.5 mM trolox secondary stock solution diluted in acetonitrile
- Trolox standards from 10 to 100 µM prepared by diluting stock solution in acetonitrile
- 1.0 mM $FeSO_4$ solution prepared in deionized water (prepared fresh daily and kept on ice under N_2)
- 1.0 mM H_2O_2 solution prepared in PBS (prepared fresh for every run)
- 1.0 mM disodium ethylenediaminetetraacetate (EDTA) solution prepared in PBS
- 250 mM 5,5-dimethyl N-oxide pyrroline (DMPO) solution prepared in PBS
- Sample extracts, diluted to be within the range of standard curve with acetonitrile
- 50 µL glass capillaries for ESR measurements

9.7.3 Procedure

- Add to a test tube in this order, 30 µL 250 mM DMPO followed by
- 30 µL 1.0 mM H_2O_2;
- 30 µL 1.0 mM EDTA;
- 22.5 µL 100 mM PBS (pH 7.4);
- 7.5 µL lipophilic antioxidant in acetonitrile, or acetonitrile for controls;
- and 30 µL 1.0 mM $FeSO_4$ to start the reaction;
- vortex mixture;
- transfer an aliquot of reaction mixture into two 50 µL glass capillaries;
- begin recording ESR spectrum using the below parameters

9.7.4 ESR Parameters

- Method developed using an X-band Varian E-109 spectrometer (Varian Inc., Palo Alto, CA)
- Scan range: 100 G
- Receiver gain: 8×10^3
- Microwave power: 15 mW
- Field modulation: 1 G
- Temperature: ambient

9.7.5 Calculations

- For sample, standards, or control, determine the ESR peak height at the second peak of the ESR spectra (Fig. 9.11) These values represent the relative concentrations of DMPO-HO˙ adducts present in reaction mixtures.
- Relative peak heights for samples and standards are calculated as control peak height minus sample or standard peak height.
- Standard curve is generated using 4–5 concentrations of a standard such as trolox, and plotting the relative peak height values of standards on the y axis versus standard concentrations in mM on the x axis (Fig. 9.12)
- Calculate mmoles/L trolox equivalents (TE) for each samples using the linear regression equation from the standard curve ($y = 4.6349x + 660.65$) as shown in Fig. 9.12. For example, if a sample relative peak height is 900.0, the TE for the sample would be 51.64 mmoles TE/L.
- Scavenging capacity values in mmoles trolox equivalents/g sample calculated as

$$\text{mmoles trolox equivalent/g} = (\text{mmoles TE/L}) \times \text{DF} \times (L_{\text{solvent}}/g_{\text{sample}})$$

where DF is the dilution factor for sample extract, and L_{solvent} and g_{sample} are the liters of solvent to dissolve grams of sample used for preparing the sample solution (or the liters of solvent used to extract the grams of sample).

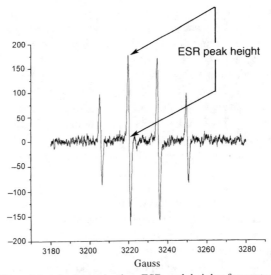

Figure 9.11   ESR peak height determination. ESR peak heights for controls, standards, and samples are determined using the second peak in the HO˙ ESR spectrum. In this example, peak height was 169.

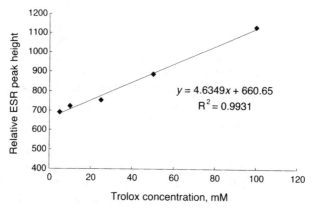

Figure 9.12 Standard curve for hydroxyl radical scavenging capacity estimation using ESR method and trolox standard. ESR stands for electron spin resonance. At least four concentrations of standard such as trolox are required to obtain a linear regression equation, which is used to calculate potential hydroxyl radical scavenging capacity for antioxidant samples. Relative ESR peak height values on y axis were calculated as control ESR peak

9.7.6 Discussion

The most critical step of this method is the addition of $FeSO_4$ to start the generation of hydroxyl radicals and the timing of subsequent steps. Following iron solution addition, consistent vortexing times should be used, as well as consistent timing to begin recording ESR spectra. Given the lifetime of the spin trapping agent, DMPO, ESR spectra must be recorded within 20 min of initiating the reaction (65). In developing this method, Cheng and coworkers (65) investigated the effect of reaction time on the precision of the protocol, and found increasing time from 1 to 20 min to increase the relative standard deviation from 2.94% to 6.21%. Shorter reaction times are therefore preferred to improve method precision. Large day-to-day variability for this method has been reported, making necessitating that all samples for comparison are completed within one day (65). Appropriate preparation and storage of the reagent solutions are also important for this assay. The $FeSO_4$ solution should be prepared fresh daily and stored on ice under nitrogen to prevent oxidation in air. DMPO solution can be made in advance and stored under freezer conditions until use. H_2O_2 solutions should be prepared fresh daily.

The major advantage of this assay versus other antioxidant capacity assays is its ability to determine scavenging capacity of lipophilic antioxidants against the highly reactive hydroxyl radical. In addition, the reaction system used for this assay has been validated to be free of interference from reagents used, such as using solubilizing agents. The major disadvantages of this assay include equipment cost and reproducibility. ESR equipment is expensive and requires a skilled operator. Large day-to-day variations observed were by Cheng and coworkers (65) when developing this method. This means that relative antioxidant capacity measurements are possible for samples, but that values cannot be quantitatively compared with those measured on different days or between labs.

9.8   TOTAL PHENOLIC CONTENTS ASSAY USING THE FOLIN–CIOCALTEU REAGENT

9.8.1   Principles and Background

The total phenolic contents (TPC) of an antioxidant sample may be estimated by measuring its reducing capacity with the Folin–Ciocalteu (FC) reagent using a spectrophotometer. This method measures the ability of samples under basic conditions to reduce the yellow FC reagent causing its color to change to dark blue. Under basic conditions, phenolic groups are deprotonated leading to a phenolate anion with reducing potential. The chemical nature of the FC reagent is not entirely understood but is thought to be a phosphomolybdate-phosphotungstate complex (64). It is believed that molybdenum in this complex, Mo(VI) has the characteristic yellow color, which upon reduction to Mo(V) by phenolate anions becomes blue (64). This method quantifies "total phenolics" or reducing capacity in reference to that of a standard phenolic acid. Gallic acid is the most commonly used TPC standard for wheat antioxidants, with results expressed as mg gallic acid equivalents per g sample.

   The development of this method dates back to 1927 when Folin and Ciocalteu reported the use of the FC reagent for tyrosine determinations in protein taking advantage of its phenol group reactivity (72). Singleton and Rossi later adapted this method for analysis of total phenols in wine including optimization studies for reagent ratios, reagent addition order, timing, and temperature (73). More recently developed versions of this method have included automation of analysis (74,75). The procedure described below is based on the procedure more recently reported by Singleton and others that closely follows their original method (76).

9.8.1.1   Folin–Ciocalteu (FC) Reagent Preparation

- Add 20 g sodium tungstate ($Na_2WO_4 \cdot 2H_2O$), 5 g sodium sodium molybdate ($Na_2MoO_4 \cdot 2H_2O$), and 140 mL distilled water in a 500 mL round-bottomed flask and mix well;
- add to flask 10 mL of 85% phosphoric acid (H_3PO_4) and 20 mL concentrated HCl;
- attach a reflux condenser to the flask, reflux for 10 h with stirring;
- cool solution, add 10 mL water, 30 g lithium sulfate ($Li_2SO_4 \cdot 4H_2O$), and stir the mixture at ambient temperature;
- add a few drops of bromine (in hood), boil for 15 min without condenser (the bromine is used to oxidize any traces of reducing agent that have blue color) in fume hood;
- color should now be yellow without any traces of green;

- cool solution and make up to 200 mL with distilled water;
- solution can be filtered if not completely clear;
- stable indefinitely when stored at ambient temperature in absence of light.

9.8.1.2 Assay Materials and Solutions Preparation

- 20% Na_2CO_3 solution
- 1 mg/mL gallic acid stock solution prepared in the same solvent as sample solutions
- Gallic acids working solutions (at least four concentrations) for preparing standard curve (5–800 µg/mL)
- FC reagent (prepared above)

9.8.2 Procedure

- Add 3 mL water to each test tube;
- add 50 µL sample, standard, or solvent (blank);
- add 250 µL FC reagent to each tube;
- vortex for 5 s;
- wait at least 1 min, but not longer than 8 min;
- add 750 µL 20% Na_2CO_3 solution to each tube;
- sealed test tubes are kept in dark at ambient temperature for 2 h;
- add blank solution to cuvette and blank spectrophotometer set at 765 nm;
- measure absorbance at 765 nm for all standard and sample solutions.

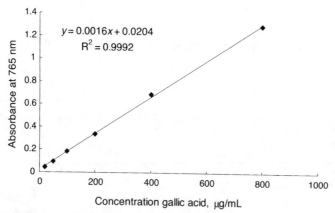

Figure 9.13 Standard curve for TPC method using gallic acid. At least four concentrations of gallic acid are required to obtain a linear regression equation, which is used to calculate TPC values for potential antioxidant samples.

9.8.3 Calculations

- Generate a standard curve by plotting the absorbance values of gallic acid standard solutions against their concentrations as shown in Fig. 9.13.
- μg gallic acid equivalent (GAE)/mL calculated for each sample using linear regression equation from standard curve ($y = 0.0016x + 0.0204$) as shown in Fig. 9.13. For example, a sample testing solution with a Abs. at 765 nm of 0.800 would have a TPC value of 488 mg GAE/mL.
- TPC value of the original sample such as wheat-based food ingredients in mg GAE/g sample may be calculated as

$$\text{mg GAE/g} = (\mu\text{g GAE/mL}) \times (1\,\text{mg}/1000\,\mu\text{g}) \times (\text{mL}_{solvent}/\text{g}_{sample})$$

where $\text{mL}_{solvent}$ and g_{sample} are the mL of solvent and grams of sample used for the sample extraction.

9.8.4 Discussion

TPC values reported for wheat grain and its fractions along with wheat-based food products are summarized in Table 9.6.

TABLE 9.6 Reported TPC Values for Wheat, Its Fractions, and Wheat-based Food Products

Wheat material	Reported TPC range	Units	References
Hard wheat brans	2.2–3.5	mg GE/g	20,32,34,36
Swiss red wheat grain	1.8	mg GE/g	8
Swiss red wheat bran	2.6	mg GE/g	8
Swiss red wheat aleurone	3–4	mg GE/g	8
Hard wheat grains	1.7–9.28	mg GE/g	31,35,77
Soft wheat grains	0.4–0.8	mg GE/g	17
Wheat-based breakfast cereals	0.203–0.35	mg GE/g	33
Hard wheat flours	0.11–0.371	mg GE/g	78
Elena, Ugo, Alba wheat germ	4.4–6.1	mg GE/g	79
Elena, Ugo, Alba wheat grains	4.4–4.9	mg GE/g	79
Soft wheat germ/bran	5.39–5.87[a]	mg GE/g	80
Soft wheat endosperm	0.33–0.37[a]	mg GE/g	80
Hard and soft wheat grains	0.23–0.87[a]	mg GE/g	81,82
Whole wheat bread	~1.2	mg GE/g	82
Hard and soft wheat germs	3.49–12.18[a]	mg FE/g	40,83
Soft and hard wheat flours	0.23–0.82[a]	mg FE/g	40,83
Soft and hard wheat grains	0.35–2.06[a]	mg FE/g	40,83
Soft and hard wheat brans	0.85–6.56[a]	mg FE/g	40,83
Soft and hard wheat shorts	3.76–4.04[a]	mg FE/g	40
Hard and soft wheat grains	0.46–0.55[a]	mg CE/g	39

[a]TPC stands for total phenolic content; GE stands for gallic acid equivalents; FE stands for ferulic acid equivalents; CE stands for catechin equivalents.
Calculated from reported data.

TABLE 9.7   Reported Interfering Compounds for the Total Phenolic Content Estimation

Compound	References
[A] Adenine, adenosine, alanine, aniline, aminobenzoic acid	30
[B] Benzaldehyde	
[C] Creatinine, cysteine, cytidine, cytosine	
[D] Dimethyanaline, diphenylamine	
[E] EDTA	
[F] Fructose	
[G] Guanine, guanosine, glycine	
[H] Histamine, histidine, hydrazine, hydroxyammonium chloride	
[I] Indole, iron ammonium sulfate, iron sulfate	
[M] Manganese sulfate, methylamine	
[N] Nitriloacetic acid	
[O] Oleic acid	
[P] Phenylthiourea, potassium nitrite, proteins, pyridoxine	
[S] Sucrose, sodium cyanide, sodium metabisulfite, sodium phosphate, sodium sulfite, sulfanilic acid	
[T] Thiourea, thymine, thymidine, tin chloride, trimethylamine, tryptophan	
[U] Uracil, uric acid	
[X] xanthine	

The TPC assay has several advantages that have made it a popular assay. This assay is easy to perform, inexpensive, and has acceptable reproducibility. It can be used to analyze large numbers of samples using even a single read spectrophotometer.

The TPC assay has several disadvantages including low specificity for phenols, lack of relevance to biological oxidative processes, and interferences. It measures reducing capacity of phenols or other reducing agents present in samples that can react with the FC reagent, and was originally developed to quantify phenols, the primary reducing agent in wines. Numerous interfering compounds for this assay have been reported and are listed in Table 9.7. In addition to these inferences, our lab has found ethanol extracts of wheat incompatible with this assay due to precipitate formation. It is recommended therefore for ethanol extracts of wheat to be redissolved in DMSO.

A recent report by Stevanato et al. (85) has described a new enzymatic total phenolic content assay that addresses some of the weaknesses of the existing FC assay. This new assay is reported to have greater specificity for phenols and fewer interferences compared to the current FC method. Use of an enzymatic reaction system for this new assay, however, may introduce new interferences through enzyme inhibition, which could lead to overestimation of TPC.

9.9   IRON(II) CHELATING CAPACITY ASSAY

9.9.1   Principles and Background

The Fe^{2+} chelating assay measure the capacity of test antioxidant samples to compete with a synthetic chelator, 2,2'-bipyridine, to form chelating complexes with iron(II)

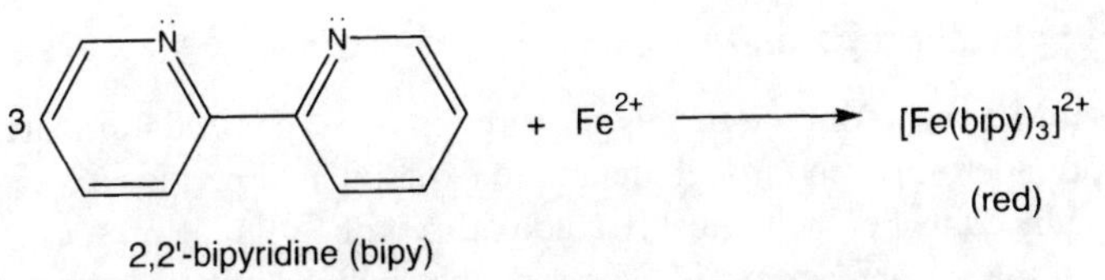

Figure 9.14 2,2'-bipyridine complex with iron(II). See References 90 and 92.

under physiological pH. The chelating complex has a red color and can be quantified at 522 nm with a spectrophotometer. The reduction of $A_{522\ nm}$ with and without the test sample reflects the presence and amount of chelating agent(s) in the sample. Ethylenediaminetetraacidic acid (EDTA), a known chelating agent, is used as a reference chelator (standard) to report the relative Fe^{2+} chelating capacity of antioxidants including wheat antioxidants (20).

Chelating agents may suppress the initiation of the free radical mediated oxidative chain reactions by reducing the availability of transition metals as the catalysts and consequently inhibiting oxidation reactions in biological and food systems. It is well accepted that initiation is the rate limiting phase of the oxidative chain reaction. Chelating agents are an important group of antioxidants and chelating potential is a very important measurement of overall antioxidant capacity for a test sample. Iron is a physiologically relevant transition metal existing as iron(II) and iron(III) and is involved in redox reactions that may lead to the formation of ROS (86). The reaction of 2,2'-bipyridine (also referred to as 2,2'-dipyridine and abbreviated as bipy) with iron(II) ions (Fig. 9.14) dates back to the late 1800s when Blau observed the intensely colored red substances formed (87–89). 2,2'-bipyridine as a bidentate ligand with a lone pair of electrons on each of the two nitrogen atoms, creates five membered rings with iron(II) and forms a colored stable complex at pH 7.4 (Fig. 9.14). Use of 2,2'-bipyridine was later described by Moss and Mellon in 1942 (90) for colorimetric iron determinations and is still widely used today for this purpose. A report by Yamaguchi and coworkers in 2000 described the use of the 2,2'-bipyridine reagent in a competitive assay to determine the chelating activity of botanical extracts (91). This method by Yamaguchi is the basis for the assay presented below.

9.9.1.1 Materials and Solutions Preparation

- 1 M Tris-HCl Buffer solution (pH 7.4)
- 0.1% (w/v) 2,2'-bipyridyl in 0.2N HCl
- 10% (w/v) hydroxylamine hydrochloride
- 400 μg/mL EDTA stock solution prepared in the same solvent used to dissolve antioxidant samples
- EDTA standard solutions (10–100 μg/mL) prepared in the same solvent
- Sample solutions may be diluted with the solvent to be within EDTA standard curve range

9.9.1.2 Special Preparation

- 1.8 mM $FeSO_4$ solution freshly prepared in deoxygenized ultrapure water and kept under nitrogen through the entire experimental period.
  - Since this Fe^{2+} will quickly autooxidize to Fe^{3+} in the presence of oxygen, a special preparation is necessary that bubbles nitrogen gas through the solution to prevent this process;
  - setup a three-neck flask as shown in Fig. 9.15. The center neck should be covered with glass stopper. Into one of the outer necks is a glass pipette

(a)

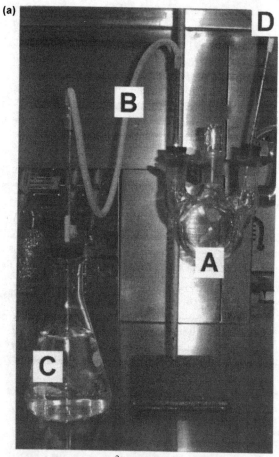

Figure 9.15 Recommended setup for Fe^{2+} solution. **A**—250 mL three neck flask attached to stand with three-finger clamp; **B**—connection between three neck flask and Erlenmeyer flask filled with water to capture gasses, using rubber tubing and disposable glass pipettes; **C**—erlenmeyer flask glass pipette submerged in water; **D**—connection between three neck flask and nitrogen gas tank using rubber tubing; **E**—glass pipette connected to nitrogen gas supply. Note that pipette tip should be at bottom of three neck flask to allow nitrogen to bubble up through Fe^{2+} solution; **F**—glass stopper in three neck flask where aliquots of Fe^{2+} solution are withdrawn from to start assay reactions. It needs to be pointed out that the Fe^{2+} solution should be stirred constantly during the entire assay.

(b)

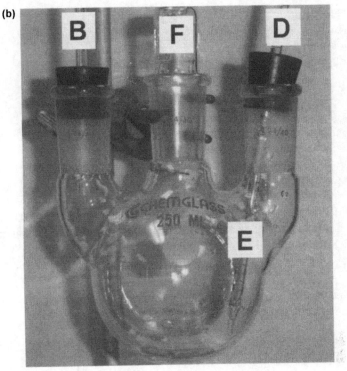

Figure 9.15   (*Continued*)

connected to a source of nitrogen gas and sealed with a rubber stopper. Into the other outer neck is a glass tube sealed with rubber stopper and connected to a flask of water to prevent entry of air into the system and allow escape of nitrogen gas;

o   add ultrapure water that has been sonicated to remove dissolved oxygen into the three-neck flask, and turn on nitrogen gas to start bubbling through water to remove any dissolved oxygen;

o   dissolve $FeSO_4$ in water in three-neck flask;

o   remove aliquots of this solution as needed for assay reactions from the center neck.

9.9.2   Procedure

- Adjust spectrophotometer to 522 nm and set zero using the reagent blank;
- add 200 µL of 1M Tris-HCI buffer to a test tube;
- add 500 µL sample solution or EDTA standard;
- add 50 µL of 0.1% 2,2′-bipyridyl solution;

- add 200 μL of 10% hydroxylamine hydrochloride;
- add 30 μL of 1.8 mM FeSO$_4$ freshly taken from three-neck flask;
- vortex for 3 s;
- transfer to cuvette and record absorbance at 522 nm.

9.9.3 Calculations

- Generate a standard curve by plotting the $A_{522\ nm}$ values of EDTA standard solutions against their concentrations as shown in Fig. 9.16.
- μg EDTA equivalents/mL can be calculated for each sample using linear regression equation from standard curve ($y = -0.0025x + 0.3751$) as shown in Fig. 9.16. For example, chelating capacity of a test sample with an $A_{522\ nm}$ value of 0.200 may would be 70.2 μg EDTA equivalents/mL from Fig. 9.16.
- Chelating capacity may be reported on a per solid sample weight basis in μg EDTA/g sample. This may be calculated as

$$\mu g\ \text{EDTA/g} = (\mu g\ \text{EDTA/mL}) \times (\text{mL}_{\text{solvent}}/g_{\text{sample}})$$

where and mL$_{\text{solvent}}$ and g$_{\text{sample}}$ are the mL of solvent and grams of sample in the sample solution or used for the sample extraction.

9.9.4 Discussion

The iron(II) chelating capacity assay presented in this section has been utilized to evaluate a range of botanical extracts including many studies on wheat. Table 9.8 summarizes the reported iron(II) chelating capacity values for wheat and its fractions along with wheat-based food products using the method described in this section. The major advantage of this assay is that it can be used to evaluate large numbers of samples

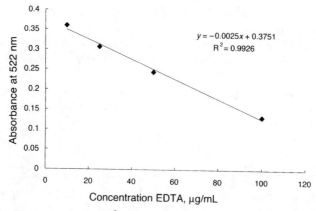

Figure 9.16 Standard curve for Fe^{2+} chelating capacity method using EDTA. At least four concentrations of EDTA are required to obtain a linear regression equation, which is used to calculate Fe^{2+} chelating capacity values for potential antioxidant samples. EDTA stands for ethylenediaminetetraacetic acid.

TABLE 9.8   Reported Fe(II) Chelating Capacity Values for Wheat, Its Fractions, and Wheat-based Food Products

Wheat material	Reported Fe(II) chelating capacity, mg EDTA equiv/g	References
Hard and soft wheat brans	0.375–2.04	20,32,34,36,38,84
Swiss red wheat grain	~0.4	8
Swiss red wheat bran	~1.4	8
Swiss red wheat aleurone	~1.25–1.75	8
Soft wheat grains	0.11–0.79	17,84
Hard wheat grains	0.629–5.4	35
Wheat based breakfast cereals	0.33–0.48	33
Hard and soft wheat flours	0.3–1.7	78,84

using a single read spectrophotometer. The major disadvantage of this assay is that it does not evaluate additional chelator properties which may be important to understand a compound's ability to inhibit iron catalyzed ROS generation including iron(III) selectivity, rigidity of ligand conformations, availability of coordination sites, affinity towards iron ions, and stoichiometry of the complex (93,94).

There are two critical points for this assay that must be followed for reproducible results. First, as measurements are made before the reaction reaches steady state, consistent timing through the assay procedure is required. The second critical point is to protect ferrous from oxidation through the assay. If iron(II) is oxidized, the resulting iron(III) can disassociate from the 2,2′-bipyridine (bipy) complex and reform a different complex with bipy with brown color, which can significantly interfere with the assay (95). In addition, solubilizing agents may be required if the test sample contains lipophilic compounds, which may form precipitate during the assay and interfere with measurements. Sodium dodecyl sulphate (SDS) has been used as solubilizing agent, with a final concentrations of 0.1% (w/v) (91). The iron(II) solutions should be prepared for each assay according to the procedure described above. 2,2′-bipyridyl solution should be prepared fresh daily, while other solutions are relatively stable and can be stored for about 1 month at ambient temperature.

9.10   COPPER(II) CHELATING CAPACITY ASSAY

9.10.1   Principles and Background

The Cu^{2+} chelating capacity assay is a qualitative assay that utilizes low-temperature electron spin resonance (ESR) to evaluate changes in cupric ion spectra as a result of its interactions with sample components. Copper is a biologically essential transition metal, in part due to its role as an enzyme cofactor in redox reactions. Similar to iron, however, the ability of copper to undergo reversible redox changes allows it to catalyze the formation of reactive oxygen species (96). Copper-induced oxidative damage has been implicated in the pathogenesis of neurodegenerative diseases such as

Alzheimer's and Parkinson's diseases (96). Copper is also thought to play a role in the formation of oxidized low-density lipoprotein, possibly involved in the pathogenesis of atherosclerosis (97). Chelation of copper ions is thought to be a potential mechanism to stabilize this reactive metal and inhibit its catalytic activity in formation of damaging free radical species.

Several qualitative spectrophotometric methods have been reported to characterize the Cu^{2+} complex forming capacity of test compounds. UV/Vis spectroscopy has been utilized by several groups, comparing the absorption spectra of sample compounds with those of sample compounds mixed with cupric ions (98). Lodge and coworkers (97) and Kong and coworkers (99) reported a UV/Vis spectroscopy assay. In this assay, colored cupric complexes were formed, and the potential decrease in absorption monitored after addition of antioxidants that may be able to competitively bind to the copper ions forming new complexes. While cupric ions do not absorb in the UV/Vis range, they do have an ESR spectra due to their paramagnetic nature. Use of ESR therefore allows direct detection of changes in cupric ion properties as a result of coordinating interactions with antioxidants. Several groups have reported use of ESR under low temperature conditions to characterize antioxidant-Cu^{2+} complexes (100,101). The below described method comes from Zhou and coworkers (38) who reported the use of low-temperature ESR to evaluate the cupric ion chelating capacity of wheat extracts.

9.10.2   Materials and Solutions Preparation

- 2 mM $CuCl_2$ solution prepared fresh daily in water
- Extracts of wheat samples
- Liquid nitrogen

9.10.3   Procedure

- Combine 150 μL 2 mM $CuCl_2$ and 150 μL wheat extracts
- Vortex
- Transfer reaction mixture into two 50 μL glass capillaries
- Place capillaries in liquid nitrogen to freeze the reaction mixture
- Begin recording ESR spectrum immediately after inserting the capillaries containing frozen reaction mixtures into ESR machine

9.10.4   ESR Parameters

- Method developed using an X-band Varian E-109 spectrometer (Varian Inc., Palo Alto, CA)
- Microwave power: 40 MW
- Field modulation: 5 G

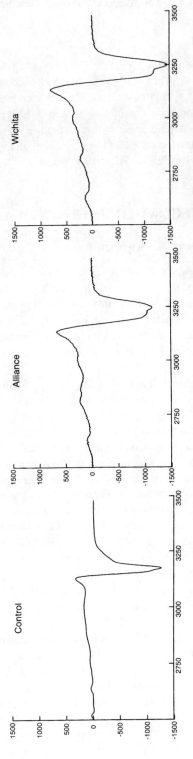

Figure 9.17 Complexes formed by Cu^{2+} and individual hard wheat bran extracts measured by ESR. The final concentrations were 1 mM for copper chloride ($CuCl_2$) and 50 mg of bran equiv/mL for bran extracts. ESR spectra were recorded at 1 min of reaction at 77K. Reproduced with permission (38).

9.10.5 Discussion

Zhou and coworkers (38) first reported the Cu^{2+} chelating capacity of wheat extract using low-temperature ESR (Fig. 9.17). Recently, Su and coworkers (102) determined the Cu^{2+} chelating capacity of 50% acetone extracts from black peppercorn, nutmeg, rosehip, cinnamon, and oregano leaf following the same ESR protocol. These previous studies demonstrate the potential use of this method in measuring the formation of antioxidant-Cu^{2+} chelating complexes. The major limitation of this method is that it cannot be used to quantify the Cu^{2+} chelating capacity.

9.11 LIPID PEROXIDATION INHIBITION ASSAY (OSI)

9.11.1 Principles and Background

Lipid oxidation is an important chain reaction process because of the damaging effects it can have in food systems and *in vivo* in biological systems. Lipid oxidation in food systems is a major cause of quality deterioration including development of rancidity and loss in nutritional value and is important to food product quality. *In vivo*, lipid oxidation and its products have been linked to a number of chronic inflammatory and neurodegenerative diseases, making it of great interest for human health (103).

Accelerated oxidation tests using increased temperatures, oxygen pressure, or catalysts have commonly been used to evaluate the oxidative stability of oil and effectiveness of antioxidants in suppressing lipid peroxidation in oils and fats (104). Determination of induction time (also referred to as induction period) for these accelerated tests is typically used to determine the oxidative stability of lipids (104). Numerous physical and chemical markers of lipid oxidation have been used to determine induction time including oxygen uptake, peroxide value, thiobarbituric acid reactive substances, direct detection of free radicals, thermal release, head space volatiles analysis, and conductivity (104,105). The capacity of a selected antioxidant to suppress to lipid oxidation in fats and oils can be evaluated using the oxidative stability index (OSI) assay (106,107). The OSI method is a standardized assay developed to evaluate the oxidative stability of fats and oils under accelerated conditions such as elevated temperature. This assay can be automated using two possible pieces of equipment, the Rancimat (Metrohm Ltd, Herisau, Switzerland) or the Oxidative Stability Instrument (Omnion Inc, Rockland, MA, USA) using a protocol outlined by the American Oil Chemists Society (106). This assay measures the secondary products of lipid peroxidation and their derivatives. Results for this assay are generally reported as "induction time," the time at which rapid formation of secondary volatile products from lipid oxidation is observed. This time point may be automatically calculated with software for these instruments. It is calculated as the inflection point of the reaction curve obtained by plotting the total amount of measurable secondary products of lipid oxidation against the reaction time as shown in Fig. 9.18. The capacity of a selected antioxidant sample in prevention of lipid oxidation in oils may be estimated by comparing the induction time of an oil with and

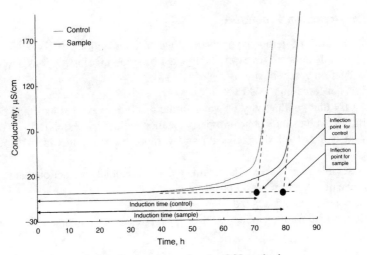

Figure 9.18   Calculation of induction time using the OSI method.

without the antioxidant. Results may be reported in hours beyond control, or protection factor or index (PF) calculated as induction time of sample divided by induction time of control (108). Results for samples measured with this assay cannot be attributed to a single antioxidant mechanism, but rather to one or a combination of mechanisms. This assay has been used to evaluate the potential of wheat antioxidants in suppressing lipid peroxidation in fish oil using the Rancimat instrument. The detailed procedure of this assay is described below.

9.11.1.1   *Materials and Solutions Preparation*

- Substrate oil of choice (stripped or antioxidant free)
- Antioxidant samples
- Deionized or distilled water with conductivity $<5\,\mu S/cm$
- Acetone (for cleaning)
- One of the following detergents (for cleaning) that does not leave contaminating residues: Liquinox, Contrad 70, Citranox RBS, or Micro

9.11.1.2   *Oil Preparation*

- Sample extract should be added to substrate oil, and solvent removed by evaporation under vacuum at 40–50°C.
- Same volume of sample extract solvent should be added to control oil samples and solvent removed under the same conditions as sample oils

9.11.1.3 *Reaction Procedure*

- Add 70 mL of deionized or distilled water to collection vessels
- Assemble collection vessels as directed by Rancimat instructions and set into collection vessel positions
- Preheat Rancimat to 80°C
- Verify that conductivity of water in the receiving vessel is less than 5 μS/cm (higher values indicate improperly cleaned collection vessels)
- Add 6 mL of oil samples with and without the test antioxidant to reaction vessels
- Assemble reaction vessels as directed by Rancimat instructions and set into preheated reaction vessel positions
- Start air-flow at 7 L/h and start timing
- Stop reactions after induction time for the reaction is obtained

9.11.2 Results

- Induction times (IT) is automatically calculated using Rancimat software or can be manually calculated as the maximum of the second derivative for conductivity versus time plot.
- Antioxidant effectiveness can be calculated as extension time or protection factor (PF) as shown below.

$$\text{Extension time} = IT_{sample} - IT_{control}$$

$$\text{Protection factor(PF)} = IT_{sample} / IT_{control}$$

where IT_{sample} and $IT_{control}$ represent the induction time of oils with and without testing antioxidant samples.

- Results should be reported as extension time or protection factor, with information given on conditions including sample concentration in oil, substrate oil, reaction temperature, and reaction air flow rate.

9.11.3 Discussion

The inhibitory capacity of wheat antioxidants on lipid oxidation in oils have been reported using the OSI assay. These data are summarized in Table 9.9. It should be noted that reaction conditions and substrate oils differed for each reported set of data that makes direct comparisons of these results difficult.

It is very critical for this assay to make sure all reaction and receiving vessels are free of contaminates, including trace metals, surface-active agents from detergents, and reaction residues from previous runs. It is recommended that reaction conditions including temperature, air flow rate, and sample extract concentrations should be adjusted so that induction times for controls and samples are between 4 and 15 h for best precision (106). Compared to other accelerated oxidative stability assays that can be used to evaluate antioxidant capacity such as peroxide value, TBARS,

TABLE 9.9   Reported Oxidative Stability Index Values for Wheat, Its Fractions, and Wheat-based Food Products

Wheat material	Reported OSI PF range	Wheat antioxidant concentration in oil	Testing conditions	References
Hard wheat grains	1.2–2.1[a]	0.06–0.09% (wheat extract in oil, w/w)	Fish oil, 80°C, air flow rate of 7 L/h	19
Hard and soft wheat flours	1.12–1.16	12.5 mg wheat equiv/mL oil	Stripped corn oil, 100°C, air flow rate of 20 L/h	84
Hard and soft wheat grains	1.19–1.21	12.5 mg wheat equiv/mL oil	Stripped corn oil, 100°C, air flow rate of 20 L/h	84
Hard and soft wheat brans	1.9–2.1	12.5 mg wheat equiv/mL oil	Stripped corn oil, 100°C, air flow rate of 20 L/h	84
Wheat bran	1.35	200 mg wheat/mL oil	Butter, 120°C, NA[b]	109
Wheat bran breakfast cereal	1.91	200 mg wheat/mL oil	Butter, 120°C, NA[b]	109

[a]PF stands for protection factor and may be calculated as induction time of sample divided by induction time of control. Method details include substrate oil type used, oil reaction temperatures, and air flow rates. Values calculated from data presented in manuscript.
[b]NA indicates that air flow rate was not reported.

and others, the OSI method has the advantage of being a simple assay and requiring a less skilled operator.

Similar to other antioxidant capacity assays, OSI does have several disadvantages. High-temperature accelerated oxidative stability tests including OSI have been criticized since changes in reaction temperature significantly alter the reaction mechanisms of lipid oxidation (104,105). Several limitations under high-temperature conditions versus normal oxidative conditions have been cited including (a) changes in oxygen availability due to its limited solubility at higher temperatures and faster consumption of oxygen leading to higher dependence of oxidation rate on local oxygen concentration, (b) possibility of side polymerization and cyclization reactions that do not occur at ambient temperature in air, and (c) possible thermal degradation of antioxidants and oil components (104,105). Compared to other assays for measuring antioxidant capacity, the OSI assay does not measure antioxidant capacity through a specific mechanism such as radical scavenging or metal chelating, but rather measures inhibition of overall lipid oxidation that could occur through multiple mechanisms. A major disadvantage of OSI compared to other assays for measuring antioxidant capacity such as ORAC, HOSC, and others is its lack of standardized conditions and an adequate quantification method. This makes comparisons of results between samples and labs difficult. Reports of natural extract antioxidant capacity using OSI have utilized various substrate oils, reaction temperatures, and air flow rates, all of which can affect results. While standardization of temperature and air flow conditions can be more easily addressed, oil substrate standardization poses a larger challenge. A recent report by Nakatani and coworkers (110) has suggested the use of methyl linoleate as a model oil substrate for OSI assays, which may make the interlaboratory comparisons easier. Lastly, inclusion of a positive control such as α-tocopherol or BHA is recommended and will help to compare the research data from different laboratories.

9.12 LOW-DENSITY LIPOPROTEIN (LDL) PEROXIDATION INHIBITION ASSAY

9.12.1 Principles and Background

This assay estimates the capacity of samples to prevent copper(II)-induced lipid oxidation to LDL by measuring the amount of secondary lipid oxidation products capable of reacting with thiobarbituric acid. These secondary lipid oxidation products, thiobarbituric acid-reactive substances (TBARS), are quantified using 1,1,3,3-tetra-ethoxypropane as a standard. Results are expressed as milligrams of TBARS reduction per gram of sample relative to a solvent control.

Oxidative modification of LDL is thought to play an important role in the pathogenesis of atherosclerosis (111). An important part of this modification is thought to be a lipid peroxidation process that may lead to protein oxidation and structural changes of LDL particles (112). The oxidized LDL cannot be recognized by liver LDL receptor and are uptaken by macrophage cells leading to lipid-laden foam cell formation, an important step in the pathogenesis of atherosclerosis (113). It has therefore become of great interest how dietary antioxidants can inhibit this oxidative process (114). While the exact mechanism of oxidation *in vivo* is still debated, cultured cell studies have shown that LDL oxidation is dependent on transition metal ions in cell media (113). Of the physiologically relevant transition metals, copper(II) has been the most commonly used inducer of oxidation for *in vitro* studies of LDL oxidation (3). Mao and coworkers in 1991 (115) were one of the first groups to report the antioxidative capacity of compounds to inhibit the Cu^{2+}-induced oxidation of LDL, by measuring the levels of TBARS formed with and without the presence of an antioxidant in the reaction suspension. Since then, numerous markers of lipid oxidation besides TBARS have been utilized in LDL oxidation experiments including conjugated dienes, total lipid peroxides, oxysterols, hydroxyl and hydroperoxy fatty acids, aldehydes, lysopho-sphatides, oxygen consumption, head space hexanal, and disappearance of polyun-saturated fatty acids (114,116). Of these, the TBARS and conjugated dienes method have been the most widely used in studies examining the antioxidant properties of compounds in the Cu^{2+} LDL system. It is possible, however, that pigments in antioxidant samples may interfere the determination of total conjugated dienes in the reaction suspensions. The Cu^{2+}-induced lipid oxidation in human LDL has been used by Yu and coworkers 2005 (37) to evaluate the inhibitory effect of wheat antioxidants on lipid peroxidation in LDL. The detailed protocol for evaluating antioxidant potential in suppressing lipid oxidation in human LDL is provided below using Cu^{2+} to induce the lipid oxidation. Additionally the laboratory protocols for LDL preparation (117), protein determination with the Bradford method (118), and the TBARS assay for measurements of lipid peroxidation (119) are provided.

9.12.1.1 Materials, LDL Preparation

- Commercial human LDL-EDTA solution (Sigma-Aldrich #L7914, or similar)
- 0.01 M phosphate buffer (PBS), pH 7.4, made oxygen free by vacuum degassing followed by nitrogen gas purging
- Dialysis tubing, 12,000–14,000 MWCO, and clamps

9.12.1.2 Materials, Bradford Protein Determination

- Bradford reagent prepared by dissolving 100 mg Coomassie brilliant blue G-250 in 50 mL 95% ethanol, adding 100 mL 85% phosphoric acid, diluting to 1 L with water, and filtering through Whatman#1 just before use (118)
- Prepare bovine serum albumin (BSA) standards containing 10–200 µg protein in 100 µL PBS
- Unknown concentration LDL in PBS, diluted to be within standard curve range

9.12.1.3 Materials, LDL Oxidation

- 150 µM $CuCl_2$ prepared in PBS
- DMSO solutions of antioxidant sample, diluted to be within the range of standard curve for TBARS if necessary
- Pure DMSO for control
- 200 µg/mL LDL protein working solution prepared in the 0.01M PBS

9.12.1.4 Materials, TBARS Method

- 0.02 M thiobarbituric acid (TBA) prepared in 90% acetic acid
- 10 mg/mL 1,1,3,3-tetraethoxypropane (TEP) primary stock solution prepared in EtOH
- 0.01 mg/mL 1,1,3,3-tetraethoxypropane (TEP) working stock solution prepared in EtOH
- TEP working standard solutions, 0.2–4 µg/mL prepared by diluting the 0.01 mg/mL TEP solution with 0.01 M pH 7.4 PBS

9.12.1.5 Procedure, EDTA Free LDL Preparation

- Dialyze LDL with 100-fold volume of oxygen-free 0.01M pH 7.4 PBS, changing buffer 4 times, with constant nitrogen purging and gentle stirring
- Determine protein content using Bradford assay (Bradford, 1976)
- Prepare 200 µg protein/mL stock solution of LDL in oxygen free 0.01M PBS
- Store solution at 4°C under nitrogen for no longer than 24 h before use

9.12.1.6 Procedure, LDL Oxidation

- Add 500 μL LDL stock solution (200 μg protein/mL) to a test tube;
- add 380 μL 0.01M pH 7.4 PBS;
- add 20 μL DMSO solution of the antioxidant sample or pure DMSO for control;
- add 100 μL CuCl$_2$ stock solution (150 μM) to start lipid peroxidation reaction;
- allow reaction to proceed for 60 min at ambient temperature.

9.12.1.7 Procedure, TBARS Assay for Oxidized LDL

- Add 1 mL of TEP standards or water (reagent blank) to test tubes;
- to each tube (LDL oxidation reaction, TEP standards, or blank) add 1 mL of TBA reagent;
- vortex for 5 s;
- keep all test tubes in boiling water bath for 30 min;
- cool the test tubes to ambient temperature;
- blank spectrophotometer at 532 nm with prepared reagent blank solution;
- measure absorbance of each LDL oxidation sample and TEP standards at 532 nm.

9.12.1.8 Calculations, Inhibition of LDL Oxidation Using TBARS

- Generate standard curve by plotting $A_{532\ nm}$ values of TEP standard solutions against TEP concentrations (μg/mL), Fig. 9.19.
- Calculate μg TBARS/mL for each sample and control using linear regression equation from the standard curve ($y = 0.2222x - 0.0141$) from Fig. 9.19. For example, if the sample $A_{532\ nm}$ value is 0.600, the sample TBARS would be 2.76 μg TEP/mL.
- Inhibition of LDL oxidation in mg TBARS reduction/g sample is calculated as

$$= (TBARS_{control}/mL - TBARS_{sample}/mL) \times DF \times (mL_{solvent}/g_{sample})$$

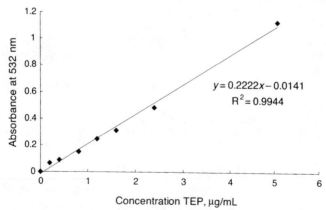

Figure 9.19 TBARS standard curve for LDL oxidation inhibition assay. At least four concentrations of TEP are required to obtain a linear regression equation, which is used to

TABLE 9.10 Inhibitory Effects of Wheat Antioxidants on Lipid Peroxidation in LDL

Wheat material	Inhibitory capacity	Units	Detection method	References
Hard wheat brans	1.03–1.56	mg TBARS reduction/g bran/100 µg protein	TBARS	37
Soft and hard wheat flours	1472–1997	µg protein protected/g defatted material[a]	CD	84
Soft and hard wheat grains	3315–3795	µg protein protected/g defatted material[a]	CD	84
Soft and hard wheat brans	4845–4978	µg protein protected/g defatted material[a]	CD	84

TBARS stands for thiobarbituric acid reactive substances and CD stands for conjugated dienes, both are secondary products of lipid peroxidation.

[a]The capacity of wheat antioxidant in preventing LDL oxidation was determined by measuring diene formation induced by copper(II) ions, calculated as percent inhibition and reported as µg protein protected/g defatted material (84).

where DF is the dilution factor for antioxidant sample (if diluted), and $mL_{solvent}$ and g_{sample} are the mL of solvent and grams of sample used to prepare the sample solution or extract.

9.12.2 Discussion

Wheat grain and its fractions have been investigated for their potential in suppressing lipid peroxidation in LDL. These previous data are summarized in Table 9.10.

The most critical points for the inhibition of LDL oxidation assay are consistent timing and reaction conditions through the assay. Also important are proper preparation and storage of solutions. Purchased EDTA-LDL solution should be stored at 4°C under nitrogen until use. Oxidation of LDL stock solutions should be minimized by dialyzing in oxygen-free PBS with nitrogen purging and by using LDL stock within 24 h of preparation. In addition, TEP and BSA stock solutions should be prepared fresh daily. Sodium dodecyl sulphate (SDS) and β-cyclodextrin have been added to reaction mixtures in our lab to solublize lipophilic antioxidants at final concentrations of 1.0% or 3.0% (w/v) respectively, although their potential interference has not been investigated.

The major advantage of the LDL oxidation inhibition assay compared to others presented in this chapter is its relevance to *in vivo* events. This method measures the degree of lipid oxidation in LDL, an important event related to the development of atherosclerosis *in vivo*, and uses copper(II), a physiologically relevant inducer of lipid oxidation. Use of antioxidants as a strategy to prevent oxidation of LDL and thereby reduce the risk of atherosclerosis has been of great interest recently (120). This assay has found wide-spread use in screening the antioxidant activity of compounds.

Similar to all antioxidant capacity assays, the LDL oxidation inhibition test has disadvantages and limitations. First, this method is quite complex, involving a number of time consuming steps, and has not been adapted for high-throughput analysis.

Second, this assay uses fixed reaction time for each steps involved for quantification that unlike area under the curve calculations, does not take into account both kinetic and thermodynamic properties of each reaction involved in the measurements. The short 1 h oxidation reaction time used in this assay likely does not allow the reaction to reach steady state and, therefore, may also cause variability in results if consistent timing is not used. A simple improvement in this assay could therefore be a kinetic study of the oxidation reactive system to determine an appropriate end time when the reaction has reached steady state, thereby reducing potential variability (114). As mentioned earlier, numerous methods have previously been used with this assay used to determine the extent of lipid oxidation for samples. The TBARS assay specifically has been criticized by several authors for its lack of specificity in evaluating lipid oxidation products (1,12,114). A review of methods to access LDL oxidation by Puhl and Esterbauer, however, has argued that the reaction system used in this assay lacks the complexity that other biological systems have that cause interference (114). This review recommends the use of the simple TBARS method for single time-point measurements of LDL oxidation (114). Use of the conjugated dienes method for estimating LDL oxidation is also recommended by Puhl and Esterbauer (114), but for kinetic measurement of LDL oxidation.

9.13 CONCLUSIONS

While all the assays presented in this chapter have been evaluated and utilized by researchers to investigate antioxidant properties, the issues and limitations described in this chapter point to the continual need for not only improved assays but also standardized assays that will enable better comparisons between laboratories. Further discussion of these and other antioxidant capacity estimation methods, and recommendations for standardization can be found in numerous recent reviews (1–3,9,12,121–123).

One notable recent area of improvements for antioxidant capacity assays has involved the use of electron spin resonance (ESR) techniques to directly detect free radicals. While some free radicals such as the DPPH and ABTS cation radicals can be directly detected spectrophotometrically due to their unique spectral properties, most physiologically relevant free radicals require more modern techniques for direct detection. ESR has been extensively utilized for direct characterization and quantification of stable free radicals, and unstable radicals with spin-labeling techniques. This has enabled more direct measurement of free radical scavenging capacity of potential antioxidative components. ESR can also be utilized to directly detect and describe metal ion complexes formed by chelating compounds. The use of ESR techniques in antioxidant capacity assays will be discussed separately in this book.

ABBREVIATIONS

AAPH   2,2′-Azobis(2-amidinopropane) dihydrochloride
ABTS$^{•+}$ 2,2′-Azinobis (3-ethylbenzothiazoline-6-sulfonic acid) cation radical

AE	Antiradical efficiency
AOC	Antioxidant capacity assay
AOCS	American Oil Chemists Society
ARP	Antiradical power
AUC	Area under the kinetic curve
bipy	$2,2'$-Bipyridine
BSA	Bovine serum albumin
DPPH$^{\bullet}$	2,2-Diphenyl-1-picrylhydrazyl radical
ESR	Electron spin resonance
ET	Electron transfer
FC	Folin and Ciocalteu
FL	Fluorescein
GAE	Gallic acid equivalents
HAT	Hydrogen atom transfer
HOSC	Hydroxyl radical scavenging capacity
HPX-XOD	Hypoxanthine-xanthine oxidase
LDL	Low-density lipoprotein
NBT	Nitroblue tetrazolium
$^{\bullet}$NO	Nitric oxide
$^{\bullet}$NO$_2$	Nitrogen dioxide
$^{\bullet}$OH	Hydroxyl radical
O$_2^{\bullet -}$	Superoxide anion radical
OONO$^-$	Peroxynitrite
ORAC	Oxygen radical absorbing capacity
OSI	Oxidative stability index
PBS	Phosphate buffer
PF	Protection factor
PV	Peroxide value
ROO$^{\bullet}$	Peroxyl radical
RAUC	Relative area under the curve
ROS	Reactive oxygen species
RSE	Radical scavenging efficiency
RDSC	Relative DPPH$^{\bullet}$ scavenging capacity
SET	Single electron transfer
SOD	Superoxide dismutase
SPLET	Sequential proton loss electron transfer
TE	Trolox equivalents
TEP	1,1,3,3-Tetraethoxypropane
TEAC	Trolox equivalent antioxidant capacity
TPC	Total phenolic contents

REFERENCES

1. Halliwell, B. Food-derived antioxidants: how to evaluate their importance in food and in-vivo. In: Cadens, E.;Packer, L. (Eds.). *Handbook of Antioxidants* 2nd edn. Marcel Dekker, New York, **2002**. pp.1–45.

2. Huang, D.; Ou, B.; Prior, R. L. The chemistry behind antioxidant capacity assay. *J. Agric. Food Chem.* **2005**, *53*, 1841–1856.

3. Frankel, E. N.; Meyer, A. S. The problems of using one-dimensional methods to evaluate multifunctional food and biological antioxidants. *J. Sci. Food Agric.* **2000**, *80*, 1925–1941.

4. Zhang, H. -Y.; Ji, H. -F. How vitamin E scavenges DPPH radicals in polar protic media. *New J. Chem.* **2006**, *30*, 503–504.

5. Leopoldini, M.; Marino, T.; Russo, N.; Toscano, M. Antioxidant properties of phenolic compounds: H-atom versus electron transfer mechanism. *J. Phys. Chem. A* **2004**, *108*, 4916–4922.

6. Nielsen, M. F.; Ingold, K. U. Kinetic solvent effects on proton and hydrogen atom transfers from phenols. Similarities and differences. *J. Am. Chem. Soc.* **2006**, *128*, 1172–1182.

7. Foti, M.; Ruberto, G. Kinetic solvent effects of phenolic antioxidants determined by spectrophotometric measurements. *J. Agric. Food Chem.* **2001**, *49*, 342–348.

8. Moore, J.; Yin, J.; Yu, L. Novel fluorometric assay for hydroxyl radical scavenging capacity (HOSC) estimation. *J. Agric. Food Chem.* **2006**, *54*, 617–626.

9. Perez-Jimenez, J.; Saura-Calixto, F. Effect of solvent and certain food constituents on different antioxidant capacity assays. *Food Res. Int.* **2006**, *39*, 791–800.

10. Blois, M. S. Antioxidant determinations by the use of a stable free radical. *Nature* **1958**, *181*, 1199–1200.

11. Brand-Williams, W.; Cuvelier, M. E.; Berset, C. Use of a free radical method to evaluate antioxidant activity. *Lebensm.-Wiss. Technol.* **1995**, *28*, 25–30.

12. Roginsky, V.; Lissi, E. A. Review of methods to determine chain-breaking antioxidant activity in food. *Food Chem.* **2005**, *92*, 235–254.

13. Cheng, Z.; Moore, J.; Yu, L. A high-throughput relative DPPH radical scavenging capacity (RDSC) assay. *J. Sci. Food Agric.* **2006**, *54*, 7429–7436.

14. Sanchez-Moreno, C.; Larrauri, J. A.; Saura-Calixto, F. A procedure to measure the antiradical efficiency of polyphenols. *J. Sci. Food Agric.* **1998**, *76*, 270–276.

15. De Beer, D.; Joubert, E.; Gelderblom, W. C. A.; Manley, M. Antioxidant activity of South African red and white cultivar wines: free radical scavenging. *J. Agric. Food Chem.* **2003**, *51*, 902–909.

16. Jimenez-Escrig, A.; Jimenez-Jimenez, I.; Sanchez-Moreno, C.; Saura-Calixto, F. Evaluation of free radical scavenging of dietary carotenoids by the stable radical 2,2-diphenyl-1-picrylhydrazyl. *J. Sci. Food Agric.* **2000**, *80*, 1686–1690.

17. Moore, J.; Hao, Z.; Zhou, K.; Luther, M.; Costa, J.; Yu, L. Carotenoid, tocopherol, phenolic acid, and antioxidant properties of Maryland-grown soft wheat. *J. Agric. Food Chem.* **2005**, *53*, 6649–6657.

18. Zhou, K.; Laux, J. J.; Yu, L. Comparison of Swiss red wheat grain, and fractions for their antioxidant properties. *J. Agric. Food Chem.* **2004**, *52*, 1118–1123.

19. Yu, L.; Haley, S.; Perret, J.; Harris, M. Antioxidant properties of hard winter wheat extracts. *Food Chem.* **2002**, *78*, 457–461.

20. Zhou, K.; Su, L.; Yu, L. Phytochemical and antioxidant properties in wheat bran. *J. Agric. Food Chem.* **2004**, *52*, 6108–6114.

21. Miller, N. J.; Rice-Evans, C.; Davies, M. J.; Gopinathan, V.; Milner, A. A novel method for measuring antioxidant capacity and its application to monitoring the antioxidant status in premature neonates. *Clin. Sci.* **1993**, *84*, 407–412.

22. Arnao, M. B.; Cano, A.; Hernandez-Ruiz, J.; Garcia-Canovas, F.; Acosta, M. Inhibition by L-ascorbic acid and other antioxidants of the 2,2'-azino-bis(3-ethylbenzthiazoline-6-sulfonic acid) oxidation catalyzed by peroxidase: a new approach for determining total antioxidant status of foods. *Anal. Biochem.* **1996**, *236*, 255–261.

23. Cano, A.; Hernandez-Ruiz, J.; Garcia-Canovas, F.; Acosta, M.; Arnao, M. B. An end-point method for estimation of the total antioxidant activity in plant material. *Phytochem. Anal.* **1998**, *9*, 196–202.

24. Miller, N. J.; Sampson, J.; Candeias, L. P.; Bramley, P. M.; Rice-Evans, C. A. Antioxidant activities of carotenes and xanthophylls. *FEBS Lett* .**1996**, *384*, 240–242.

25. Van den Berg, R.; Haenen, G. R. M. M.; Van den Berg, H.; Bast, A. Applicability of an improved trolox equivalent antioxidant capacity (TEAC) assay for evaluation of antioxidant capacity measurements of mixtures. *Food Chem.* **1999**, *66*, 511–517.

26. Cano, A.; Acosta, M.; Arnao, M. B. A method to measure antioxidant activity in organic media: application to lipophilic vitamins. *Redox Report* **2000**, *5*, 365–370.

27. Arnao, M. B.; Cano, A.; Acosta, M. The hydrophilic and lipophilic contribution to total antioxidant activity. *Food Chem.* **2001**, *73*, 239–244.

28. Chen, I. -C.; Chang, H. -C.; Yang, H. -W.; Chen, G. -L. Evaluation of total antioxidant activity of several popular vegetables and Chinese herbs: A fast approach with ABTS/H₂O₂/HRP system in microplates. *J. Food Drug Anal.* **2004**, *12*, 29–33.

29. Erel, O. A novel automated direct measurement method for total antioxidant capacity using a new generation, more stable ABTS radical cation. *Clin. Biochem.* **2004**, *37*, 277–285.

30. Prior, R. L.; Wu, X.; Schaich, K. Standarized methods for determination of antioxidant capacity and phenolics in foods and dietary supplements. *J. Agric. Food Chem.* **2005**, *53*, 4290–4302.

31. Yu, L.; Haley, S.; Perret, J.; Harris, M.; Wilson, J.; Qian, M. Free radical scavenging properties of wheat extracts. *J. Agric. Food Chem.* **2002**, *50*, 1619–1624.

32. Yu, L.; Zhou, K. Antioxidant properties of bran extracts from 'Platte' wheat grown at different locations. *Food Chem.* **2004**, *90*, 311–316.

33. Yu, L.; Perret, J.; Davy, D.; Wilson, J.; Melby, C. L. Antioxidant properties of cereal products. *J. Food Sci.* **2002**, *67*, 2600–2603.

34. Yu, L.; Perret, J.; Harris, M.; Wilson, J.; Haley, S. Antioxidant properties of bran extracts from "Akron" wheat grown at different locations. *J. Agric. Food Chem.* **2003**, *51*, 1566–1570.

35. Zhou, K.; Yu, L. Antioxidant properties of bran extracts from Trego wheat grown at different locations. *J. Agric. Food Chem.* **2004**, *52*, 1112–1117.

36. Moore, J.; Liu, J.; Zhou, K.; Yu, L. Effects of genotype and environment on the antioxidant properties of hard winter wheat bran. *J. Agric. Food Chem.* **2006**, *54*, 5313–5322.

37. Yu, L.; Zhou, K.; Parry, J. W. Inhibitory effects of wheat bran extracts on human LDL oxidation and free radicals. *Lebensm.-Wiss. Technol.* **2005**, *38*, 463–470.

38. Zhou, K.; Yin, J.; Yu, L. Phenolic acid, tocopherol and carotenoid compositions, and antioxidant functions of hard red winter wheat bran. *J. Agric. Food Chem.* **2005**, *53*, 3916–3922.

39. Zielinski, H.; Kozlowska, H. Antioxidant activity and total phenolics in selected cereal grains and their different morphological fractions. *J. Agric. Food Chem.* **2000**, *48*, 2008–2016.

40. Liyana-Pathirana, C. M.; Shahidi, F. Antioxidant properties of commercial soft and hard wheats (Triticum aestivum L.) and their milling fractions. *J. Sci. Food Agr.* **2006**, *86*, 477–485.

41. Brand, M. D.; Affourtit, C.; Esteves, T. C.; Green, K.; Lambert, A. J.; Miwa, S.; Pakay, J. L.; Parker, N. Mitochondrial superoxide: production, biological effects, and activation of uncoupling proteins. *Free Radical Bio. Med.* **2004**, *37*, 755–767.

42. McCord, J. M.; Fridovich, I. Superoxide Dismutase: an enzymatic function for erythrocuprein (hemocuprein). *J. Biol. Chem.* **1969**, *244*, 6049–6055.

43. Archibald, F. S.; Fridovich, I. The scavenging of superoxide radical by manganous complexes—*in vitro*. *Arch. Biochem. Biophys.* **1982**, *214*, 452–463.

44. Halliwell, B. How to characterize a biological antioxidant. *Free Rad. Res. Comm.* **1990**, *9*, 1–32.

45. Robak, J.; Gryglewski, R. J. Flavonoids are scavengers of superoxide anions. *Biochem. Pharmacol.* **1988**, *37*, 837–841.

46. Olojo, R. O.; Xia, R. H.; Abramson, J. J. Spectrophotometric and fluorometric assay of superoxide ion using 4-chloro-7-nitrobenzo-2-oxa-1,3-diazole. *Anal. Biochem.* **2005**, *339*, 338–344.

47. Quick, K. L.; Hardt, J. I.; Dugan, L. L. Rapid microplate assay for superoxide scavenging efficiency. *J. Neurosci. Meth.* **2000**, *97*, 139–144.

48. Paya, M.; Halliwell, B.; Hoult, J. R. S. Interactions of a series of coumarins with reactive oxygen species. *Biochem. Pharmacol.* **1992**, *44*, 205–214.

49. Pascual, C.; Delcastillo, M. D.; Romay, C. A new luminal sensitized chemiluminescence method for determination of superoxide-dismutase. *Anal. Lett.* **1992**, *25*, 837–849.

50. Flohe, L.; Otting, F. Superoxide dismutase assays. *Method. Enzymol.* **1984**, *105*, 93–104.

51. Beauchamp, C.; Fridovich, I. Superoxide dismutase: improved assays and an assay applicable to acrylamide gels. *Anal. Biochem.* **1971**, *44*, 276–287.

52. Wang, S. Y.; Jiao, H. Scavenging capacity of berry crops on superoxide radicals, hydrogen peroxide, hydrogen radicals, and singlet oxygen. *J. Agric. Food Chem.* **2000**, *48*, 5677–5684.

53. Gaulejac, N. S. -C.; Provost, C.; Vivas, N. Comparative study of polyphenol scavenging activities assessed by different methods. *J. Agric. Food Chem.* **1999**, *47*, 425–431.

54. Ukeda, H.; Kawana, D.; Maeda, S.; Sawamura, M. Spectrophotometric assay for superoxide dismutase based on the reduction of highly water–soluble tetrazolium salts by xanthine–xanthine oxidase. *Biosci. Biotechnol. Biochem.* **1999**, *63*, 485–488.

55. Madesh, M.; Balasubramanian, K. A. A microtiter plate assay for superoxide using MTT reduction method. *Indian J. Biochem Bio.* **1997**, *34*, 535–539.

56. Lenaerts, I.; Braeckman, B. P.; Matthijssens, F.; Vanfletern, J. R. A high-throughput microtiter plate assay for superoxide dismutase based on lucifenin chemiluminescence. *Anal. Biochem.* **2002**, *311*, 90–92.

57. Calzuola, I.; Marsili, V.; Gianfranceschi, G. L. Synthesis of antioxidants in wheat sprouts. *J. Agric. Food Chem.* **2004**, *52*, 5201–5206.

58. Cao, G. H.; Alessio, H. M.; Cutler, R. G. Oxygen-radical absorbance capacity assay for antioxidants. *Free Radical Biol. Med.* **1993**, *14*, 303–311.

59. Cao, G.; Wu, A. H.; Wang, H.; Prior, R. L. Automated assay of oxygen radical absorbance capacity with the cobas fara II. *Clin. Chem.* **1995**, *41*, 1738–1744.

60. Ou, B.; Hampsch-Woodill, M.; Prior, R. L. Development and validation of an improved oxygen radical absorbance capacity assay using fluorescein as the fluorescent probe. *J. Agric. Food Chem.* **2001**, *49*, 4619–4626.

61. Huang, D.; Ou, B.; Hampsch-Woodill, M.; Flanagan, J. A.; Prior, R. L. High-throughput assay of oxygen radical absorbance capacity (ORAC) using a multichannel liquid handling system coupled with a microplate fluorescence reader in 96-well format. *J. Agric. Food Chem.* **2002**, *50*, 4437–4444.

62. Huang, D.; Ou, B.; Hampsch-Woodill, M.; Flanagan, Deemer E. K. Development and validation of oxygen radical absorbance capacity assay for lipophilic antioxidants using randomly methylated β-cyclodextrin as the solubility enhancer. *J. Agric. Food Chem.* **2002**, *50*, 1815–1821.

63. Miller, H. E.; Rigelhof, F.; Marquart, L.; Prakash, A.; Kanter, M. Antioxidant contents of whole grain breakfast cereals, fruits and vegetables. *J. Am. Coll. Nutr.* **2000**, *19*, 312S–319S.

64. Huang, D.; Boxin, O.; Prior, R. L. The chemistry behind antioxidant capacity assays. *J. Agric. Food Chem.* **2005**, *53*, 1841–1856.

65. Cheng, Z.; Zhou, H.; Yin, J. -J.; Yu, L. ESR estimation of hydroxyl radical scavenging capacity for lipophilic antioxidants. *J. Agric. Food Chem.* **2007**, *55*, 3325–3333.

66. Halliwell, B.; Gutteridge, J. M. C. An introduction to oxygen toxicity and free radicals. Free Radicals in Biology and Medicine. Oxford Press, New York, NY, **1989**. pp.1–543.

67. Aruoma, O. I. Deoxyribose assay for detecting hydroxyl radicals. In: Packer, L. (Ed.).1st edn.Oxygen Radicals in Biological Systems, Part C. Methods in Enzymology, Vol. 233, Academic Press, San Diego, **1994**. pp. 57–66.

68. Ou, B.; Hampsch-Woodill, M.; Flanagan, J.; Deemer, E. K.; Prior, R. L.; Huang, D. J. Novel fluorometric assay for hydroxyl radical prevention capacity using fluorescein as the probe. *J. Agric. Food Chem.* **2002**, *50*, 2772–2777.

69. Li, B.; Gutierrez, P. L.; Blough, N. V. Trace determination of hydroxyl radical in biological systems. *Anal. Chem.* **1997**, *69*, 4295–4302.

70. Yang, X.; Guo, X. Fe(II)-EDTA chelate-induced aromatic hydroxylation of terephthalate as a new method for the evelution of hydroxyl radical-scavenging ability. *Analyst* **2001**, *126*, 928–932.

71. Tobin, D.; Arvanitidis, M.; Bisby, R. H. One-electron oxidation of "photo-Fenton" reagents and inhibition of lipid peroxidation. *Biochem. Bioph. Res Co.* **2002**, *299*, 155–159.

72. Folin, O.; Ciocalteu, V. On tyrosine and tryptophane determinations in proteins. *J. Biol. Chem.* **1927**, *73*, 627–650.

73. Singleton, V. L.; Rossi, J. A. Colorimetry of total phenolics with phosphomolybdic-phosphotungstic acid reagents. *Am. J. Enol. Vitic.* **1965**, *16*, 144–158.

74. Slinkard, K.; Singleton, V. L. Total phenol analysis: automation and comparison with manual methods. *Am. J. Enol. Vitic.* **1977**, *28*, 49–55.

75. Magalhaes, L. M.; Segundo, M. A.; Reis, S.; Lima, J. L. F. C.; Rangel, A. O. S. S. Automatic method for the determination of Folin-Ciocalteu reducing capacity in food products. *J. Agric. Food Chem.* **2006**, *54*, 5241–5246.

76. Singleton, V. L.; Orthofer, R.; Lamuela-Raventos, R. M. Analysis of total phenols and other oxidation substrates and antioxidants by means of folin-ciocalteau reagent. *Method. Enzmol.* **1999**, *299*, 152–179.

77. Mpofu, A.; Sapirstein, H. D.; Beta, T. Genotype and Environmental Variation in Phenolic Content, Phenolic Acid Composition, and Antioxidant Activity of Hard Spring Wheat. *J. Agric. Food Chem.* **2006**, *54*, 1265–1270.

78. Yu, L.; Haley, S.; Perret, J.; Harris, M. Comparison of wheat flours grown at different locations for their antioxidant properties. *Food Chem.* **2004**, *86*, 11–16.

79. Amarowicz, R.; Maramac?, M. Antioxidant activity of wheat caryopses and embryo extracts. *J. Food Lipids.* **2002**, *9*, 201–210.

80. Adom, K. K.; Sorrells, M. E.; Liu, R. H. Phytochemicals and antioxidant activity of milled fractions of different wheat varieties. *J. Agric. Food Chem.* **2005**, *53*, 2297–2306.

81. Adom, K. K.; Sorrells, M. E.; Liu, R. H. Phytochemical profiles and antioxidant activity of wheat varieties. *J. Agric. Food Chem.* **2003**, *51*, 7825–7834.

82. Gelinas, P.; McKinnon, C. M. Effect of wheat variety, farming site, and bread-baking on total phenolics. *Int. J. Food Sci. Tech.* **2006**, *41*, 329–332.

83. Velioglu, Y. S.; Mazza, G.; Gao, L.; Oomah, B. D. Antioxidant activity and total phenolics in selected fruits, vegetables, and grain products. *J. Agric. Food Chem.* **1998**, *46*, 4113–4117.

84. Liyana-Pathirana, C. M.; Shahidi, F. Importance of insoluble-bound phenolics to antioxidant properties of wheat. *J. Agric. Food Chem.* **2006**, *54*, 1256–1264.

85. Stevanato, R.; Fabris, S.; Momo, F. New enzymatic method for the determination of total phenolic content in tea and wine. *J. Agric. Food Chem.* **2004**, *52*, 6287–6293.

86. Qian, S. Y.; Buettner, G. R. Iron and dioxygen chemistry is an important route to initiation of biological free radical oxidations: an electron paramagnetic resonance spin trapping study. *Free Rad. Med. Biol.* **1999**, *26*, 1447–1456.

87. Blau, F. Die destillation pyridinmonocarbonsaurer salze (The distillation of pyridinmonocarbonsaurer salts). *Chemische Berichte (now Eur. J. Inorg. Chem.)* **1888**, *21*, 1077–1078.

88. Blau, F. U?ber neue organische Metallverbindungen (Over new organic metal connections). *Monatsh. Chem.* **1898**, *19*, 647–689.

89. Brandt, W. W.; Dwyer, F. P.; Gyarfas, E. C. Chelate complexes of 1,10-phenanthroline and related compounds. *Chem. Rev.* **1954**, *54*, 959–1017.

90. Moss, M. L.; Mellon, M. G. Colorimetric determination of iron with 2,2′-bipyridyl and with 2,2,′2″-terpyridyl. *Ind. Eng. Chem.* **1942**, *14*, 862–865.

91. Yamaguchi, F.; Ariga, T.; Yoshimura, Y.; Nakazawa, H. Antioxidative and anti-glycation activity of garcinol from *Garcinia indica* fruit rind. *J. Agric. Food Chem.* **2000**, *48*, 180–185.

92. Baxendale, J. H.; George, P. A fourth order reaction. *Nature* **1948**, *162*, 777–778.

93. Buss, J. L.; Torti, F. M.; Torti, S. V. The role of iron chelation in cancer therapy. *Curr. Med. Chem.* **2003**, *10*, 1021–1034.

94. Liu, Z. D.; Hider, R. C. Design of iron chelators with therapeutic applications. *Coordin. Chem. Rev.* **2002**, *232*, 151–171.

95. Ehman, D. L.; Sawyer, D. T. Electrochemistry and hydrolysis kinetics of 2,2′-bipyridine complexes of iron(III) and iron(II). *Inorg. Chem.* **1969**, *8*, 900–903.

96. Offen, D.; Gilgun-Sherki, Y.; Barhum, Y.; Benhar, M.; Grinberg, L.; Reich, R.; Melamed, E.; Atlas, D. A low weight molecular weight copper chelator crossese the blood-brain barrier and attenuates experimental autoimmune encephalomyelitis. *J. Neurochem.* **2004**, *89*, 1241–1251.

97. Lodge, J. K.; Traber, M. G.; Packer, L. Thiol chelation of Cu^{2+} by dihydrolipoic acid prevents human low density lipoprotein peroxidation. *Free Rad. Med. Biol.* **1998**, *25*, 287–297.

98. Briante, R.; Febbraio, F.; Nucci, R. Antioxidant properties of low molecular weight phenols present in Mediterranean diet. *J. Agric. Food Chem.* **2003**, *51*, 6575–6981.

99. Kong, B.; Xiong, Y. L. Antioxidant activity of zein hydrolysates in a liposome system and the possible mode of action. *J. Agric. Food Chem.* **2006**, *54*, 6059–6068.

100. Antholine, W. E.; Basosi, R.; Hyde, J. S.; Lyman, S.; Petering, D. H. Immobile- and mobile-phase ESR spectroscopy of copper complexes: studies on biologically interesting bis(thiosemicarbazonato) copper(II) chelates. *Inorg. Chem.* **1984**, *23*, 3543–3548.

101. Krishnankutty, K.; John, V. D. Synthesis, characterization, and antitumour studies of metal chelates of some synthetic curcuminoids. *Transit. Metal Chem.* **2005**, *30*, 229–233.

102. Su, L.; Yin, J. -J.; Charles, D.; Zhou, K.; Moore, J.; Yu, L. Total phenolic contents, chelating capacities, and radical-scavenging properties of black peppercorn, nutmeg, rosehip, cinnamon, and oregano leaf. *Food Chem.* **2007**, *100*, 990–997.

103. Spiteller, G. Peroxyl radicals: inductors of neurodegenerative and other inflammatory diseases. Their origin and how they transform cholesterol, phospholipids, plasmalogens, polyunsaturated fatty acids, sugars, and proteins into deleterious products. *Free Rad. Biol. Med.* **2006**, *41*, 362–387.

104. Frankel, E. N. In search of better methods to evaluate natural antioxidants and oxidative stability in food lipids. *Trends Food Sci. Tech.* **1993**, *4*, 220–225.

105. Velasco, J.; Andersen, M. L.; Skibsted, L. H. Evaluation of oxidative stability of vegetable oils by monitoring the tendency to radical formation. A comparison of electron spin resonance spectroscopy with the Rancimat method and differential scanning calorimetry. *Food Chem.* **2004**, *85*, 623–632.

106. AOCS. Method Cd 12b-92. *Official Methods and Recommended Practices of the American Oil Chemists' Society.* American Oil Chemists' Society, 5th edn, Champaign, IL, **1992**.

107. Akoh, C. C. Oxidative stability of fat substitutes and vegetable-oils by the oxidative stability index method. *J. Am. Oil Chem. Soc.* **1994**, *71*, 211–216.

108. Liang, C.; Schwarzer, K. Comparison of four accelerated stability methods for lard and tallow with and without antioxidants. *J. Am. Oil Chem. Soc.* **1998**, *75*, 1441–1443.

109. Martinez-Tome, M.; Murcia, M. A.; Frega, N.; Ruggieri, S.; Jimenez, A. M.; Roses, F.; Parras, P. Evaluation of antioxidant capacity of cereal brans. *J. Agric. Food Chem.* **2004**, *52*, 4690–4699.

110. Nakatani, N.; Tachibana, Y.; Kikuzaki, H. Establishment of a model substrate oil for antioxidant activity assessment by Oil Stability Index method. *J. Am. Oil Chem. Soc.* **2001**, *78*, 19–23.

111. Berliner, J. A.; Heinecke, J. W. The role of oxidized lipoproteins in atherogenesis. *Free Rad. Bio. Med.* **1996**, *5*, 707–727.

112. Esterbauer, H.; Striegl, G.; Puhl, H.; Rotheneder, M. Continuous monitoring of in vitro oxidation of human low density lipoprotein. *Free Rad. Res. Comm.* **1989**, *6*, 67–75.

113. Retsky, K. L.; Chen, K.; Zeind, J.; Frei, B. Inhibition of copper-induced LDL oxidation by vitamin C is associated with decreased copper-binding to LDL and 2-oxo-histidine formation. *Free Rad. Med. Biol.* **1999**, *26*, 90–98.

114. Puhl, H.; Waeg, G.; Esterbauer, H. Methods to determine oxidation of low-density lipoproteins. *Methods Enzymol.* **1994**, *233*, 425–441.

115. Mao, S. J. T.; Yates, M. T.; Rechtin, A. E.; Jackson, R. L.; Sickle, W. A. V. Antioxidant activity of probucol and its analogues in hypercholesterolemic watanabe rabbits. *J. Med. Chem.* **1991**, *34*, 298–302.

116. Frankel, E. N.; German, J. B.; Davis, P. A. Headspace gas chromatography to determine human low density lipoprotein oxidation. *Lipids* **1992**, *27*, 1047–1051.

117. Ohta, T.; Semboku, N.; Kuchii, A.; Egashira, Y.; Sanada, H. Antioxidant activity of corn bran cell-wall fragments in the LDL oxidation system. *J. Agric. Food Chem.* **1997**, *45*, 1644–1648.

118. Bradford, M. M. A rapid and sensitive method for the quantization of microgram quantities of protein utilizing the principle of protein-dye binding. *Anal. Biochem.* **1976**, *72*, 248–254.

119. Tarladgis, B. G.; Pearson, A. M.; Dugan, L. R. Chemistry of the 2-thiobarbituric acid test for determination of oxidative rancidity in foods. *J. Sci. Food Agric.* **1964**, *15*, 602–607.

120. Steinberg, D. Low density lipoprotein oxidation and its pathobioliogical significance. *J. Biol. Chem.* **1997**, *272*, 20963–20966.

121. Becker, E. M.; Nissen, L. R.; Skibsted, L. H. Antioxidant evaluation protocols: food quality or health effects. *Eur. Food Res. Technol.* **2004**, *219*, 561–571.

122. Arnao, M. B.; Cano, A.; Acosta, M. Methods to measure the antioxidant activity in plant material. A comparative discussion. *Free Rad. Res.* **1999**, *31*, S89–S96.

123. Sanchez-Moreno, C. Review: methods used to evaluate the free radical scavenging activity in foods and biological systems. *Food Sci. Tech. Int.* **2002**, *8*, 121–137.

CHAPTER 10

APPLICATION OF ESR IN WHEAT ANTIOXIDANT DETERMINATION

Zhihong Cheng
Liangping Yu
JunJie Yin

10.1 INTRODUCTION

Electron spin resonance (ESR) spectroscopy has been widely used in the detection of free radicals, transition-metal ions, and the irradiation history of foodstuffs. ESR is very specific in its applications. Only the systems that contain an electron whose spin is not paired with the oppositely directed spin of another electron will give an ESR signal (1). Although many methods have been used in studying the free radical scavenging capacities of antioxidants, such as spectrometric, fluorometric, and electrochemical methods, ESR by far is considered to be the most direct and reliable method for detection of free radicals (2,3).

Wheat is a major agricultural commodity and dietary component across the world. It is one of the most important cereals in view of nutritional values. It serves as a significant source for carbohydrates, proteins, and fibers in human diets. Wheat is morphologically composed of endosperm and embryo enclosed by bran layers. The endosperm portion (also known as flour) is primarily starch, and the bran is gradually revealed to contain plenty of natural antioxidants, mainly phenolics. Recently, several comprehensive researches on wheat and wheat fractions have been carried out (4–9). These studies have confirmed that wheat, especially wheat bran, possesses significant antioxidant activities in addition to general nutritional values, which implies the potential for utilizing and promoting the health benefits of wheat and wheat-based food products. Phenolics including free-, bound-, and conjugated phenolic acids; tocopherols (α-, δ-, and γ-); and carotenoids are among the main groups of components that are believed to be responsible for the overall antioxidant properties in wheat (4,10). The spectrometric or fluorometric free radical scavenging assays such as 2,2-diphenyl-1-picrylhydrazyl (DPPH$^\bullet$) radical, oxygen radical absorption capacity (ORAC), and cation 2,2′-azino-di (3-ethylbenzthiazoline sulfonate) (ABTS$^{\bullet+}$) radical assays, are the most frequently used methods for estimating the antioxidant activities

Wheat Antioxidants, Edited by Liangli Yu
Copyright © 2008 John Wiley & Sons, Inc.

of food-related systems. These methods are simple and easy to perform in any laboratory, but most of them do not provide clear information on the mechanisms of antioxidant-radical reactions. Such information could be important, especially in the development of functional foods or nutraceuticals to target the prevention of certain diseases. ESR, on the contrary, could be a useful tool in investigating the free radical scavenging mechanisms and the effectiveness of various antioxidants in their reactions with different types of free radicals, especially in biological systems such as cultured cells. Note that the nature of the free radicals can also be elucidated by ESR method, which is a unique merit of ESR compared to other methods. The application of ESR in antioxidant research is attracting more and more interests.

10.2 THE PRINCIPLES OF ESR

ESR is also called EPR (electron paramagnetic resonance). It is such a spectrometric technique that allows direct detection of species that have one or more unpaired electrons. Such an atom or a molecule with an unpaired electron is called a free radical. Therefore, ESR is a direct and unambiguous method that is highly specific for free radicals and other paramagnetic species such as transition metals (e.g., Fe^{2+}, Cu^{2+}) (11). It is very sensitive and able to detect 2×10^{11} spins (ca. 3×10^{-9} M) at the lower limit (12). A typical continuous wave (CW) X-band ESR instrument includes four main components: (i) a magnet that generates and modulates a uniform magnetic field of several thousand Gauss, (ii) a microwave system that includes an electromagnetic radiation source and a detector to generate and detect microwave power, (iii) a sample cavity, and (iv) a console that processes signals and controls electronics (Fig. 10.1).

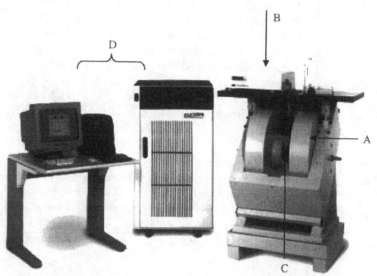

Figure 10.1 A colorful picture of Bruker EMX continuous wave (CW) ESR instrument. A, B, C, and D represent magnet, microwave system, sample cavity, and a console, respectively, of typical ESR instruments.

There are two major ESR methods in biological fields: ESR spin-trapping method and ESR spin-labeling method. The ESR spin-trapping method is used to detect low-level and short-lived (LLSL) radicals. In this method, the exogenous spin-trapping molecule is added, which intercepts LLSL radicals to give more stable secondary radical adducts that can easily be detected by ESR. 5,5-Dimethyl-1-pyrroline *N*-oxide (DMPO), 5-*tert*-butoxycarbonyl 5-methyl-1-pyrroline *N*-oxide (BMPO), 5-ethoxycarbonyl-5-methyl-1-pyrroline *N*-oxide (EMPO), and *N*-*tert*-butyl-α-phenylnitrone (PBN) (Fig. 10.2) are a few examples of spin-trapping agents that are widely used *in vitro* and *in vivo* studies to estimate the free radical scavenging capacities of antioxidants and the biological molecular damages, and to better understand the role of free radicals in many pathologies, toxicities, and diseases. ESR spin-labeling method uses a paramagnetic molecule (the spin label) to "tag" macromolecules in specific regions and to determine the type of environment in which the spin label is located. This spin label gives an ESR signal; it can be incorporated into membrane lipids or attached to proteins to enable otherwise inaccessible systems to be studied. As a result, the information of motion or orientation of biological molecules can be obtained through studying the ESR spectra of a spin-labeled molecule. Recent applications of spin labels in proteins and membranes were reviewed by Borbat et al. (13).

ESR method can analyze both solid and liquid samples *in vivo* or *in vitro*. It is noteworthy that ESR is a nondestructive analytical technique, that is able to detect the dynamics of reaction. Sample preparation of most of the food systems for ESR assay is relatively simple regardless of the selection of models. Samples rich in antioxidants are mixed with free radical producing systems (with or without the addition of a spin trap depending on the stability of free radicals). Generally, 50 μL of reaction mixture is transferred to a glass capillary tube. After the sample is loaded, the capillary tube is sealed and inserted into ESR cavity for ESR determination. In order to keep reaction systems in a biological temperature of 37°C or a temperature lower than ambient temperature, some thermal control compartments are equipped in ESR.

Free radical scavenging activities of antioxidants can be quantified through reactions between antioxidants and free radicals. The free radical scavengers react with free radicals by donating an electron, which is associated with the reduction of the ESR signal intensity. The lower the ESR signal intensity, the higher the free radical scavenging activity. In addition, ESR could be a useful tool for finding out information, such as what kinds of free radicals are present, how many they are, and how fast they react.

The ESR spectra can be intercepted by two important parameters: *g* value and hyperfine splitting, which correspond to chemical shift and coupling constant in nuclear magnetic resonance (NMR). The first parameter is to measure how far the

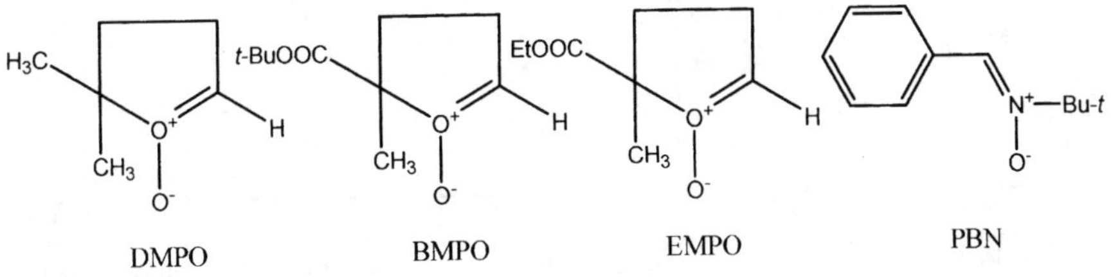

Figure 10.2   The structure of four commonly used ESR spin traps.

magnetic environment of the unpaired electrons differs from that of a free, gas-phase electron. The hyperfine splitting is the result of the interaction energy between the electron spin and a magnetic nucleus, which results in the splitting of lines in EPR spectra. One species demonstrates only one ESR signal. The hyperfine splitting from paramagnetic complexes having numerous magnetic nuclei may give rise to complicated spectra with a large number of overlapping hyperfine lines (14). Computer simulation can facilitate the elucidation of the nature of radicals by ESR simulation software.

10.3   THE APPLICATION OF ESR IN FOOD SYSTEMS

Many direct and indirect methods such as gas chromatography–mass spectrometry (GC–MS) and high performance liquid chromatography (HPLC) have been used to assess free radical scavenging activities of antioxidants. More recently, Kopani et al. published a brief review to compare traditional methods with ESR methods for the estimation of antioxidant activities (15). Most of the traditional methods showed good correlations with ESR technology, although different mechanisms were revealed.

There are four major applications of ESR in food systems: (i) the evaluation of free radical scavenging capacity of antioxidants, (ii) the determination of oxygen consumption, (iii) the detection of an irradiation history of certain foods, and (iv) the determination of oxidation stability and shelf life of foods. Most of the previous studies were focused on using ESR to estimate free radical scavenging capacity of a potential antioxidant as a result of a growing interest in searching for natural antioxidants. Various food systems including wheat, barley, fats and oils, beverages, spices, and fruit juices have been studied for their free radical scavenging capacities against biologically related radicals including superoxide radicals ($O_2^{\cdot-}$) (16,17), hydroxyl radicals (HO$^{\cdot}$) (17–19), peroxide radicals ($O_2^{\cdot-2}$) (20), nitric oxide (NO), and even carbon-centered radicals (21) and nonbiologically related radicals such as DPPH$^{\cdot}$ (19,22,23) and ABTS$^{\cdot+}$ (19). For example, free radicals scavenging capacities of conjugated linoleic acids (CLA) were evaluated against DPPH$^{\cdot}$ by ESR spectrometric method (Fig. 10.3) for the first time by Yu (22). Results from this study clearly demonstrated that CLA isomers could directly react with DPPH$^{\cdot}$, and thus provided evidence of antioxidant activity of CLA. In 1996, the antioxidant activities of nine different spices were determined by using ESR spin-trapping method and by electrochemical measurement of oxygen depletion (18).

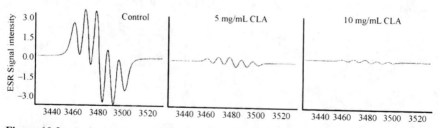

Figure 10.3   Antioxidant activities of CLA on DPPH$^{\cdot}$ assayed by ESR. (Redrawn from Reference 22.)

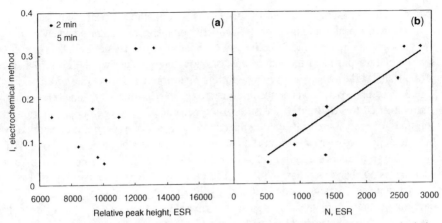

Figure 10.4 Antioxidative index of nine spice extracts evaluated from initial oxygen depletion rate versus (**a**) relative peak height obtained from ESR method detected at 2 and 5 min, and (**b**) the difference of the relative peak height detected at 5 and 2 min: ph_5–ph_2. (Redrawn from Reference 18.)

Clearly, different methods demonstrated different mechanisms of antioxidative effects of different spices. Figure 10.4 demonstrated that no correlations were found between the electrochemical method and the ESR method at 2 and 5 min determination. However, a better correlation was found between the electrochemical method and the ESR method when the differences of values between 2 and 5 min determination were used. The authors concluded that ESR method was related to the early stage of oxidation in which radicals were generated, while the electrochemical method was a direct measurement of the effect of antioxidants on the oxygen depletion in the propagation stage of oxidation. Most of the previous studies were not conducted quantitatively. The results were frequently expressed as the percent of inhibition of ESR signal intensity compared to the controls, and no quantitative information of radicals was determined. Further studies are needed to investigate the possibility of quantitative determination. Certain standards, such as 6-hydroxy-2,5,7,8-tetramethylchroman-2-carboxylic acid (trolox) or α-tocopherol, could be used to express the results in order to make the data comparable from one study to another.

The second popular application of ESR is to determine oxygen consumption inside biological systems by ESR oximetry method. Liposomes or liver cells are normally used as model systems for this type of study. It is well known that oxygen can easily cross cell membranes, and the reduction rate of oxygen inside biological microenvironment is proportional to the rate of lipid peroxidation. In addition, oxygen is a unique resource for converting reactive oxygen species (ROS), such as $HO^\bullet$, H_2O_2, $O_2^{\bullet-}$, and $RO_2^\bullet$, which react with almost all the cell components including proteins, lipids, and DNA to cause oxidative damage. Therefore, the determination of oxygen consumption could be a useful tool for evaluation of the extent of lipid peroxidation.

Oxygen molecule has two unpaired electrons with paramagnetic property in the ground state, and its ESR line is so broad due to its fast relaxation. However, it can be detected indirectly through the oxygen-induced ESR line broadening of other stable paramagnetic substances, termed as spin labels. The method is commonly known as ESR oximetry, which is based on Heisenberg spin–spin exchange between oxygen molecule and a spin label (24,25). The presence of oxygen molecules broadens the ESR signals of the spin label in such a way that a change in line width Δw is linearly proportional to the local oxygen concentration. Therefore, a narrower line width of ESR signal is correlated to a lower concentration of oxygen. This line width can be measured and converted to pO_2 (partial pressure of oxygen) or $[O_2]$ (oxygen concentration in solution) using an appropriate calibration curve. Note that no oxygen was consumed by the interaction between oxygen and spin labels, since this is a physical interaction (26).

The spin labels are mainly classified into two groups: water soluble nitroxides such as ^{15}N-PDT and 3-carbamoyl-2,2,5,5-tetramethyl-3-pyrroline-1-yloxyl (CTPO), and particularly, paramagnetic materials, such as lithium phthalocyanine, lithium naphthalocyanine, and inks. An extensive comparison of these two types of spin labels was previously reviewed by Dunn and Swartz (27). Here, we just focus on applications of soluble spin labels, though some limitations were found for this kind of labels due to the susceptibility to cellular reduction over time (28). A more detailed description of this oximetry method will be given in Section 10.4.

An example of determination of oxygen consumption by ESR was recently reported by Yin et al. Synthetic phosphatidylcholines (PC) with either of the two CLA isomers (c9, t11-CLA or t10, c12-CLA) or linoleic acid (LA) at the position sn-2 were studied with or without four types of natural liposomes (soybean, egg yolk, rat brain, and rat heart) (29). On the incorporation of 5 mol % LA or CLA into the phospholipids, all exerted significant antioxidant activities in ethanol solutions in which the PC molecules were randomly dispersed. However, when the measurements were conducted in liposomes, 5 mol % PC (CLA) showed different effects on oxygen consumption. PC (CLA) acted either as an antioxidant or pro-oxidant, while 5 mol % PC (LA) demonstrated antioxidant activity consistently (Fig. 10.5). These observations showed that, unlike the "linear" structure of LA, a "bend" configuration of CLA affected the integrity of the alignments of natural PC, which probably affected the transportation property of

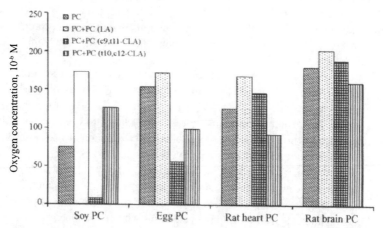

Figure 10.5   Effect of CLA isomers and LA on oxygen consumption in different liposome models. (Redrawn from Reference 29.)

oxygen and led to the increase in the collision rate of oxygen including reactive oxygen species and biomolecules of the membrane. Therefore, CLA can be either an antioxidant or a prooxidant depending on different systems and the result of the combination of two effects mentioned above (29,30). However, the applications of ESR oximetry in food systems are limited. More researches are needed to fully demonstrate the benefits or the possible limitations of the method.

The third application of ESR in food systems is to detect the irradiation history of certain foods. Irradiation has been well accepted and widely used as a method for food preservation to reduce pathogenic microorganisms and extend the shelf life of foods (31). Irradiation is carried out by exposing the foods to a γ-emitting source, such as radioactive isotopes of cobalt or cesium, or to a linear accelerated electron beam. The loss of nutrients and the generation of free radicals often occur in the process of irradiation. Although the free radicals generated in the irradiation process are not of safety concerns, it does provide a reliable way for the detection of irradiated foods. Goodman et al. (32) demonstrated that ESR could be used in determining the irradiation history from bone in meat and fish or fruit seeds. Dodd et al. (33) also reported that ESR could be a valid method for detection of irradiation history in certain foods containing bone or other calcified tissue, or dry foodstuffs such as seeds and spices.

Another application of ESR in food systems is to evaluate the oxidation stability and predict the shelf life of foods. Kristensen and Skibsted (34) successfully used three ESR methods (direct, spin trapping, and spin labeling) to obtain the information on free radical development in processed cheese when stored under different temperature and light conditions. Their study clearly demonstrated that oxidation in processed cheese could be monitored by using ESR spectrometry. ESR could be a useful tool in prediction of shelf life of foods.

10.4 ESR DETERMINATION OF WHEAT ANTIOXIDANTS

Although ESR method has been widely used to estimate the free radical scavenging capacities of natural antioxidants, few applications were successful for assessing wheat antioxidants. To the best of our knowledge, ESR so far has been applied to wheat antioxidants for estimation of free radical scavenging capacities, chelating activities against Cu^{2+} and Fe^{2+}, and lipid peroxidation retardation in liposomes.

10.4.1 Free Radical Scavenging Capacities of Wheat Antioxidants

DPPH˙ is a stable nitro radical that is used as a standard to tune ESR instrument. Spectrometric DPPH˙ assay is widely employed in the estimation or screening of antioxidants due to its characteristics of easy performance and clear reaction mechanism, though this radical is not biologically related. It can accept an electron or hydrogen radical to become a stable, diamagnetic molecule and lose its deep violet color measured at 515–517 nm. Therefore, DPPH˙ can be adapted to ESR method for determination of free radical scavenging capacities without using any spin-trapping agent. As one of the earliest studies on the wheat antioxidants using ESR method, Sripriya and coworkers studied DPPH˙ quenching action of finger millet comparing with an unidentified wheat species that was used as one of the positive controls. In their study, wheat extracts showed less activity than that of finger millet (35). The typical six peaks and special hyperfine splitting of DPPH˙-ESR spectra were observed in wheat extracts although wheat was not the focus of the study. In 2002, a study focused on antioxidant activities of hard wheat was performed by Yu et al. (5). The results showed that three wheat extracts (Trego, Akron, and Platte) differed in their capacities to quench DPPH˙. Among the three wheat species studied, Akron demonstrated the highest activity in a time-dependent manner (Fig. 10.6). The data from DPPH˙-ESR method can give confirmable information on

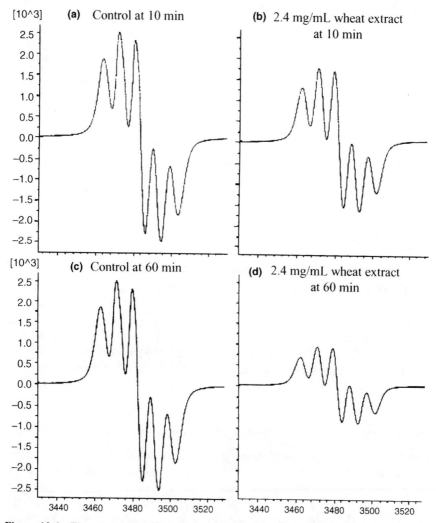

Figure 10.6 Time-dependent effect of Akron wheat extracts on DPPH˙ determined by ESR. (**a**) and (**b**) represent the measurement at 10 min after the reaction was initiated without and with 2.4 mg/mL Akron extract, respectively, while (**c**) and (**d**) represented the detection at 60 min for (**a**) and (**b**), respectively. (Redrawn from Reference 5.)

direct interaction between wheat antioxidants and free radicals. This is an important advantage of using DPPH˙-ESR method compared to the spectrometric method (5). In other words, no impure radicals other than DPPH˙ -antioxidants were produced in the system. This is of utmost significance, and details will be given in ESR-HO˙ section. As a consecutive contribution to explore the antioxidant benefits of hard winter wheat, the bran (the part rich in antioxidants of wheat) of two different species (Alliance and Wichita) was determined by the same group using DPPH˙–ESR method. The results were expressed as the percentage of DPPH˙ radicals quenched (10). The free radical scavenging capacities of the two varieties were compared and quantified. The ESR-$O_2^{\cdot-}$ and ESR-HO˙ models were also used to estimate the wheat antioxidant effects, and the details will be discussed later. In order to better understand the antioxidant action of wheat, radical scavenging capacities of some pure phenolic acids (4-coumaric, ferulic, 4-hydroxylbenzoic, vanillic, and syringic acids) naturally present in wheat were investigated and compared by using ESR-DPPH˙ method as well as ESR-$O_2^{\cdot-}$ and ESR-HO˙ assays (36). The free radical scavenging capacities of these phenolic acids were strongly depended on their structures. For example, the substitutes on phenyl ring, the conjugated carbon skeleton, and the presence of an additional methoxyl group in the *ortho* position of the hydroxyl group showed strong influences on the antioxidant activities.

In addition to DPPH˙, HO˙ production models were also applied for evaluating antioxidant activities using ESR (36–38). As it is known, HO˙ is a highly reactive species in organisms, with a very short *in vivo* half-life (ca. 10^{-9} s). In contrast to DPPH˙, spin traps must be used in ESR-HO˙ study to get a stable secondary radical for ESR detection. Screening HO˙ radical scavengers and developing methods for assessing HO˙ radicals are of high interests. As a successful model for production of HO˙, Fenton reaction ($Fe^{2+} + H_2O_2 \rightarrow Fe^{3+} + HO˙ + HO^-$) is commonly used in most of the literatures. Fe^{2+} is liable to be autooxidized, and so carefully controlled experiment conditions are necessary. ESR spectra of DMPO-HO˙ typically include four peaks with the peak height ratio of $1:2:2:1$. An earlier study on comparison of antioxidants among an unidentified wheat bran, alfalfa, and *Ginkgo biloba* showed that wheat bran at a concentration of 4 mg/mL did not show any antioxidant activities against HO˙ (39). However, a recent study by Zhou and coworkers found significant HO˙ scavenging activity with a time response in two hard winter wheat bran samples (Alliance and Wichita) at a concentration of 100 mg/mL (10). Different wheat varieties concentration used and no quantitative measurement of HO˙ production in the reaction systems were believed to cause the discrepancies between the two studies. Afterward, a study conducted by the same group revealed that 2 mM phenolic acids such as 4-coumaric, ferulic, and vanillic acids in wheat bran demonstrated hydroxyl radical scavenging capacities (36) (Fig. 10.7). The study confirmed that phenolic acids in wheat contributed partially to its overall antioxidant activities.

As mentioned in ESR-DPPH˙ section, ESR plays an important role in confirming the direct reaction between free radicals and antioxidants in complicated reaction systems (22). The following example will give a better understanding of this point. More recently, a new fluorometric assay for determining HO˙ scavenging capacities (HOSC) was successfully developed in a Fenton-like reaction system with the help of ESR (40). Several commonly used solvents for extracting or assessing natural

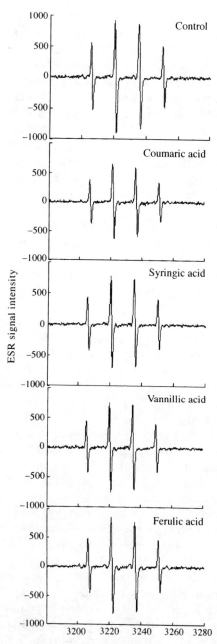

Figure 10.7 Hydroxyl radical scavenging capacities of four main phenolic acids in wheat on production of DMPO-HO˙ by ESR detected at 20 min of the reaction. The reaction mixture contained 10 μL of freshly made 3.00 mM FeSO$_4$, 80 μL of 0.75 mM EDTA, 15 μL of 1.00 M DMPO, 15 μL of 0.50 mM H$_2$O$_2$, and 30 μL of 2.00 mM phenolic acids. (Redrawn from Reference 36.)

antioxidants were investigated for their potential effects on $HO^•$. As a result, some solvents, such as ethanol, methanol, and DMSO, acted as $HO^•$ scavengers and carbon-centered radicals were produced; while acetone showed no obvious effect on the production of DMPO-$HO^•$ (Fig. 10.8). Results from this study indicated again that ESR acted as a "mirror" that facilitated the observation of what reactions or side reactions occurred in the complicated systems assayed by spectrometric or fluorometric methods. It can also give a better understanding of the possible reaction mechanism and the nature of the radicals.

Other free radicals such as $O_2^{•-}$ were also studied for their potential applications in wheat antioxidants. Similar to the results of ESR-$DPPH^•$ and ESR-$HO^•$ reported by Zhou et al., wheat bran and some of the phenolic acids present in wheat showed different $O_2^{•-}$ scavenging capacities depending on wheat varieties and chemical structures (10,36). Interestingly, a study of the effects of heat stress and postharvest treatments on the availability of wheat antioxidants showed opposite results by using ESR-$O_2^{•-}$ method to those obtained from $DPPH^•$, $ABTS^{+•}$, and total phenolic content (TPC) by using spectrometric methods (16).

Differences from ESR-$DPPH^•$, ESR-$HO^•$, and ESR-$O_2^{•-}$ methods could not be easily quantified due to the difficulties in quantifying $HO^•$ or $O_2^{•-}$. Further studies on

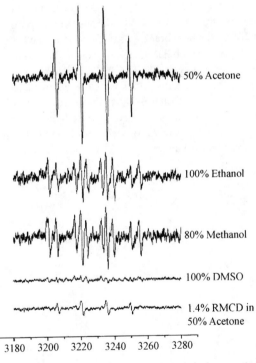

Figure 10.8   Effects of several commonly used solvents on production of DMPO-$HO^•$ by ESR. The reaction mixture contained 42.50 mM phosphate buffer (pH 7.4), 0.10M DMPO, 0.69 mM Fe(III), 26.50 mM H_2O_2, and 10% (v/v) of the solvents. DMSO and RMCD represented dimethyl sulfoxide and β-cyclodextrin, respectively. (Redrawn from Reference 40.)

ESR-DPPH$^\bullet$/HO$^\bullet$/O$_2^{\bullet-}$ methods are needed using a standard, such as trolox or α-tocopherol, to express the results and make data comparable from one study to another.

10.4.2 Chelating Activity Against Cu^{2+}

Transition metals such as Fe^{2+}, Cu^{2+}, Zn^{2+}, and Co^{2+} play an important role in biological systems, although there are only trace amounts present in organisms. For instance, cycling in redox states of metals may cause oxidative damage to cells through catalyzing Fenton or Fenton-like reactions, in which one of the highly reactive radicals HO$^\bullet$ was produced causing the damage to almost every type of biomolecules in living cells, such as DNA, proteins, and membranes (41). Therefore, antioxidants such as flavonoids were screened for chelating metal ions. The chelation of metal ions may contribute to antioxidant activities by preventing redox-active metals from catalyzing decomposition of hydrogen peroxides and further formation of free radicals (42). A well-established spectrometric chelating assay using 2,2′-bipyridyl is used for this purpose (4). A competition mechanism is involved between 2,2′-bipyridyl and antioxidants against metal ions. Similar to the spectrometric DPPH$^\bullet$ assay, qualitative and quantitative determination of antioxidants can be achieved by this spectrometric chelating assay. However, since the reaction system is quite complicated, possible side reactions could occur resulting in incorrect estimation of antioxidants. Recently, Zhou and coworkers demonstrated for the first time the unique profile of ESR spectra of chelating activity of hard wheat bran against Cu^{2+} (10). The ESR spectra obtained in the study clearly showed that chelating complexes of Cu^{2+} were formed, although no quantitative information was available (Fig. 10.9). In a consecutive study, the chelating activities of five pure phenolic acids, which were present in wheat, were determined by ESR against Fe^{2+} and Cu^{2+} (36). Interestingly, no chelating activities against Fe^{2+} were found, while vanillic and ferulic acids demonstrated significant chelating activities against Cu^{2+}. Importantly, this study provided evidence that the chelating activities of vanillic and ferulic acids (36) against Cu^{2+} contributed to the chelating activities of

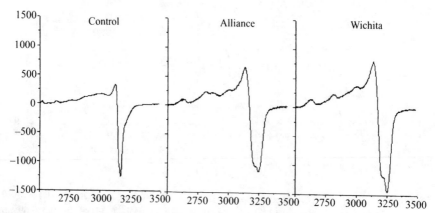

Figure 10.9 Interaction between Cu^{2+} and extracts of wheat bran measured by ESR detected at 1 min of the reaction. (Redrawn from Reference 10.)

wheat (10). Further studies are needed to investigate the quantitative protocols and the nature of the complexes of metal ions and antioxidants.

10.4.3   Effects of Wheat Antioxidants on Lipid Peroxidation in Liposomes

Lipid peroxidation is one of the most important biologically relevant free radical chain reactions that contributes mainly to the damage or nonintegrity of lipid-containing tissues (43). Therefore, accessing the extent of lipid peroxidation is important for better understanding the status of oxidation stress in biological systems. Generally, the methods used to investigate lipid peroxidation monitor changes of either the principal reactants such as unsaturated fatty acids, hydroperoxides, and oxygen molecules or one of the reaction products including thiobarbituric acid, fluorescent compounds, liberated hydrocarbons, and other putative end products (44) by ESR, spectrometric, fluorometric, and some special methods. In contrast to other methods that determine the breakdown products of lipid peroxidation, ESR intercepts the intermediate radicals by using a spin trap. In addition, the early stage of lipid peroxidation can also be studied by ESR oximetry method through measuring oxygen uptake (43). Therefore, in combination with the optimum lipid models, such as liposomes, lipoproteins, and isolated biological membranes, ESR method will facilitate the discovery of the possible pathogenesis of the oxidation-related diseases. In 1998, a study on natural antioxidants against lipid radicals was performed in rat liver microsomes. The extracts of wheat bran, alfalfa, and *G. biloba* were used in the study (39). A 26% reduction of the ESR signal was observed by using wheat bran extracts at a concentration of 4 mg/mL. The wheat bran extract was less effective compared to that of *G. biloba*, which resulted in 76% reduction of the ESR signal. Although many methods were employed to evaluate the effects of wheat extracts on lipid peroxidation using PC liposomes, no similar data were obtained by using ESR spin label method (9).

Recently, we conducted a study on the effects of soft wheat (Choptank) extracts on lipid peroxidation by ESR method using two types of liposomes (rat liver PC and egg PC) (unpublished data). In this study, oxygen consumption in liposomes was determined by an ESR oximetry method. This method is based on the broadening effect of oxygen molecules on the ESR spectra of a spin label via Heisenberg spin exchange as mentioned previously. In a closed reaction system, with the oxygen consumed by the lipid peroxidation initiated by AAPH or AMVN, the interaction between the oxygen and the spin label became weaker, resulting in the decrease in the line width of ESR-spin label signal (Δw), as shown in Fig. 10.10. However, when antioxidants were introduced into the system, the lipid peroxidation was somewhat retarded. In other words, the oxygen consumption in lipid peroxidation in the local environment slowed down. On the basis of this principle, the effects of water- and lipo-soluble extracts of Choptank wheat were determined and compared using AAPH and AMVN as water and lipo-soluble free radical initiators, respectively. As a result, water-soluble extract of Choptank wheat reduced oxygen consumption (Fig. 10.11), while incorporation of the same concentrations of the lipo-soluble extract slightly increased the oxygen consumption. The lipo-soluble extract actually acted as a pro-oxidant. Liposomes with similar constituents as to the living cell membranes

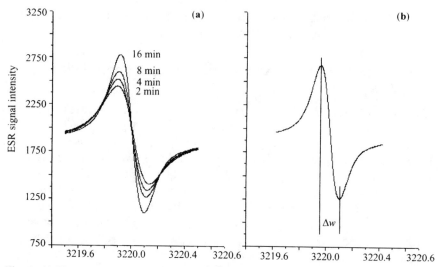

Figure 10.10 Typical schemes for determination of oxygen consumption using ^{15}N-PDT as a spin labeling in liver liposome initiated with AAPH (**a**) and calculating the ESR line width Δw (**b**).

are proven to be good models for evaluation of the biological activity. A better prediction can be achieved.

As a conclusion, wheat antioxidants so far are mainly evaluated by free radical scavenging activities (including DPPH$^{\bullet}$, ABTS$^{\bullet+}$, and O$_2^{\bullet-}$) and chelating activities against Cu^{2+} and Fe^{2+} using spectrometric and fluorometric methods *in vitro*. Some biological models are gradually adapted in wheat antioxidants researches. Although ESR methods have been widely used in evaluating antioxidants from various natural sources, limited literatures are available for the applications of ESR on wheat

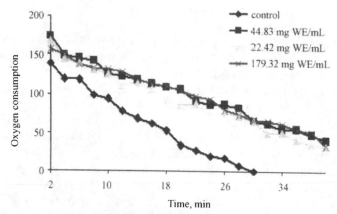

Figure 10.11 Effect of Choptsan wheat on oxygen consumption in egg PC. The reaction mixture contained 30 mg/mL of egg PC, 0.04 mM of ^{15}N-PDT, 0.02 mM of AAPH, and different concentrations of water-soluble extract of Choptsan wheat. ESR signals were recorded within 45 min with the interval of 2 min at 37°C.

antioxidants. This is probably due to the relatively high cost of the ESR instruments. However, the unique information obtained from the ESR method, such as the nature of complexes between antioxidants and free radicals or metal ions and the possible side reactions, and using ESR as a confirmation tool for spectrometric and fluorometric methods make ESR methods superior to the conventional spectrometric and fluorometric methods. A careful consideration of combining the ESR and spectrometric or fluorometric methods will facilitate a better understanding of the possible mechanisms of antioxidant action of wheat and wheat related foodstuffs, and the development of new methods for evaluating antioxidants.

ABBREVIATIONS

AAPH	2,2′-Azobis (2-amidinopropane) hydrochloride
ABTS$^{\cdot+}$	Cation 2,2′-azino-di (3-ethylbenzthiazoline sulfonate) radicals
AMVN	2,2′-Azobis (2,4-dimethylvaleronitrile)
BMPO	5-*tert*-Butoxycarbonyl 5-methyl-1-pyrroline *N*-oxide
CLA	Conjugated linoleic acids
CW	Continuous wave
CTPO	3-Carbamoyl-2,2,5,5-tetramethyl-3-pyrroline-1-yloxyl
DMPO	5,5-Dimethyl-1-pyrroline *N*-oxide
DPPH$^{\cdot}$	2,2-Diphenyl-1-picryhydrazyl radicals
EMPO	5-Ethoxycarbonyl-5-methyl-1-pyrroline *N*-oxide
EPR	Electron paramagnetic resonance
ESR	Electron spin resonance
GC-MS	Gas chromatography–mass spectrometry
HOSC	hydroxyl radical scavenging capacity
HPLC	High performance liquid chromatography
LA	Linoleic acid
LLSL	Low-level and short-lived
NMR	Nuclear magnetic resonance
^{15}N-PDT	4-Oxo-2,2,6,6-tetramethylpiperidine-d16-1-oxyl
ORAC	Oxygen radical absorption capacity
PBN	*N-tert*-Butyl-α-phenylnitrone
PC	Phosphatidylcholines
RMCD	Randomly methylated β-cyclodextrin
ROS	Reactive oxygen species
TPC	Total phenolic content
Trolox	6-Hydroxy-2,5,7,8-tetramethylchroman-2-carboxylic acid

REFERENCES

1. Knowles, P. F.; Marsh, D.; Rattle, H. W. E. *Magnetic, Resonance of Biomolecules.* Wiley, New York, **1976**.
2. Tarpey, M. M.; Wink, D. A.; Grisham, M. B. Methods for detection of reactive metabolites of oxygen and nitrogen: *in vitro* and *in vivo* considerations. *Am. J. Physiol. Regul. Integr. Comp. Physiol.* **2004**, *286*, 431–444.

3. Takeshita, K.; Ozawa, T. Recent progress in *in vivo* ESR spectroscopy. *J. Radiat. Res.* **2004**, *45*, 373–384.

4. Zhou, K.; Laux, J. J.; Yu, L. Comparison of Swiss red wheat grain and fractions for their antioxidant properties. *J. Agric. Food Chem.* **2004**, *52*, 1118–1123.

5. Yu, L.; Haley, S.; Perret, J.; Harris, M.; Wilson, J.; Qian, M. Free radical scavenging properties of wheat extracts. *J. Agric. Food Chem.* **2002**, *50*, 1619–1624.

6. Yu, L.; Haley, S.; Perret, J.; Harris, M. Antioxidant properties of hard winter wheat extracts. *Food Chem.* **2002**, *78*, 457–461.

7. Yu, L.; Perret, J.; Harris, M.; Wilson, J.; Haley, S. Antioxidant properties of bran extracts from "Akron" wheat grown at different locations. *J. Agric. Food Chem.* **2003**, *51*, 1566–1570.

8. Onyeneho, S. N.; Hettiarachchy, N. S. Antioxidant activity of durum wheat bran. *J. Agric. Food Chem.* **1992**, *40*, 1496–1500.

9. Zielinski, H.; Kozlowska, H. Antioxidant activity and total phenolics in selected cereal grains and their different morphological fractions. *J. Agric. Food Chem.* **2000**, *48*, 2008–2016.

10. Zhou, K.; Yin, J. J.; Yu, L. Phenolic acid, tocopherol and carotenoid compositions, and antioxidant functions of hard red winter wheat bran. *J. Agric. Food Chem.* **2005**, *53*, 3916–3922.

11. Schaich, K. M. EPR methods for studying free radicals in foods. In: Morello, M.J.; Shahidi, F.; Ho, C.T. (Eds.). *Free Radical in Food: Chemisty, Nutrition, and Health Effects.* American Chemical Society, Washington, DC, **2002**. pp.12–34.

12. Swart, H. M.; Bolton, J. R.; Borg, D. C. *Biological Applications of Electron Spin Resonance.* Wiley, New York, **1972**.

13. Borbat, P. P.; Costa-Filho, A. J.; Earle, K. A.; Moscicki, J. K.; Freed, J. H. Electron spin resonance in studies of membranes and proteins. *Science* **2001**, *291*, 266–269.

14. Jones, M. T.; Phillips, W. D. Electron spin resonance. *Ann. Rev. Phys. Chem.* **1966**, *17*, 323–348.

15. Kopani, M.; Celec, P.; Danixovi, L.; Michalka, P.; Biro, C. Oxidative stress and electron spin resonance. *Clin. Chim. Acta* **2006**, *364*, 61–66.

16. Cheng, Z. H.; Su, L.; Moore, J.; Zhou, K. Q.; Luther, M.; Yin, J. J.; Yu, L. L. Effects of post-harvesting treatment and heat stress on availability of wheat antioxidants. *J. Agric. Food Chem.* **2006**, *54*, 5623–5629.

17. Leonard, S. S.; Cutler, D.; Ding, M.; Vallyathan, V.; Castranova, V.; Shi, X. Antioxidant properties of fruit and vegetable juices: more to the story than ascorbic acid. *Ann. Clin. Lab. Sci.* **2002**, *32*, 193–200.

18. Madsen, H. L.; Nielsen, B. R.; Bertelsen, G.; Skibsted, L. H. Screening of antioxidative activity of spices. A comparison between assays based on ESR spin trapping and electrochemical measurement of oxygen consumption. *Food Chem.* **1996**, *57*, 331–337.

19. Zhao, H.; Dong, J.; Lu, J.; Chen, J.; Li, Y.; Shan, L.; Lin, Y.; Fan, W.; Gu, G. Effects of extraction solvent mixtures on antioxidant activity evaluation and their extraction capacity and selectivity for free phenolic compounds in barley (*Hordeum vulgare* L.). *J. Agric. Food Chem.* **2006**, *54*, 7277–7286.

20. Parry, J.; Su, L.; Moore, J.; Cheng, Z. H.; Luther, M.; Rao, J. N.; Wang, J. Y.; Yu, L. L. Chemical compositions, antioxidant capacities, and antiproliferative activities of selected fruit seed flours. *J. Agric. Food Chem.* **2006**, *54*, 3773–3778.

21. Yoshimura, Y.; Nakazawa, H.; Yamaguchi, F. Evaluation of the NO scavenging activity of procyanidin in grape seed by use of the TMA-PTIO/NOC 7 ESR system. *J. Agric. Food Chem.* **2003**, *51*, 6409–6412.

22. Yu, L. Free radical scavenging properties of conjugated linoleic acids. *J. Agric. Food Chem.* **2001**, *49*, 3452–3456.

23. Chen, C.; Tang, H. R.; Sutcliffe, L. H.; Belton, P. S. Green tea polyphenols react with 1,1-diphenyl-2-picrylhydrazyl free radicals in the bilayer of liposomes: direct evidence from electron spin resonance studies. *J. Agric. Food Chem.* **2000**, *48*, 5710–5714.

24. Ilangovan, G.; Zweier, J. L.; Kuppusamy, P. Electrochemical preparation and EPR studies of lithium phthalocyanine. Part 2: particle-size-dependent line broadening by molecular oxygen and its implications as an oximetry probe. *J. Phys. Chem. B* **2000**, *104*, 9404–9410.

25. Andersen, A. B.; Risbo, J.; Andersen, M. L.; Skibsted, L. H. Oxygen permeation through an oil-encapsulating glassy food matrix studied by ESR line broadening using a nitroxyl spin probe. *Food Chem.* **2000**, *70*, 499–508.

26. Shen, J. G.; Khan, N.; Lewis, L. D.; Armand, R.; Grinberg, O.; Demidenko, E.; Swartz, H. Oxygen consumption rates and oxygen concentration in Molt-4 cells and their mtDNA depleted ($\rho°$) mutants. *Biophys. J.* **2003**, *84*, 1291–1298.

27. Dunn, J. F.; Swartz, H. M. *in vivo* electron paramagnetic resonance oximetry with particulate materials. *Methods* **2003**, *30*, 159–166.

28. Presley, T.; Kuppusamy, P.; Zweier, J. L.; Ilangovan, G. EPR oximetry as a quantitative method to measure cellular respiration: a consideration of oxygen diffusion interference. *Biophys. J.* **2006**. doi:10.1529/biophysj.106.090175.

29. Yin, J. J.; Kramer, J. K. G.; Yurawecz, M. P.; Eynard, A. R.; Mossoba, M. M.; Yu, L. L. Effects of conjugated linoleic acid (CLA) isomers on oxygen diffusion-concentration products in liposomes and phospholipid solutions. *J. Agric. Food Chem.* **2006**, *54*, 7287–7293.

30. Yin, J. J.; Mossoba, M. M.; Kramer, J. K.; Yurawecz, M. P.; Eulitz, K.; Morehouse, K. M.; Ku, Y. Effects of conjugated linoleic acid on oxygen diffusion-concentration product and depletion in membranes by using electron spin resonance spin-label oximetry. *Lipids* **1999**, *34*, 1017–1023.

31. Shea, K. M. Technical report: irradiation of food. Committee on Environmental Health. *Pediatrics.* **2000**, *106*, 1505–1510.

32. Goodman, B. A.; McPhail, D. B.; Duthie, D. M. L. Electron spin resonance spectroscopy of some irradiated foodstuffs. *J. Sci. Food Agric.* **1989**, *47*, 101–111.

33. Dodd, N. J. F.; Lea, J. S.; Swallow, A. J. ESR detection of irradiated food. *Nature* **1988**, *334*, 387.

34. Kristensen, D.; Skibsted, L. H. Comparison of three methods based on electron spin resonance spectrometry for evaluation of oxidative stability of processed cheese. *J. Agric. Food Chem.* **1999**, *47*, 3099–3104.

35. Sripriya, G.; Chandrasekharan, K.; Murty, V. S.; Chandra, T. S. ESR spectrometric studies on free radical quenching action of finger millet (*Eleusine coracana*). *Food Chem.* **1996**, *57*, 537–540.

36. Zhou, K. Q.; Yin, J. J.; Yu, L. L. ESR determination of the reactions between selected phenolic acids and free radicals or transition metals. *Food Chem.* **2006**, *95*, 446–457.

37. Stajner, D.; Milic-Demarino, N.; Canadanovic-Brunet, J.; Stajner, M.; Popovic, B. M. Screening for antioxidant properties of *Allium giganteum*. *Fitoterapia* **2006**, *77*, 268–270.

38. Chou, D. S.; Hsiao, G.; Shen, M. Y.; Fong, T. H.; Lin, C., H.; Chen, T. F.; Sheu, J. R. Low concentration of oxidized low density lipoprotein suppresses platelet reactivity *in vitro*: an intracellular study. *Lipids* **2004**, *39*, 433–440.

39. Boveris, A. D.; Puntarulo, S. Free-radical scavenging action of natural antioxidants. *Nutr. Res.* **1998**, *18*, 1545–1557.

40. Moore, J.; Yin, J. J.; Yu, L. Novel fluorometric assay for hydroxyl radical scavenging capacity (HOSC) estimation. *J. Agric. Food Chem.* **2006**, *54*, 617–626.

41. Kaur, A.; Pan, M.; Meislin, M.; Facciotti, M. T.; El-Gewely, R.; Baliga, N. S. A systems view of haloarchaeal strategies to withstand stress from transition metals. *Genome Res.* **2006**, *16*, 841–854.

42. Frei, B.; Higdon, J. V. Antioxidant activity of tea polyphenols *in vivo*: evidence from animal studies. *J. Nutr.* **2003**, *133*, 3275S–3284S.

43. Halliwell, B.; Chirico, S. Lipid peroxidation: its mechanism, measurement, and significance. *Am. J. Clin. Nutr.* **1993**, *57* (suppl.), 715S–725S.

44. Yin, J. J.; Smith, M. J.; Eppley, R. M.; Page, S. W.; Sphon, J. A. Effect of fumonisin B_1 on lipid peroxidation in membranes. *Biochim. Biophy. Acta* **1998**, *1371*, 134–142.

CHAPTER 11

ANALYSIS OF TOCOPHEROLS AND CAROTENOIDS IN WHEAT MATERIALS USING LIQUID CHROMATOGRAPHY–MASS SPECTROMETRY TECHNOLOGY

Zhigang Hao
Denys Charles

11.1 INTRODUCTION

Carotenoids (provitamin A) and tocopherols (vitamin E) are fat-soluble antioxidants associated with decreased risk of several degenerative diseases (1). Both antioxidants occur naturally in wheat. They are widely present in different wheat tissues from the seven outer layers of bran like a protective shell or bran envelope to the inner tiny germ and grain "heart" endosperm. The endosperm is the source of white flour after the bran and germ have been sifted out. In 1912, Monier-Williams (2) assigned the yellow color of wheat flour to the presence of carotenes by comparing the 440-nm light absorbance of wheat extracts to that of carotene from carrots. In 1940, Zechmeister and Cholnoky (3) confirmed the preponderance of xanthophylls, mainly lutein, to the yellow pigments in wheat flour. In 1968, Lepage and Sims (4) found that free lutein accounts for 84.8% of the yellow pigments in wheat. Another important class of lipid antioxidants in wheat is tocopherols that are also known as vitamin E. Vitamin E was first discovered in 1922 when Evans and Bishop (5) described a "substance X" that was essential for maintaining rat fertility. After obtaining similar results, Sure (6) called the substance X as "vitamin E" because vitamins A, B, C, and D were already known at that time. Pure α-tocopherol was first isolated from wheat-germ oil in 1936 by Evans's group (7) and they called this isolated substance X (vitamin E) as "tocopherol," a name derived from the ancient Greek word *phero*, "to bring" and the word *tocos* meaning "childbirth."

Tocopherols and carotenoids are widely present in food and nutraceutical products and have been extensively studied for disease prevention and health promotion. However, there is still no official and standardized chromatographic method for qualitative and quantitative analyses of these antioxidants. An AACC method (8) has been established for analyzing wheat yellow pigments by using pure β-carotene as standard at 440 nm wavelength. However, this method does not provide the individual carotenoid content information.

In 1964, Pennock and coworkers (9) first reported the existence of eight toco-derived compounds, α-, β-, γ-, and δ-tocopherols of $2R$, $4'R$, $8'R$-configuration along with the corresponding 2R-trans/trans-tocotrienols in nature (Fig. 11.1). The first chromatographic separation of tocopherol isomers was reported in 1973 (10). Since then, both normal and reverse phases of HPLC technologies have been extensively used for the analysis of antioxidant mixtures in various sample matrices, which have been reviewed by Abidi (11) and Ruperez et al. (12). Despite obvious shortcomings of long equilibration times inherent with normal phase systems and the employment of hazardous volatile organic solvents, most researchers have still preferred normal-phase HPLC techniques for tocopherol isomer separation, especially between β- and γ-tocopherols, since the unusual selectivity of silica-based stationary phases has been demonstrated for the differentiation of the aromatic ring (chromanol head) positional isomers of tocopherol antioxidants.

Different from tocopherol isomers, which have different aromatic ring position substituents, carotenoids have an important structural feature, which is the existence of a highly conjugated system of double bonds resulting in *cis* and *trans* isomers (Fig. 11.2). The separation of either different carotenoids or carotenoid isomers is therefore of common interest. The separation of carotenoids has been previously achieved by using either normal or reverse phases of HPLC methods. Most of the chromatographic methods for carotenoid separation have been reviewed by de Quiros

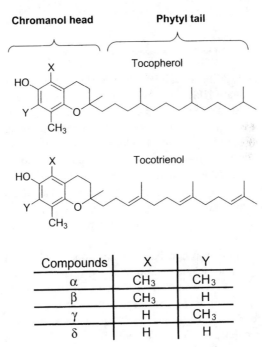

Compounds	X	Y
α	CH_3	CH_3
β	CH_3	H
γ	H	CH_3
δ	H	H

Figure 11.1   Selected structures of tocopherols and tocotrienols.

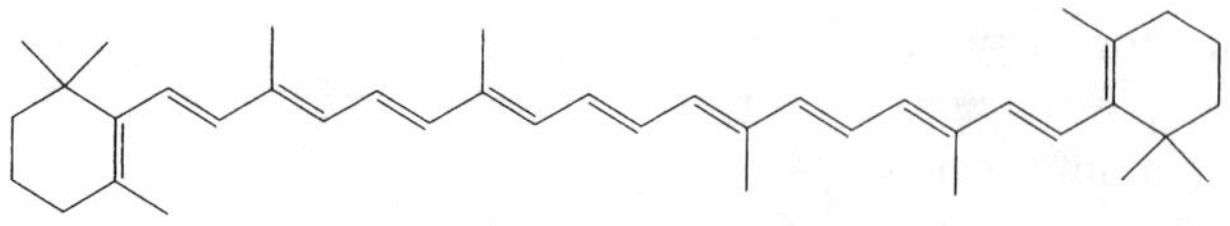

All *trans*-β-carotene

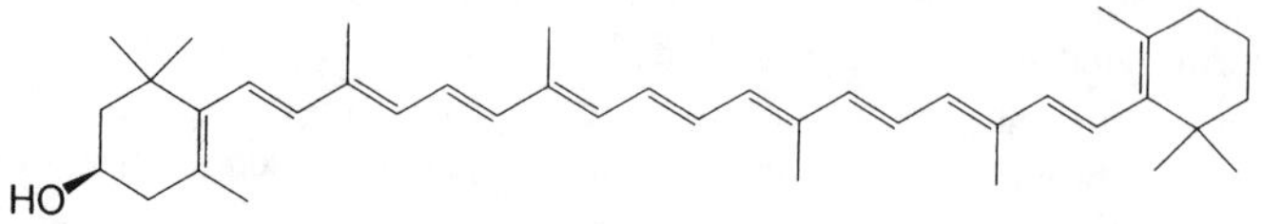

All *trans*-β-crypoxanthin

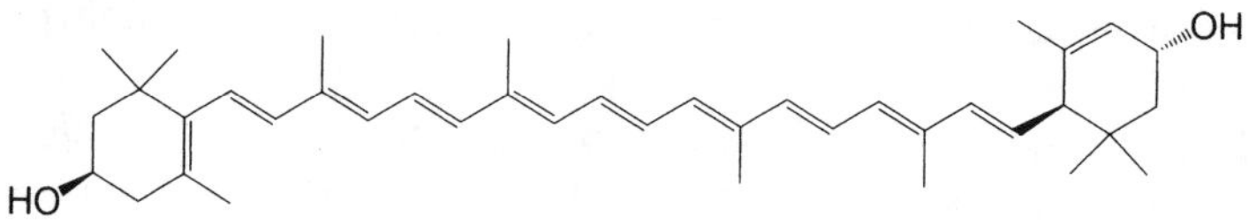

All *trans*-lutein

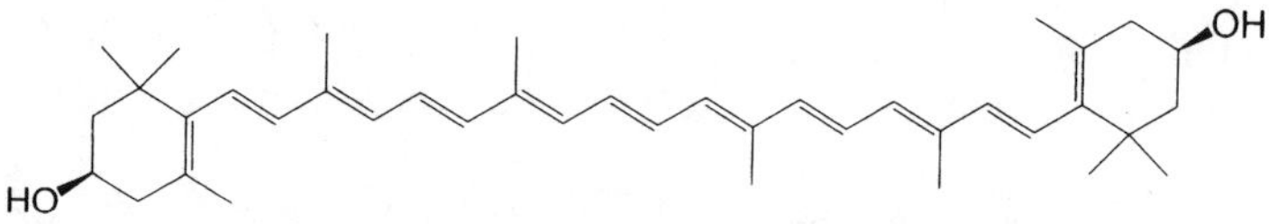

All *trans*-zeaxanthin

Figure 11.2 Selected structures of carotenoids present in wheat grain.

(13) and Tsao (14). Just like tocopherol, the lipophilic characteristics of carotenoids have made normal-phase HPLC a more favorable choice for the separation of these antioxidants. To avoid hazardous organic solvents, C_8 and C_{18} columns with reverse-phase HPLC have been proven to be well suited for routine separation of different carotenoids. Especially, Tsao et al. developed a LC–DAD–MS method with C_{18} column to separate several *cis* isomers of lutein diesters (15). However, for separation of more complicated geometric isomers of carotenoids containing 40 carbons (tetra-terpenoids), the longer C_{30} and C_{34} columns seem to have better separation and selectivity than the conventional C_8 and C_{18} columns. Sander et al. (16,17) developed a polymeric C_{30} liquid chromatographic column, especially designed for the separation of carotenoid isomers.

Wheat is a major crop and basic human food staple, supplying significant amounts of dietary carbohydrate and protein, and it is also a useful source of anti-oxidant compounds (18). Carotenoids and tocopherols have contributed mainly to the antioxidant property of wheat and its products such as cereals (19–23). For better understanding the chemical profiles of wheat antioxidants, a new simultaneous quantitation of tocopherol and carotenoids by LC–MS was established, which provided a better structural identity and detection sensitivity (24).

MS instrument can be coupled with variable interface options such as electrospray ionization (ESI) and atmospheric pressure chemical ionization (APCI) as a newer LC system detector. These interface options have to be fully understood to meet the requirements for LC mobile phase composition. Combined with specific physiochemical characters of carotenoid and tocopherol structures in wheat, the compatibility among the extraction solvents, chromatographic mobile phases, and LC–MS interfaces has become a very critical factor for column performance, method reproducibility, and detection sensitivity. However, this compatibility has been ignored in many instances. For example, extraction solvent, hexane or dichloromethane, was injected into reverse-phase C_{18} or C_{30} column (18,25). These solvents can be adhered into C_{18} or C_{30} materials and change the column properties during the sample running.

This chapter will cover some analytical views from the three stages in tocopherol and carotenoid analysis, namely sample extraction, chromatographic separation, and MS detection, to stress the importance of method compatibility. Most common extraction solvents will be discussed for their extraction efficiency and their impact on the mobile-phase selection during HPLC method development. Different HPLC columns from normal-phase bare silica to reversed-phase C_{18}, C_{30}, and C_{34}, corresponding temperature effects, and method resolution will be compared. In addition, the selection of ESI and APCI interfaces for MS detection will be addressed based on the physiochemical properties of tocopherols and carotenoids.

11.2 TERMINOLOGY

Tocopherols and carotenoids are natural lipid-soluble antioxidants. More than 700 compounds that produce the red, yellow, and orange colors in fruits, vegetables, and other plants have been reported. Their natural sources, chemical structures, physiochemical properties, biological functions, and health benefits have been extensively described. The terminology for this class of compounds is very confusing in the literature. Tocopherols are collectively known as vitamin E, and carotenoids are called terpenoids or yellow pigments, and both can be described as lipophilic, lipid, or hydrophobic antioxidants. Numerous analytical method reviews have been written recently for the different applications (12,13,26–28). To choose an analytical method for the individual targeted projects and ingredients, a comprehensive literature search using all potential names for tocopherols and carotenoids is essential. The objective for tocopherol and carotenoid study in botanical or food product could be completely different to that in clinical research, but a good chromatographic method can be shared in all the areas of investigation.

11.3 ANALYSIS OF TOCOPHEROLS AND CAROTENOIDS BY LC–MS TECHNOLOGY

For different research and manufacturing purposes, wheat materials containing tocopherols and carotenoids can be sampled at any stage from its initial growth to the final food products. After harvest, wheat grain can be dried and stored under certain controlled conditions. The bran and germ can be separated from the endosperm that is the source of wheat flour. The wheat germ can be used for preparing germ oil, and wheat flour can be further applied for the manufacture of food products such as pizza and wheat bread. The water content and structure of the different wheat materials can be dramatically changed between different stages described above. To obtain the most reliable and consistent results for variable wheat materials, the following factors need to be considered before the experiments:

(a) The solubility of tocopherols and carotenoids in extraction solvent. Only soluble ingredients can be extracted from wheat tissues.
(b) Water content in wheat samples. Hydrophobic solvents can only be used for materials containing less water because water usually blocks the hydrophobic solvent from penetrating into wheat tissues.
(c) Agreement between extraction solvents and LC mobile-phase composition. Injecting extraction solvents with strong elution power into the LC system will change column consistency.
(d) Compatibility between LC mobile-phase composition and LC–MS interfaces (ESI or APCI). ESI likes solvents with free hydrogen atom while APCI likes solvents with high evaporation pressure.

To date, no single solvent has been found for the optimal extraction of both tocopherols and carotenoids from botanical or food samples, and no single column from normal phase to reverse phase can separate all the isomers for all tocopherols and carotenoids. Both ESI and APCI LC–MS interfaces have been applied to the analysis of tocopherols and carotenoids. As a general rule, a better understanding of the properties of the sample materials and a thorough literature search for the available methods are essential before embarking on the research, especially for a new project.

11.3.1 Liquid–Liquid Extraction

Tocopherols and carotenoids are strong reducing ingredients. As a general rule, the extraction of carotenoids must be carried out very quickly, avoiding exposure to light, oxygen, high temperatures, and to prooxidant metals, such as iron or copper, in order to minimize auto-oxidation and *cis–trans* isomerization (29,30). In addition, THF and ethyl ether have been reported to develop peroxides that can rapidly degrade β-carotene, and may contribute to the production of artifacts (30,31). Hence, the addition of antioxidants such as ascorbic acid and pyrogallol has been recommended to prevent carotenoid losses during the extraction procedure (32–35).

In addition, for a successful liquid–liquid extraction of tocopherols and carotenoids, the similarity in polarity between extracting solvents and the extracted ingredients is a key factor. Free tocopherols and carotenoids without glucose or protein attachment are nonpolar. As such, nonpolar solvents such as hexane or chloroform should be the ideal solvents for extraction. The solubility of both β-carotene and lutein in different organic solvents was examined by Craft and Soares (36). β-carotene was least soluble in methanol and acetonitrile (solubility for both is 10 µg/mL) while lutein was least soluble in hexane (only 20 µg/mL) and both carotenoids were most soluble in THF (10,000 µg/mL for carotene and 8000 µg/mL for lutein). Although THF is subject to peroxide formation, it is increasingly used for tocopherol and carotenoid extraction due to the high solubility of a wide polarity range of carotenoids. The presence of water in most botanicals and food materials will block the physical contact between extraction solvents and tocopherols and carotenoids inside the materials. In order to minimize this kind of blocking effect, more polar solvents need to be used. Because of their complex structure and wide variety of botanicals and food materials, there is not one method to apply for the analysis of all of them, and as such, various organic solvents have been tried for tocopherol and carotenoid extraction for different research objectives. For botanical materials containing a significant amount of water, it is desirable to use an organic solvent that is miscible with water in an attempt to penetrate into the inside of the material tissues and prevent the formation of an emulsion (37). To improve extraction efficiency, special solvent solubility is required in material extraction. For example, acetone and THF can be employed for most tocopherol and carotenoid extractions. However, lycopene has a higher solubility in CH_2Cl_2 and $CHCl_3$ and poor solubility in other organic solvents. Therefore, lycopene-rich materials, such as tomatoes, need to be extracted with a solvent containing CH_2Cl_2 or $CHCl_3$. For the above reasons, mixed solvents such as CH_2Cl_2:MeOH were also applied for tocopherol and carotenoid extractions. To better understand solvent extraction of tocopherols and carotenoids, different solvents and their combinations have been evaluated by two groups (38,39). Both groups concluded that the best extraction efficiency was achieved with solvent EtOH:hexane mixture. In addition, sequential extraction procedures have been reported to avoid water interference. Deli et al. (40) used MeOH followed by diethyl ether to extract carotenoids from fruits of pepper, and Gandul-Rojas et al. (41) used DMF followed by hexane to extract carotenoids from olive fruits.

For wheat materials, dried grain flour powder or wheat oil can be extracted directly with hexane because the latter can easily penetrate into the powder or oil materials and dissolve tocopherols and carotenoids. If wheat grain flour powder has been formulated into aqueous liquid solution, tocopherols and carotenoids can be partitioned into hexane. It is very critical to completely remove water from the hexane extract prior to injection into normal-phase LC system because water will change the silica surface property and affect the reproducibility of the HPLC method. However, for reverse-phase LC system, hexane has to be removed and the extract reconstituted into methanol or other reverse-phase LC system friendly solvents before the extract could be delivered into the reverse-phase LC system because hexane will be strongly adhered into the stationary phase and affect the method reproducibility. Wheat bran is difficult to be directly extracted by hexane because of its surface structure, moisture, and tissue membrane structure. Methanol/tetrahydrofuran (1 : 1. v/v) should be a good extraction solvent because methanol not only is considered as the most powerful penetrating solvent for botanical materials but also mixes well with the moisture on the surface of wet hard wheat bran tissues. Tetrahydrofuran can enhance the solubility of tocopherols and carotenoids during extraction. The methanol/tetrahydrofuran extract solution can be directly injected into the reverse-phase LC system, but it is not suitable for normal-phase LC system. For normal-phase LC, the methanol/tetrahydrofuran has to be completely removed and the dried extract reconstituted into hexane before the sample solution is injected into the bare silica column to retain the reproducibility of the stationary phase. The detailed extraction schemes for tocopherols and carotenoids from wheat materials for normal- and reverse-phase LC systems, respectively, are presented in Fig. 11.3a and b.

The influence of polar solvents from sample extracts on normal-phase LC performance has been previously investigated in our lab (24). Methanol, tetrahydrofuran, or water from the sample extracts can dramatically alter the retention time of the interested compounds on bare-silica column. If methanol and tetrahydrofuran are needed during the sampling processes, they should be removed under nitrogen gas prior to sample injection into the LC system. Moisture also significantly alters the retention performance of the silica column and has to be removed by using anhydrous Na_2SO_4 before sample injection. This not only prevents the extended reequilibration time required to condition the column after running each sample, but also provides

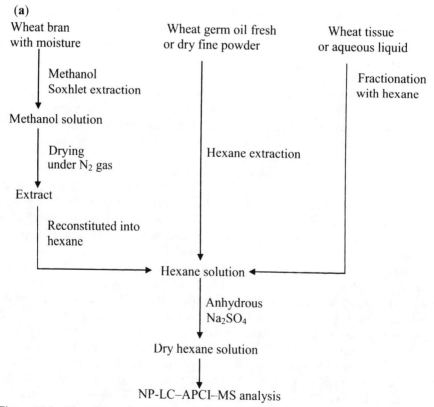

Figure 11.3 The schemes for (a) normal-phase LC–APCI–MS and (b) reverse-phase LC–ESI–MS analyses of wheat tocopherols and carotenoids.

(b)

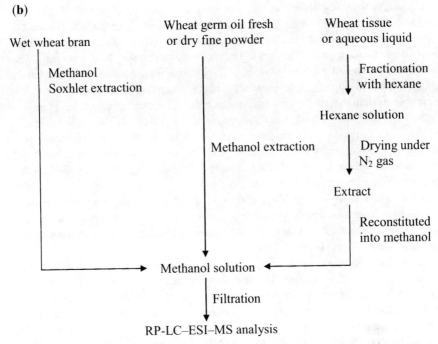

Figure 11.3   (*Continued*)

better reproducibility on retention time. In fact, the effects from polar solvent like water or methanol in silica column is just like injecting hexane or ethyl acetate solvent into C_{18} or C_{30} reverse-phase column (16). Generally, reequilibration in the reverse-phase LC does not take long because nonpolar organic solvents are not commonly used in the sample extraction for reverse-phase chromatography.

11.3.2   Chromatographic Separation

Main lipid antioxidants in wheat products include carotenoids and tocopherols. Two classes of carotenoids are recognized in wheat materials: the carotenes, which are tetraterpenoid hydrocarbons, and xanthophylls, which are carotenoids with one or more oxygenated functional groups present in the molecule (18). Five representative lipid antioxidants in wheat materials are α-tocopherol, β-carotene, β-cryptoxanthin, lutein, and zeaxanthin (18,23,42,43). Lutein and zeaxanthin in the four common carotenoids have the same molecular weight and very similar conjugated system in the structures (Fig. 11.2), and therefore, their resolution on the common reverse-phase column is poor (44). Furthermore, UV and even tandem MS cannot provide the identity information during the quantitation without a good chromatographic separation.

Some common chemical properties such as long terpenoid-conjugated double bonds, stereoisomers and geoisomers and strong hydrophobicity can be found for carotenoid and tocopherol structures. These chemical characteristics can cause some special problems for the reverse-phase chromatographic separation such as permanently adhering to the stationary phase that will result in sample loss and low recovery. Sample losses for tocopherols and carotenoids on reverse-phase columns are very common phenomena because of their extremely strong hydrophobic property and permanent sticking on C_{18} materials. The recovery can be very low, especially when acetonitrile-based solvents or polymeric C_{18} columns are used. The use of methanol or methanol-based solvents can provide a higher recovery of the tocopherols and carotenoids than the use of acetonitrile or acetonitrile-based solvents. Acid contamination in chlorinated solvents has also been blamed for carotenoid losses (45). Because hydrolysis of ethyl acetate produces acetic acid, the acidity may also cause some tocopherol and carotenoid losses. Kamber and Pfander (46) found it is necessary to add N,N-diisopropylethylamine to silica column to obtain a good carotenoid recovery because diisopropylethylamine is a nucleophilic base (46). Epler et al. (44) from NIST also found that triethylamine (TEA) and ammonium acetate can improve the recovery. Ammonium acetate is poorly soluble in acetonitrile but 0.05M ammonium acetate can be dissolved in methanol, which can provide enough buffer capacity for acidity from silica and organic modifiers.

The acetonitrile- or methanol-based eluents combined with the various organic modifiers such as tetrahydrofuran, diisopropyl ether, chloroform, dichloromethane, ethyl acetate, and even hexane have been tried for carotenoid and tocopherol chromatographic separations (47–49). The organic modifiers can increase the mobile-phase elution power and decrease the running time but they cannot significantly affect the selectivity for lutein and zeaxanthin (44). The organic modifiers will affect the protonization process of the ESI and decrease the ESI–MS detection sensitivity.

The NIST has done extensive work on tocopherol and carotenoid separation using C_{18} column. C_{18} columns can be classified into monomeric, polymeric, or intermediate based on their selectivity for a test mixture of polycyclic aromatic hydrocarbons (PAHs), Standard Reference Materials (SRM) 869 (50). This classification is strongly related to the procedures used in the bonded-phase synthesis, that is, monomeric phases prepared by the reaction of silica with monofunctional silanes (usually in the absence of water) have properties that differ from polymeric phases prepared by reaction of silica with trifunctional silanes in the presence of water. The third group is an arbitrary classification for which the phase chemistry is less certain. This group has properties that are intermediate to the monomeric and polymeric classes. On the basis of this classification, the polymeric phase C_{18} columns exhibit enhanced shape selectivity toward isomers and other groups of compounds with similar structures (51) and have the potential to be able to partially separate lutein and zeaxanthin (selectivity, α value can reach 1.32). In addition, column pore size does not affect the selectivity of lutein and zeaxanthin because of their similar molecular size. The only change is that more modifier needs to be added as the pore size decreases (44). This is as expected since the greater surface area of the base silica of the narrow-pore columns results in a higher carbon load.

Methanol-based solvent system and polymeric phase column have been proven to be better conditions for method resolution and recovery. However, for separation of the isomer pair of lutein and zeaxanthin, no significant improvement was made. To enhance the discrimination of geometric differences of tocopherols and carotenoids in reverse phase LC system, engineered stationary phases (C_{30} and C_{34}) were developed (16,32). It is informative to consider the size of carotenoid molecules for comparison with typical bonded phase dimensions. Thickness for various monomeric and polymeric alkyl phases has been measured using small-angle neutron scattering (SANS) (52) in a reverse-phase environment (methanol and deuterated methanol). It is clear that carotenoids are significantly larger than either monomeric or polymeric C_{18} phases and closer to either monomeric or polymeric C_{30} phases. To further address the separation of lutein and zeaxanthin isomers, temperature effects on the retention of carotenoids on C_{18}, C_{30}, and C_{34} columns during isocratic elution were studied (53). For the C_{30} and C_{34} stationary phases, temperature has a distinctly different effect on the retention of lutein and zeaxanthin, the van't Hoff plots $\ln k' = -\Delta H/RT + \Delta S/R + \ln \Phi$ are nonlinear and appear to flatten out and diverge from each other as the temperature is decreased. It was believed that the retention of these polar carotenoids involves interaction with silanols on or near the surface of the silica, and that retention results from a combination of mechanisms (17). The

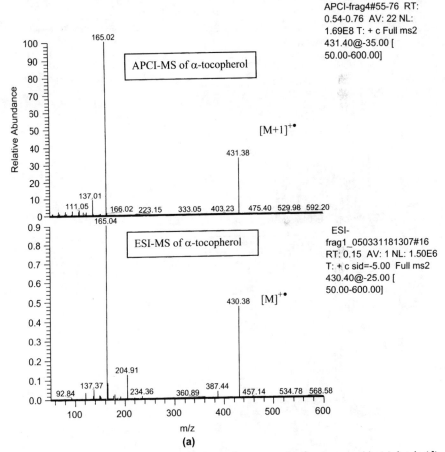

Figure 11.4 Mass spectra of α-tocopherol (**a**), β-carotene (**b**), β-cryptoxanthin (**c**), lutein (**d**), and zeaxanthin (**e**) via ESI and APCI interfaces. (Adapted from Reference 24.)

nonlinear nature of the van't Hoff plots on the C_{30} and C_{34} stationary phases indicates temperature-dependent changes in the analyte–stationary-phase interactions. The increased rigidity and order induced by lower temperature may prevent the analytes from penetrating all the way to the surface and interacting with silanol, thus changing the relative contributions of analyte–silanol and solute-bonded phase interactions to retention. In this case, each contributing mechanism would be governed by one of van't Hoff equations, but the overall effect would not be so correlated. For the shorter C_{18} phase, this effect is not apparent in the practical temperature range. However, on the C_{30} and C_{34} phases, the van't Hoff plots for lutein and zeaxanthin were significantly different, which provides a very good opportunity for chromatographic separation of tocopherols and carotenoids under different temperature conditions (54). The requirements of gradient elution with long running time and C_{30} column with low temperature facility are still problems for the routine analysis of tocopherols and carotenoids. For these reasons, a newer simplified normal-phase LC system has been developed for the analysis of carotenoids and tocopherols in wheat materials under

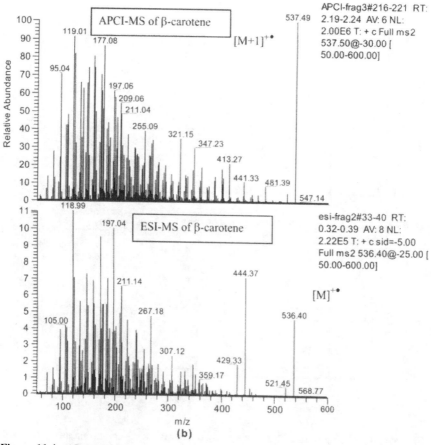

Figure 11.4 (*Continued*)

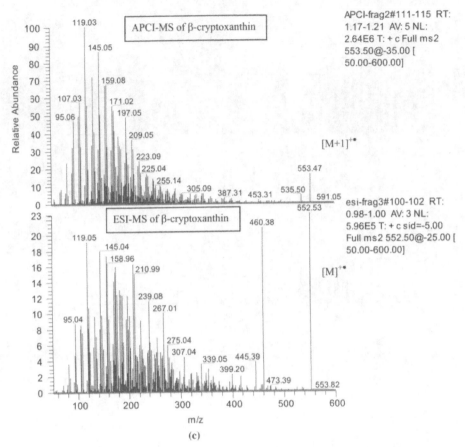

Figure 11.4 (*Continued*)

isocratic elution, short running time, and ambient column temperature conditions (24). And most importantly, this method can be compatible with hexane solvent extraction and APCI interface for MS detection.

11.3.3 LC–MS Interfaces and MS Detection

Normal- and reverse-phase LC systems can be coupled to VIS–UV, fluorescence, electrochemical detectors for the qualitative and quantitative analyses of tocopherols and carotenoids. However, the structural information obtained by those detectors is very limited. NMR and MS detectors can provide more information on functional groups and molecular weight but the mass amount requirement for NMR makes MS detector more practical and useful for routine chromatographic analysis, especially for low concentration of samples. Mass detectors measure mass-to-charge ratios (*m/z*) of ionized molecules and the ionized molecules can be formed under a variety of ionization techniques, which include EI (electronic impact), CI (chemical ionization), FAB (fast atom bombardment), and API (atmospheric pressure ionization). Both ESI

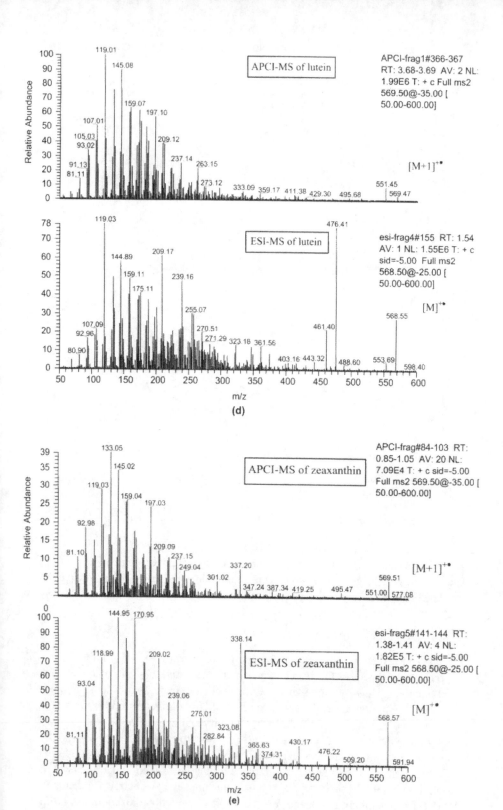

Figure 11.4 (*Continued*)

and APCI techniques in API category involve molecular ionization at atmospheric pressure instead of vacuum pressure used in other techniques and so they are often called as soft-ionization techniques because proton transfer between donor and acceptor need much less energy than electron-capture or electron-loss process in the normal EI processes, even the extended polyene chain in carotenoids can stabilize either a negative or positive charge in their molecular ions $[M^{\cdot+}$ or $M^{\cdot-}]$ (55). ESI and APCI interfaces are ionization devices that were initially developed to enable the effluent from an LC column to be passed directly into a mass spectrometer. It is necessary to remove the solvent from the LC column effluent without removing too much of the analyte itself and without applying too much heat. When samples are acquired using the ESI interface, the effluent is sprayed at atmospheric pressure into a desolvation region, where most of the solvent evaporates, and the residual sample (analyte) molecules pass into a mass analyzer, usually as protonated molecules. The sample does not need to be heated and the protonated (quasi) molecular ions are thermally equilibrated by collisional processes during the evaporation of solvent. The major difference between ESI and APCI is the formation of protonated molecular ions. In ESI, ions are preformed in the liquid phase before the analyte even reaches the source probe. Most commonly, this can be accomplished by adding a proton donor such as formic acid to the polar group such as OH, NH_2 or C=O in the analyte structures or removing a proton from the polar group such as COOH to the LC eluent. However, fat-soluble tocopherols and carotenoids lack such polar protonation or deprotonation sites in the liquid phase and their proton transfers become harder to occur. Competing with these proton transfer processes, the molecular ion $[M]^{\cdot+}$ (electron loss) instead of $[M+1]^{\cdot+}$ formation becomes predominant via ESI interface (56). However, the sensitivity obtained under this unusual phenomenon by ESI is sacrificed to a certain degree. To improve the detection sensitivity of the fat-soluble tocopherols and carotenoids, the MS detection via APCI interface after LC separation was applied. When samples are acquired using APCI interface, tocopherols and carotenoids become gaseous in the vaporizer before they reach the corona discharge needle. All the mass spectra of tocopherols and carotenoids return to the normal $[M+1]^{\cdot+}$ phenomenon. The protonated $[M+1]^{\cdot+}$ molecular ions from gas-phase ion-molecular reactions inside the APCI interface might be considered as either a straightforward CI process or as a result of the heating energy from the APCI vaporizer, which promotes tocopherols and carotenoids to various excited states that exhibit unusually high proton affinity compared with the ground-state molecules (55). The mass spectra of α-tocopherol and β-carotene, β-cryptoxanthin, lutein, and zeaxanthin via ESI and APCI are displayed in Fig. 11.4a–e. Our experiments indicate that detection using APCI mode usually can be 100 times more sensitive that with an ESI mode.

The thermal stability is a major concern for the APCI approach because the APCI vaporizer is usually heated to around 550°C in order to convert the analytes into a gas phase. During our experiments, the thermal dehydration of lutein was found to be significant when the temperature was greater than 450°C. The formation of dehydrolutein is shown in Fig. 11.5. This dehydration happened postcolumn and pre-ionization at the corona discharge needle in the APCI interface because lutein still displayed m/z at 569 $[M+1]^{\cdot+}$ under lower APCI vaporizer temperature and the

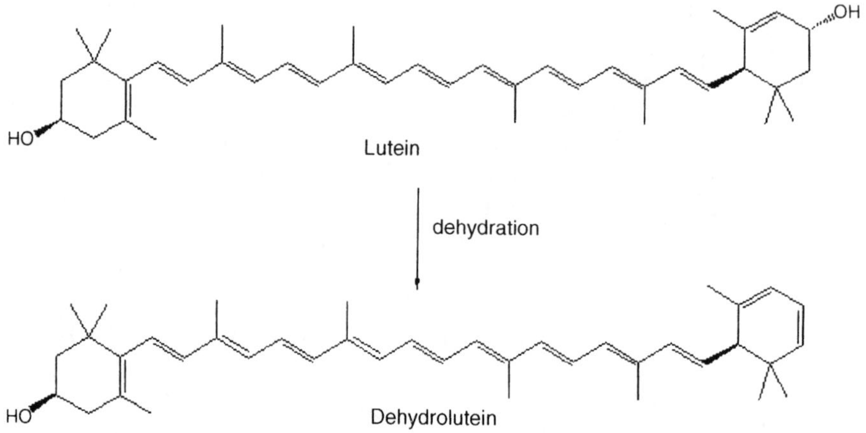

Figure 11.5   Dehydrolutein formation from lutein under high temperature in APCI vaporizer. (Adapted from Reference 24.)

dehydration ion is 551 [568 − H_2O + 1] instead of 550 [568 − H_2O]. The proton was added to the dehydrated molecule [568 − H_2O] before softionization occurred. The SRM from 551.4 to 119.2 instead of from 569.5 to 119.2 was used for lutein quantitation because of this postcolumn dehydration. Zeaxanthin can also be dehydrated at high temperatures but the ratio between dehydrated and nondehydrated molecules is much lower than lutein because it lacks the allyl alcohol skeleton (CH=CHCHOH) in the structure, which is displayed in Fig. 11.2. Therefore, the SRM from 569.5 to 119.2 was used to quantify the zeaxanthin in the samples.

11.4   SUMMARY

Tocopherols and carotenoids can be extracted by most organic solvents. Both normal and reversed phase can be applied for chromatographic separation of tocopherols and carotenoids. Either ESI or APCI interface has been reported for MS detection. The most critical point for successfully developing a reliable analytical method for tocopherols and carotenoids is the compatibility among the sampling process, the chromatographic separation, and the interfaces to MS analyzer. An ESI interface is more suitable for reverse-phase mobile phase such as methanol or acetonitrile with very low concentration of proton donor such as formic acid. For such an LC–MS system, hexane should be avoided at the end of the extraction procedures. However, if hexane is necessary for extraction, it should be completely removed and the dried sample reconstituted into a highly polar organic solvent such as methanol for injection. The hexane solution without moisture is better in the normal-phase LC system. The mobile phase such as high ratio of hexane or ethyl acetate in normal-phase LC is a perfect match for the APCI interface because higher evaporation pressure can easily be achieved in the vaporizer in order to promote the formation of analyte gas phase.

ACKNOWLEDGMENT

We like to thank Professor Chi-Tang Ho from Rutgers University for reviewing this submission and providing his insightful comments.

REFERENCES

1. Winklhofer-Roob, B. M.; Rock, E.; Ribalta, J.; Shmerling, D. H.; Roob, J. M. Effects of vitamin E and carotenoid status on oxidative stress in health and disease. Evidence obtained from human intervention studies (REVIEW). *Mol. Aspect. Med.* **2003**, *24*, 391–402.

2. Monier-Williams, G. W. Reports to the local government board on the nature of flour and its relationship to the process of natural and artificial bleaching. Reports, Local government board (Great Britain). *Public Health and Medical Subjects, New Series.* **1912**, 73.

3. Zechmeister, L.; Cholnoky, L. Carotenoids of Hungarian wheat flour. *J. Biol. Chem.* **1940**, *135*, 31–36.

4. Lepage, M.; Sims, R. P. A. Caroteinoids of wheat flour: their identification and composition. *Cereal Chem* .**1968**, *45*, 600–604.

5. Evans, H. M.; Bishop, K. S. On the existence of a hitherto unrecognized dietary factor essential for reproduction. *Science* **1922**, *56*, 650–651.

6. Sure, B. Dietary requirements for reproduction. II. The existence of a specific vitamin for reproduction. *J. Biol. Chem.* **1924**, *58*, 693–709.

7. Evans, H. M.; Emerson, O. H.; Emerson, G. A. The isolation from wheat germ oil of an alcohol, α-tocopherol, having the properties of vitamin E. *J. Biol. Chem.* **1936**, *113*, 319–332.

8. AACC (American Association of Cereal Chemists). AACC official method 14–50. 1961. (revised). In: Approved Methods of the American Association of Cereal Chemists, Minneapolis, MN, USA,**1994**.

9. Pennock, J. F.; Hemming, F. W.; Kerr, J. D. *Biochem. Biophys. Res. Commun.* **1964**, *17*, 542.

10. Van Niekerk, P. J. The direct determination of free tocopherols in plant oils by liquid-solid chromatography. *Anal. Biochem.* **1973**, *52*, 533–537.

11. Abidi, S. L. Chromatographic analysis of toco-derived lipid antioxidants. *J. Chromatogr. A* **2000**, *881*, 197–216.

12. Ruperez, F. J.; Herrera, M. E.; Barbas, H. C. Chromatographic analysis of α-tocopherol and related compounds in various matrices. *J. Chromatogr. A* **2001**, *935*, 45–69.

13. de Quiros, A. R. B.; Costa, H. S. Analysis of carotenoids in vegetable and plasma samples: A review. *J. Food Comp. Anal.* **2006**, *19*, 97–111.

14. Tsao, R.; Deng, Z. Separation procedures for naturally occurring antioxidant phytochemicals. *J. Chromatogr. A* **2004**, *812*, 85–99.

15. Tsao, R.; Yang, R.; Young, C.; Zhu, H.; Manolis, T. Separation of geometric isomers of native lutein diesters in marigold (*Tagetes eracta* L.) by high performance liquid chromatography–mass spectrometry. *J. Chromatogr. A* **2004**, *1045*, 65–70.

16. Sander, L. C.; Sharples, K. E.; Pursch, M. C$_{30}$ stationary phases for the analysis of food by liquid chromatography (review). *J. Chromatogr. A* **2000**, *880*, 189–202.

17. Sander, L. C.; Sharpless, K. E.; Wise, S. A. Development of engineered stationary phases for the separation of carotenoid isomers. *Anal. Chem.* **1994**, *66*, 1667–1674.

18. Hidalgo, A.; Brandolini, A.; Pompei, C.; Piscozzi, R. Carotenoids and tocols of einkorn wheat (*Triticum monococcum* ssp. *monococcum* L.). *J. Cereal Sci.* **2006**, *44*, 182–193.

19. Adom, K. K.; Sorrells, M. E.; Liu, R. H. Phytochemical profiles and antioxidant activity of wheat varieties. *J. Agric. Food Chem.* **2003**, *51*, 7825–7834.

20. Adom, K. K.; Sorrells, M. E.; Liu R. H. Phytochemicals and antioxidant activity of milled fractions of different wheat varieties. *J. Agric. Food Chem.* **2005**, *53*, 2297–2306.

21. Panfili, G.; Fratianni, A.; Irano, M. Normal-phase high performance liquid chromatography method for the determination of tocopherols and tocotrienols in cereals. *J. Agric. Food Chem.* **2003**, *51*, 3940–3944.

22. Panfili, G.; Fratianni, A.; Irano, M. Improved normal-phase high performance liquid chromatography procedure for the determination of carotenoids in cereals. *J. Agric. Food Chem.* **2004**, *52*, 6373–6377.

23. Moore, J.; Hao, Z.; Zhou, K.; Luther, M.; Costa, J.; Yu, L. L. Carotenoid, tocopherol, phenolic acid, and antioxidant properties of Maryland-grown soft wheat. *J. Agric. Food Chem.* **2005**, *53*, 6649–6657.

24. Hao, Z.; Parker, B.; Knapp, M.; Yu, L. Simultaneous quantification of α-tocopherol and fours major carotenoids in botanical materials by normal phase liquid chromatography-atmospheric pressure chemical ionization-tandem mass spectrometry. *J. Chromatogr. A* **2005**, *1094*, 83–90.

25. Rentel, C.; Strohschein, S.; Albert, K.; Bayer, E. Silver-plated vitamins: a method of detecting tocopherols and carotenoids in LC/ESI–MS coupling. *Anal. Chem.* **1998**, *70*, 4394–4400.

26. Schoefs, B. Determination of pigments in vegetables (review). *J. Chromatogr. A* **2004**, *1054*, 217–226.

27. Cserhati, T.; Forgacs, E. Liquid chromatographic separation of terpenoid pigments in food and food products. *J. Chromatogr. A* **2001**, *936*, 119–137.

28. Aust, O.; Sies, H.; Stahl, W.; Polidori, M. C. Analysis of lipophilic antioxidants in human serum and tissues: tocopherols and carotenoids (Review). *J. Chromatogr. A* **2001**, *936*, 83–93.

29. Van den Berg, H.; Faulks, R.; Granado, H. F.; Hirschberg, J.; Olmedilla, B.; Sandmann, G.; Southon, S.; Stahl, W. The potential for the improvement of carotenoid levels in food and likely systemic effects. *J. Sci. Food Agric.* **2000**, *80*, 880–912.

30. Marsili, M.; Callahn, J. Comparison of a liquid solvent extraction technique and supercritical fluid extraction for the determination of α- and β-carotene in vegetables. *J. Chromatogr. Sci.* **1993**, *31*, 422–428.

31. Khachik, F.; Beecher, G. R.; Whittaker, N. F. Separation, identification and quantification of the major carotenoid and chlorophyll constituents in extracts of several green vegetables by liquid chromatography. *J. Agric. Food Chem.* **1986**, *34*, 603–616.

32. Sharpless, K. E. Value assignment of retinal, retinyl palmitate, tocopherol and carotenoid concentrations in standard reference material 2383 (Baby Food Composite). *J. AOAC Int.* **1999**, *82*, 288–296.

33. Nierenberg, F.; Nann, F. A method for determining concentration of retinol, tocopherol and five carotenoids in human plasma and tissue samples. *Am. J. Clin. Nutr.* **1992**, *56*, 417–426.

34. Hagg, M.; Ylikoski, S.; Kumpulainen, J. Vitamin C and α- and β-carotene contents in vegetables consumed in Finland during 1988–1989 and 1992–1993. *J. Food Compos. Anal.* **1994**, *7*, 252–259.

35. Granelli, J.; Helmorsson, M. Rapid high-performance liquid chromatographic method for determination of beta-carotene in milk. *J. Chromatogr. A* **1996**, *721*, 355–358.

36. Craft, N. E.; Soares, J. H. Relative solubility, stability and absorptivity of lutein and β-carotene in organic solvents. *J. Agric. Food Chem.* **1992**, *40*, 431–434.

37. Khachik, F.; Goli, M. B.; Beecher, G. R.; Holden, J.; Lusby, W. R.; Tenorio, M. D.; Barrera, M. R. Effect of food preparation on qualitative and quantitative distribution of major carotenoid constituents of tomatoes and several green vegetables. *J. Agric. Food Chem.* **1992**, *40*, 390–398.

38. Taungbodhitham, A. K.; Jones, G. P.; Wahlqvist, M. L.; Briggs, D. R. Evaluation of extraction method for the analysis of carotenoids in fruits and vegetables. *Food Chem* .**1998**, *63*, 577–584.

39. Lin, C. H.; Chen, B. H. Determination of carotenoids in tomato juice by liquid chromatography. *J. Chromatogr. A* **2003**, *1012*, 103–109.

40. Deli, J.; Molnar, P.; Matus, Z.; Toth, G. Carotenoid composition in the fruits of red paprika (*Capsicum annuum* var. *lycopersiciforme rubrum*) during ripening; biosynthesis of carotenoids in red paprika. *J. Agric. Food Chem.* **2001**, *49*, 1517–1523.

41. Gandul-Rojas, B.; Cepero, M. R. L.; Minguez-Mosquera, M. I. Chlorophyll and carotenoid patterns in olive fruits, *Olea europaea* Cv. Arbequina. *J. Agric. Food Chem.* **1999**, *47*, 2207–2212.

42. Wennermark, B.; Ahlmein, H.; Jagerstad, M. Improved vitamin E retention by using freshly milled whole-meal wheat four during drum-drying. *J. Agric. Food Chem.* **1994**, *42*, 1348–1351.

43. Zhou, K.; Yin, J. J.; Yu, L. Phenolic acid, tocopherol and carotenoid compositions, and antioxidant functions of hard red winter wheat bran. *J. Agric. Food Chem.* **2005**, *53*, 3916–3922.

44. Epler, K. S.; Sander, L. C.; Ziegler, R. G.; Wise, S. A.; Craft, N. E. Evaluation of reversed-phase liquid chromatographic column for recovery and selectivity of selected carotenoids. *J. Chromatogr. A* **1992**, *595*, 89–101.

45. Britton, G. *Methods Enzymol* .**1985**, *111*, 113–116.

46. Kamber, M.; Pfander, H. Separation carotenoids by high-performance liquid chromatography: III. 1, 2-epoxycarotenoids. *J. Chromatogr.* **1984**, *295*, 295.

47. Fisher, J. F.; Rouseff, R. L. Solid-phase extraction and HPLC determination of β-cryptoxanthin and α- and β-carotene in orange juice. *J. Agric. Food Chem.* **1986**, *34*, 985–989.

48. Heinonen, M. I. Carotenoids and provitamin A activity of carrot (*Daucus carota* L.) cultivars. *J. Agric. Food Chem.* **1990**, *38*, 609–612.

49. Khachik, F.; Beecher, G. R.; Lusby, W. R. Separation, identification and quantitation of the major carotenoid in extracts of apricots, peaches, cantaloupe and pink grapefruit by liquid chromatography. *J. Agric. Food Chem.* **1989**, *37*, 1465–1473.

50. Certificate of Analysis.SRM 869,column selectivity test mixture for liquid chromatography..,**1990**,

51. Sander, L. C.; Parris, R. M.; Wise, S. A.; Garrigues, P. Shape discrimination in liquid chromatography using charge-transfer phases. *Anal. Chem.* **1991**, *63*, 2589–2597.

52. Sander, L. C.; Glinka, C.; Wise, S. A. Determination of bonded phase thickness in liquid chromatography by small angle neutron scatting. *Anal. Chem.* **1990**, *62*, 1099–1101.

53. Bell, C. M.; Sander, L. C.; Wise, S. A. Temperature dependent of carotenoids on C_{18}, C_{30} and C_{34} bonded stationary phase. *J. Chromatogr. A* **1997**, *757*, 29–39.

54. Hao, Z.; Lu, C.; Xiao, B.; Weng, N.; Parker, B.; Knapp, M.; Ho C. Separation of amino acids, peptides and corresponding Amadori compounds on a silica column at elevated temperature. *J. Chromatogr. A* **2007**, (Forthcoming).

55. Breemen, R. B.; Huang, C. R.; Tan, Y.; Sander, L. C.; Schilling, A. B. Liquid chromatography/mass spectrometry of carotenoids using atmospheric pressure chemical ionization. *J. Mass Spectr.* **1996**, *31*, 975–979.

56. Breemen, R. B. Electrospray liquid chromatography–mass spectrometry of carotenoids. *Anal. Chem.* **1995**, *67*, 2004–2009.

CHAPTER 12

QUANTIFICATION OF PHENOLIC ACIDS IN WHEAT AND WHEAT-BASED PRODUCTS

Kequan Zhou

12.1 INTRODUCTION

Phenolic acids have received considerable attention as potentially protective factors against human chronic diseases, which is in part related to their potent antioxidant activities. Research has shown that phenolic acids could protect low density lipoproteins from oxidative damage (1), improve anti-inflammatory response of cells (2), inhibit tumor cell proliferation and carcinogenesis in animals (3), and prevent atherosclerotic development (4). Wheat may provide a significant amount of dietary phenolic acids and, it has been estimated that one gram of whole grain may contain about 0.5 milligram of total phenolic acids (5–7). However, the phenolic acid contents in wheat are often underestimated since the majority of the phenolic acids are bound to cell wall and unable to be directly extracted without specific treatments. Those insoluble bound phenolic acids primarily exist as ester linkages to cell wall polysaccharides that constitute up to 90% of the total phenolic acids in wheat (5,8,9). The current detectable phenolic acids in wheat consist of two groups, hydroxycinnamic acids and hydroxybenzoic acids, depending on their discrepancy in chemical structures (10,11). The former includes ferulic, caffeic, and *p*-coumaric acid, while the later includes vanillic, chlorogenic, gentisic, syringic, and *p*-hydroxybenzoic acid. Recent research has shown that phenolic acids with different aglycones and formulations may have quite different responses in terms of their bioavailability and potential health effects (12–14). Thus, knowledge regarding the concentrations and daily intake of specific phenolic acids is essential. It is important to have an optimized and validated method for the determination of a variety of phenolic acids and their formulations in wheat and wheat-based products.

Wheat Antioxidants, Edited by Liangli Yu

12.2 BACKGROUND

Phenolic acids (PAs) were identified in wheat as early as the 1960s. Seven phenolic acids were determined from ethanolic extracts of wheat including *p*-coumaric, ferulic, caffeic, sinapic, *p*-hydroxybenzoic, syringic, and vanillic acids (15). However, the earlier studies on wheat phenolic compounds failed to include the bound forms of phenolic acids in wheat. Sosulski et al. was the first to report the existence of insoluble bound phenolic acids in wheat flour. The bound phenolic acids were released by alkaline hydrolysis and accounted for 83% of total phenolic acids in wheat flour. The hydrolysis was conducted by 2N NaOH at room temperature under nitrogen in order to minimize the deterioration of liberated phenolic acids (16). Such a sample preparation procedure has been widely adopted in many subsequent studies. Wheat phenolic acids were usually separated and determined by chromatography technologies including thin layer chromatography (TLC), gas liquid chromatography (GLC), and high performance liquid chromatography (HPLC). Earlier, separation of wheat phenolic acids was conducted by TLC with semiquantization until GLC technology emerged (17). GLC dramatically improved the sensitivity and accuracy in phenolic acid determination, but GLC methods needed Me_3Si ether derivatization, which may result in the underestimation of and significant variation in measurements. Recently, HPLC became more popular due to its unique analytical capacity and reliability. Reverse-phase HPLC with gradient eluent could separate all wheat phenolic acids in a single run (8,11,18–20). The method presented below described a complete sample preparation and HPLC procedure for determination of free, soluble, and insoluble phenolic acids in wheat samples or wheat-based products.

12.3 CHEMICALS AND EQUIPMENTS

- Phenolic acid standards: gallic, protocatechuic, *p*-hydroxybenzoic, chlorogenic, vanillic, caffeic, syringic, *p*-coumaric, and ferulic acids (Sigma, St. Louis, MO)
- Methanol/acetone/water (7/7/6, v/v/v): freshly prepared in fume hood before sample extraction
- 2 N NaOH: freshly prepared
- 2 and 6 N HCl
- Ethyl ether/ethyl acetate (1/1, v/v): freshly prepared in fume hood
- Glass tubes (16×125 mm and 25×200 mm with screw caps) and flasks (25 mL, 125 mL, 1500 mL) (Fisher Scientific, Pittsburgh, PA)
- Water bath shaker (New Brunswick Scientific, Edison, NJ)
- Nitrogen evaporator (Labconco, Kansas City, MO)
- Rotary evaporator (Buchi, New Castle, DE)
- Freeze-dry system (Labconco, Kansas City, MO)
- Vortex (VMR, West Chester, PA)
- Filter papers (Whatman, Florham Park, NJ)
- Glass vacuum filter unit (Whatman, Florham Park, NJ)
- 0.45 µm filtration membrane (Whatman, Florham Park, NJ)
- Syringe (1 ml with needle) and 0.45 µm syringe filter (Fisher scientific, Pittsburgh, PA)

- Waters HPLC System with 600 pump and controller, 996 PDA detector, 717 autosampler, Millennium 32.2 software (Waters, Dublin, CA)
- Phenomenex C18 column (250 × 4.6 mm)

12.4   METHODS

12.4.1   Sample Preparation 1

Phenolic acids in wheat were prepared and extracted as three phases: free, total soluble, and insoluble bound (Fig. 12.1). The procedures were described as follows:

1. Add 1 g ground wheat sample to 25-mL flask.
2. Add 10 mL methanol/acetone/water (7/7/6, v/v/v) to the flask and seal the flask.
3. Shake the mixture in a water bath shaker for 12 h for extraction.
4. Stop the shaker and rest the mixture until the extraction become clear.
5. Withdraw 6 mL extraction solution and transfer 3 mL to a glass tube (16 × 200 mm) for free phenolic acid determination; add another 3 ml to a glass tube (25 × 200 mm) for total soluble phenolic acid determination; and evaporate the organic solvents in the sample residues in a nitrogen evaporator for insoluble phenolic acid determination.

For free PA in wheat sample:
6. Evaporate the organic solvents in 3 ml phenolic acid extraction solution under nitrogen.
7. Acidify the extraction in the tube to pH 2–4 with 2N HCl.
8. Add 10 mL ethyl acetate/ethyl ether (1:1, v/v) to the tube and vortex for 10 min to extract free PA from aqueous extraction, rest the mixture for 15–30 min until the two phases separate, then transfer the organic phase to another glass tube (25 × 200 mm).
9. Repeat step 8 twice and combine organic phases.
10. Evaporate the collected organic phases under nitrogen and immediately reconstitute the residue with 1 mL of methanol.
11. Withdraw the methanol solution with a 1-mL syringe.
12. Filter the solution with 0.45-μm syringe filter to a glass vial for HPLC analysis.

For total soluble PA in wheat sample:
13. Evaporate the organic solvents in 3 ml phenolic acid extraction under nitrogen.
14. Add 10 mL of 2 N NaOH, flush the tube with nitrogen and seal the tube, and then shake the mixture at 45°C in a water bath for 4 h.
15. Repeat step 7–12.

For insoluble bound PA in wheat sample:
16. Add 20 mL 2N NaOH to the flask after the organic solvents have been completely evaporated under nitrogen.

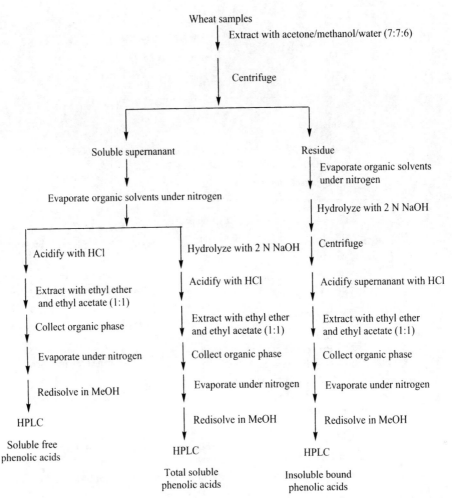

Figure 12.1 Sample preparations for the determination of wheat phenolic acids. Modified from Reference 6.

17. Flush the flask with nitrogen and seal the flash, shake the mixture in a water bath under 45°C for 12 h.
18. Stop the shaker and rest the mixture for 1 h, and transfer 3 mL to a glass tube (25 × 200 mm).
19. Repeat step 7–12.

12.4.2   Sample Preparation 2 According to the Protocol Reported by Kim et al. (Fig. 12.2):

1. Add 200 g wheat flour/ground wheat sample to an flask.
2. Add 800 mL hexane to the flask and shake the mixture for 1 h.
3. Filter the mixture through a Whatman No. 1 filter paper.

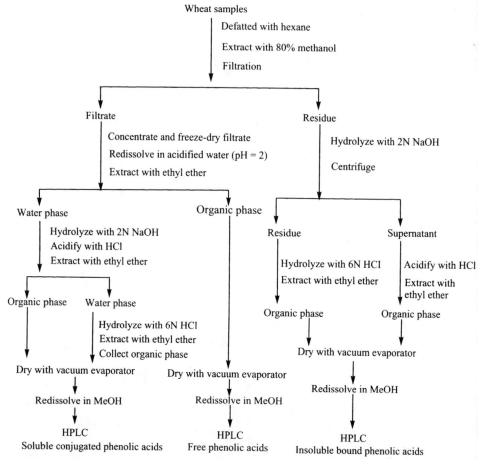

Figure 12.2 Sample preparations for the determination of wheat phenolic acids according to Reference 22.

4. Repeat step 2–3; dry the defatted samples in a hood.
5. Extract the defatted samples twice with 80% methanol at a 5 : 1 ratio (v/w) for 1 h at room temperature.
6. Filter the mixture, concentrate the collected supernatant in a rotary evaporator at 40°C, and further freeze-dry the resulting extract.

For free and soluble conjugated PA in wheat sample:
7. Redissolve the freeze-dried extract (equivalent to 1 g original sample) in 4 mL acidified water (pH 2 with HCl).
8. Add 4 mL ethyl ether to extract free PA from aqueous phase; repeat three times.
9. Combine organic phases and further dry the extract in a rotary evaporator.
10. Reconstitute the residue in methanol, filter through a 0.45-μm syringe filter for free PA determination.

11. Neutralize the remaining water phase from step 8 to pH 7 with 2N NaOH and dry the solution in a rotary evaporator.
12. Dissolve the residue in 4 mL of 2N NaOH and stir the mixture for 4 h at room temperature.
13. Acidify the mixture to pH 2 with HCl.
14. Repeat step 8–10 for the determination of the soluble conjugated PA after alkaline hydrolysis.
15. Add 3 mL of 6N HCl to the remaining water phase (from step 14) and heat the solution at 95°C for 20 min.
16. Repeat step 8–10 for the determination of the soluble conjugated PA after subsequent acidic hydrolysis.

For insoluble bound PA in wheat sample:
17. Hydrolyze 1 g residue from methanolic extract (step 6) in 40 mL of 2N NaOH for 4 h at room temperature.
18. Filter the mixture, acidify the filtrate to pH 2, and extract with ethyl ether (40 mL × 3).
19. Repeat step 9–10 for the determination of the insoluble bound PA after alkaline hydrolysis.
20. Subsequently, hydrolyze the residue from step 18 with 6N HCl at 95°C for 1 h.
21. Filter the mixture and extract the filtrate with ethyl ether (40 mL × 3).
22. Repeat step 9–10 for the determination of the insoluble bound PA after subsequent acid hydrolysis.

12.4.3   HPLC Separation and Determination

1. Mobile phase: solvent A: acetic acid/H_2O (2 : 98, v/v); solvent B: acetic acid/acetonitrile/H_2O (2 : 30 : 68, v/v/v); prior to use, vacuum filter the mobile phase through a 0.45-μm membrane.
2. Gradient program: solvent B from 10% to 100% within 42 min and allow 10 min of postrun for reconditioning.
3. Flow rate: 1.0 mL/min.
4. Identification of PA: compare the retention time of peaks from sample solutions with corresponding standards at 280 nm.
5. Quantification of PA: $C_s = (C_r/M_r) \times M_s$ where C_s and C_r stand for the concentration of individual phenolic acid in sample and standard solutions, respectively. M_s and M_r represent the total area under the corresponding peak of PA in sample and standard solution.

12.5   DISCUSSION

Phenolic acids in wheat grains exist in free, soluble conjugated, and insoluble bound forms with the majority of the phenolic acids linking to cell-wall materials such as polysaccharides and lignins (21). This brings a challenge for the determination of the different types of phenolic acids and their existing forms in wheat and wheat-based products. As phenolic acids differ in their chemical nature and their sensitivities to the conditions of extraction and hydrolysis, sample preparations, particularly hydrolysis procedures, may significantly affect the yield and profile of phenolic acids in wheat (11,18,22,23). Table 12.1 summarized a number of extraction and hydrolysis procedures for the determination of phenolic acids in wheat and wheat-based products.

Soluble phenolic acids in wheat have been extracted using mixtures of methanol, acetone, and water (6,7,16,24), aqueous ethanol/methanol/acetone (10,11,17,18,22,25), absolute methanol (26,27), or mixtures of water and sulfuric acid (20). Previous studies have investigated the extraction efficiency of different solvent systems for soluble phenolic acids in wheat (16,25,28). No single solvent including methanol, ethanol, or acetone can completely extract soluble phenolic acids in wheat (28). Eight percent ethanol was shown to be a better solvent than absolute methanol or ethanol for wheat phenolic acid extraction (25). The extraction could be improved by a solvent system consisting of three different solvents rather than just two (16). We preliminarily observed that the mixture of methanol, acetone, and water (7:7:6) showed better extraction capability than 50% acetone, 70% methanol, and 70% ethanol for wheat bran samples. Indeed, mixtures of methanol, acetone, and water have been used in a number of studies for phenolic acid extractions (6,7,16). Following extraction, soluble conjugated phenolic acids have to be further hydrolyzed before loading for chromatographic analysis. The reactions are often conducted using alkaline hydrolysis (10,11,24) or acidic hydrolysis (17) or both (22,27), where two factors are critical: hydrolytic efficiency and phenolic acid degradation during the reaction. In this regard, alkaline hydrolysis is better than acidic hydrolysis due to its higher efficiency in hydrolyzing ester bonds of phenolic acids. Moreover, the acidic hydrolysis necessitated boiling condition, which resulted in significant loss of phenolic acids (22,25). A previous study showed that the harsh conditions of acidic hydrolysis caused 73% and 78% loss of p-coumaric and ferulic acids, respectively (25). Kim et al. recently investigated the effects of hydrolysis conditions on the release of extractable phenolic acids in wheat bran (22). Alkaline hydrolysis of wheat bran extracts yielded twice higher amounts of free phenolic acids than acidic hydrolysis did. In addition, alkaline hydrolysis worked particularly well on hydrolyzing ferulic acid conjugates with over 89% of extractable ferulic acid released by 2 N NaOH (22). Also noted is that the reaction has preference to be carried out under nitrogen conditions, which helps to prevent phenolic acids degradation.

As for insoluble phenolic acids in wheat, alkaline hydrolysis is also preferable with or without subsequent acidic hydrolysis (6,7,10,11,18,22,24–27). Alkaline hydrolysis (with 2 N NaOH) was reported to release about 91% of bound phenolic acids from red and white wheat bran (22). Alkaline hydrolysis is excellent for liberating bound ferulic acid, the major phenolic acid in wheat. Kim et al. reported that ferulic acid released by acidic hydrolysis was only 1.7% of that by alkaline hydrolysis (22). It should also be noted that maintaining the concentration of NaOH at about 2 N was important as higher concentrations of NaOH may result in starch gelatinization in residue.

TABLE 12.1 Phenolic Acid Preparation and Determination in Different Wheat Samples and Products

Wheat samples	Extraction solvents	Hydrolysis	Separation and determination[a]	Content range[a]	References
Spring/winter wheat	80% ethanol	6 N HCl	GLC	Total PA: 0.19–0.22 mg/g grain	17
Neepawa wheat flour	Methanol/acetone water (7/7/6, v/v/v)	2–4N NaOH (4 h) under nitrogen	GLC/GLC-MS	Total PA: 0.71 mg/g wheat flour	16
wheat grain	0.1 M sulfuric acid	Fungal α-amylase treatment	HPLC-12% methanol–citrate buffer (pH 5.4)	Total FA: 0.49–0.52 mg/g whole grain	20
Canadian wheat flour	80% acetone followed by 80% methanol	2N NaOH (4 h)	HPLC-0.1% TFA (A), 0.1% TFA in acetonitrile (B)	Total soluble PA: 49.2–70.8 μg/g wheat flour	24
Wheat endosperms, bran/germ fractions	80% ethanol	2N NaOH (1 h)	HPLC: 20% acetonitrile adjusted to pH 2 with TFA	Total FA: 0.29–0.65 mg/g whole grain; 1.1–2.2 mg/g bran; 30–40 μg/g endosperm	18,25
Wheat grains	Methanol/acetone/water (7/7/6,v/v/v)	2N NaOH (4 h)	GLC analysis	Total FA: 0.49–0.78 mg/g whole grain	7
Hard winter wheat bran	50% acetone	4N NaOH (4 h)	HPLC: 2% HAc (A), HAc/acetonitrile/H$_2$O (2/30/68) (B)	Total soluble PA: 154.2–185.0 μg/g wheat bran	10,11
China-grown wheat bran	Methanol	4N NaOH (4 h)	HPLC: 1% HAc (A), methanol (B)	Total PA: 2.0–3.1 mg/g wheat bran	26
Wheat grain, bran, flours, wheat bread	Methanol	2.5N NaOH (16 h), then 1.2N HCl (30 min)	HPLC: 50 mM H$_3$PO$_4$, pH 2.5 (A), acetonitrile (B)	Total PA: 1.3 mg/g grain; 4.5 mg/g bran; 0.15–0.17 mg/g flour; 0.11 mg/g wheat bread	27
Soft red wheat grains	Methanol/acetone water (7/7/6, v/v/v)	2N NaOH (12 h)	HPLC: 2% HAc (A), HAc/acetonitrile/H$_2$O (2/30/68) (B)	Total PA: 0.49–0.65 mg/g wheat grain	6
Hard red and soft white wheat bran	80% methanol, after defatted with hexane,	2N NaOH (4 h), then 6N HCl (1 h)	HPLC: acetonitrile (A), 2% acetic acid (B)	Total PA: 1.6 mg/g white wheat bran; 2.3 mg/g red wheat bran	22

[a]GLC-Gas Liquid Chromatography; MS-Mass spectrometry; HPLC-high performance liquid chromatography; HA-acetic acid; TFA-triflouroacetic acid; PA-phenolic acid; FA-ferulic acid.

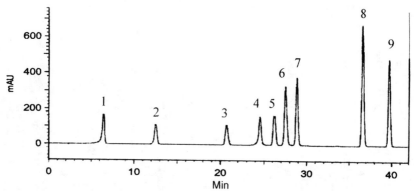

Figure 12.3 Wheat phenolic acid standards separated by reverse-phase HPLC. Wheat phenolic acid standards: 1. gallic acid, 2. protocatechuic acid, 3. 4-OH benzoic acid, 4. chlorogenic acid, 5. vanillic acid, 6. caffeic acid, 7. syringic acid, 8. coumaric acid, and 9. ferulic acid.

After hydrolysis, the reaction mixture needs to be acidified to make sure that all released phenolic acids have been protonated before further purification by liquid–liquid extraction with diethyl ether, ethyl acetate, or mixtures of diethyl ether and ethyl acetate. A mixture of diethyl ether and ethyl acetate (1 : 1, v/v) proved to be an excellent solvent for purifying free phenolic acids with recoveries ranging from 87% to 112% (29). Three 10-min extractions with thoroughly mixing could transfer virtually all of the free phenolic acids into the organic solvents. The separation and quantification of individual wheat phenolic acids are usually performed with reverse-phase HPLC (6,10,11,18,22,24–27). The mobile phases commonly consisted of aqueous methanol or acetonitrile and were often modified with acetic acid or triflouroacetic acid. The gradient program described in the Section 12.4 could separate phenolic acids in wheat extracts within 42 min (Figs. 12.3 and 12.4). Identification of individual phenolic acids

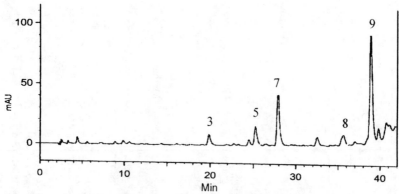

Figure 12.4 Phenolic acids in wheat sample separated by reverse-phase HPLC. Wheat phenolic acids: 3, 4-OH benzoic acid; 5, vanillic acid; 7, syringic acid; 8, coumaric acid; 9, ferulic acid.

could be accomplished by comparing their relative retention time and absorption spectra with corresponding standards detected at 280 nm.

REFERENCES

1. Castelluccio, C.; Bolwell, G. P.; Gerrish, C.; Rice-Evans, C. Differential distribution of ferulic acid to the major plasma constituents in relation to its potential as an antioxidant. *Biochem. J.* **1996**, *316*, 691–694.
2. Hellberg, M. R.; Namil, A.; Delgado, P.; David, K. C.; Kessler, T. L.; Graff, G.; Haggard, K. S.; Nixon, J. C. Novel esters and amides of nonsteroidal antiinflammatory carboxylic acids as antioxidants and antiproliferative agents. *J. Med. Chem.* **1999**, *42*, 267–276.
3. Soleas, G. J.; Grass, L.; Josephy, P. D.; Goldberg, D. M.; Diamandis, E. P. A comparison of the anticarcinogenic properties of four red wine polyphenols. *Clin. Biochem.* **2006**, *39*, 492–497.
4. Hernandez, A.; Martin, A.; Aranda, E.; Bartolome, T.; deGuiaCordoba, M. Detection of smoked paprika "Pimentón de La Vera" adulteration by free zone capillary electrophoresis (FZCE). *J. Agric. Food Chem.* **2006**, *54*, 4141–4147.
5. Zhou, K.; Laux, J. J.; Yu, L. Comparison of Swiss red wheat grain and fractions for their antioxidant properties. *J. Agric. Food Chem.* **2004**, *52*, 1118–1123.
6. Moore, J.; Hao, Z.; Zhou, K.; Luther, M.; Costa, J.; Yu, L. L. Carotenoid, tocopherol, phenolic acid, and antioxidant properties of Maryland-grown soft wheat. *J. Agric. Food Chem.* **2005**, *53*, 6649–6657.
7. Abdel-Aal, E. S.; Hucl, P.; Sosulski, F. W.; Graf, R.; Gillott, C.; Pietrzak, L. Screening spring wheat for midge resistance in relation to ferulic acid content. *J. Agric. Food Chem.* **2001**, *49*, 3559–3566.
8. Onyeneho, S. N.; Hettiarachchy, N. S. Antioxidant activity of durum wheat bran. *J. Agric. Food Chem.* **1992**, *40*, 1496–1500.
9. Mpofu, A.; Sapirstein, H. D.; Beta, T. Genotype and environmental variation in phenolic content, phenolic acid composition, and antioxidant activity of hard spring wheat. *J. Agric. Food Chem.* **2006**, *54*, 1265–1270.
10. Zhou, K.; Su, L.; Yu, L. L. Phytochemicals and antioxidant properties in wheat bran. *J. Agric. Food Chem.* **2004**, *52*, 6108–6114.
11. Zhou, K.; Yin, J. J.; Yu, L. L. Phenolic acid, tocopherol and carotenoid compositions, and antioxidant functions of hard red winter wheat bran. *J. Agric. Food Chem.* **2005**, *53*, 3916–3922.
12. Zhou, K.; Yin, J. J.; Yu, L. L. ESR determination of the reactions between wheat phenolic acids and free radicals or transition metals. *Food Chem.* **2006**, *95*, 446–457.
13. Yeh, C. T.; Yen, G. C. Effects of phenolic acids on human phenolsulfotransferases in relation to their antioxidant activity. *J. Agric. Food Chem.* **2003**, *51*, 1474–1479.
14. Mateos, R.; Goya, L.; Bravo, L. Uptake and metabolism of hydroxycinnamic acids (chlorogenic, caffeic, and ferulic acids) by HepG2 cells as a model of the human liver. *J. Agric. Food Chem.* **2006**, *54*, 8724–8732.
15. El-Basyouni, S.; Towers, G. H. The phenolic acids in wheat. Ii. natural occurrence of orthoferulic acid (2-hydroxy-3-methoxycinnamic acid). *Can. J. Biochem. Physiol.* **1964**, *42*, 493–497.
16. Sosulski, F.; Krygier, K.; Hogge, L. Free, esterified, and insoluble-bound phenolic acids. III. Composition of phenolic acids in cereal and potato flours. *J. Agric. Food Chem.* **1982**, *30*, 337–340.
17. Maga, J. A.; Lorenz, K. Phenolic acid composition and distribution in wheat flours and various triticale milling fractions. *Lebensm. Wiss. Technol.* **1974**, *7*, 273–278.
18. Adom, K. K.; Sorrells, M. E.; Liu, R. H. Phytochemical profiles and antioxidant activity of wheat varieties. *J. Agric. Food Chem.* **2003**, *51*, 7825–7834.
19. Sun, R. C.; Sun, X. F.; Zhang, S. H. Quantitative determination of hydroxycinnamic acids in wheat, rice, rye, and barley straws, maize stems, oil palm frond fiber, and fast-growing poplar wood. *J.Agric. Food Chem.* **2001**, *49*, 5122–5129.
20. Pussayanawin, V.; Wetzel, D. L. High-performance liquid chromatographic determination of ferulic acid in wheat milling fractions as a measure of bran contamination. *J. Chromatogr.* **1987**, *391*, 243–255.
21. Mattila, P.; Pihlava, J. M.; Hellstrom, J. Contents of phenolic acids, alkyl- and alkenylresorcinols, and avenanthramides in commercial grain products. *J. Agric. Food Chem.* **2005**, *53*, 8290–8295.

22. Kim, K. H.; Tsao, R.; Yang, R.; Cui, S. W. Phenolic acid profiles and antioxidant activities of wheat bran extracts and the effect of hydrolysis conditions. *Food Chem* .**2006**, *95*, 466–473.

23. Saadi, A.; Lempereur, I.; Sharonov, S.; Autran, J. C.; Manfait, M. Spatial distribution of phenolic materials in Durum wheat grain as probed by confocal fluorescence spectral imaging. *J. Cereal Sci.* **1998**, *28*, 107–114.

24. Hatcher, D. W.; Kruger, J. E. Simple phenol acids in flours prepared from Canadian wheat: relationship to ash content, color, and polyphenol oxidase activity. *Cereal Chem* .**1997**, *74*, 337–343.

25. Adom, K. K.; Liu, R. H. Antioxidant activity of grains. *J. Agric. Food Chem.* **2002**, *50*, 6182–6187.

26. Li, W.; Shan, F.; Shun, S.; Corke, H.; Beta, T. Free radical scavenging properties and phenolic content of Chinese black-grained wheat. *J. Agric. Food Chem.* **2005**, *53*, 8533–8536.

27. Mattila, P.; Pihlava, J. M.; Hellstrom, J. Contents of phenolic acids, alkyl- and alkenylresorcinols, and avenanthramides in commercial grain products. *J. Agric. Food Chem.* **2005**, *53*, 8290–8295.

28. Krygier, K.; Sosulski, F.; Hogge, L. Free, esterified, and insoluble-bound phenolic acids. I. Extraction and purification procedure. *J. Agric. Food Chem.* **1982**, *30*, 330–334.

29. Mattila, P.; Kumpulainen, J. Determination of free and total phenolic acids in plant-derived foods by HPLC with Diode-Array Detection. *J. Agric. Food Chem.* **2002**, *50*, 3660–3667.

CHAPTER 13

EFFECTS OF WHEAT ON NORMAL INTESTINE

John Parry

13.1 INTRODUCTION

Wheat is one of the most important sources of food on the planet. Human beings consume wheat more than any other grain, and over the last few years, the global harvest has been surpassing 600 million metric tonnes. Even though wheat is the most popular grain eaten worldwide, its effects on normal gastrointestinal (GI) tract in humans have only been minimally investigated. Most studies of wheat on the human GI tract have examined negative health afflictions such as celiac disease, irritable bowel syndrome, diverticulosis, and cancer. Previous studies of wheat effects on normal intestinal tissues have primarily involved animal models. The animal studies mainly investigated morphological changes in the intestine, transit-time effects of particle size, and GI protective effects of whole wheat and wheat components.

Wheat is known to contain significant levels of phenolic acids and primarily consists of ferulic acid. Phenolics have been described as antinutrients, which are compounds in foods that may reduce the bioavailability of dietary nutritional components in the intestines. Phenolic compounds are thought to have antinutritional activity by coordinating with minerals leading to a reduction in their absorption. However, wheat phenolic compounds have demonstrated significant antioxidant activities and may provide protection from free radical damage in the GI tract and the body. Phenolic compounds may also have other activities in the GI tract that affect the physiology of normal intestinal epithelial cells resulting in a reduction in cell proliferation. We examined this last hypothesis by testing the effect of two wheat bran extracts and ferulic acid for antiproliferative activity on normal intestinal IEC-6 cells. Mechanisms that may be directly related to antiproliferation include apoptosis and cell cycle arrest. These were examined by testing cellular caspase-3 expression and flow cytometry analysis, respectively.

13.2 WHEAT COMPONENT EFFECTS ON NORMAL INTESTINAL EPITHELIAL CELLS *IN VITRO*

13.2.1 Background Information

13.2.1.1 IEC-6 Cells The small intestine is comprised of several differentiated types of cells. Four of these cell types include absorptive, goblet, enteroendocrine, and paneth cells. Absorptive cells are involved in nutrient transport, goblet cells produce mucus, enteroendocrine cells secrete several different hormones, and paneth cells provide defense against microorganisms. All of these cells originate from the same primary undifferentiated intestinal epithelial cells that are located in the crypts of Leiberkuhn. IEC-6 cells are normal primary intestinal epithelial crypt cells from rat (Rattus norvegicus), which have previously been characterized (1).

13.2.1.2 Ferulic Acid: The Primary Phenolic Acid in Wheat Ferulic acid is a phenolic acid classified as a hydroxycinnamic acid. It is the most abundant phenolic acid found in wheat, usually greater than 80% of the total phenolic acids (2), and its highest concentration is found in the bran. Approximately 1% of the ferulic acid in wheat is free while the other 99% are mainly bound to sugar molecules. Two of the ferulic acid glycosides are 5-O-feruloyl-L-arabinofuranose (FAF) and feruloyl-arabinoxylan (FAXn). Several studies have demonstrated that ferulic acid is a strong antioxidant, and it has similar antioxidant activities compared to the other phenolic acids found in wheat including *p*-coumaric, caffeic, 4-hydroxybenzoic, and syringic acids (3).

13.2.1.3 Ferulic Acid Metabolism and Absorption Several researches have investigated and supported the intestinal metabolism and the absorption of ferulic acid. In 1997, Kroon and coworkers (4) tested the release of ferulic acid from its glycosidic bound forms in coarse wheat bran and fine wheat bran using a human gut fermentation model following 24 h of incubation. The results showed that the coarse wheat bran sample had released 81% ferulic acid compared to that of 57% from the fine bran.

In 1998, Day and coworkers (5) investigated a broad-spectrum β-glucosidase, which is an enzyme present in the small intestine and liver of humans that functions to catalyze the hydrolysis of β-glycosidic bonds. Results showed that β-glucosidase purified from human small intestine and liver was capable of hydrolyzing naturally occurring flavonoids and isoflavonoids from their glycosidic linkages.

In 2001, Andreasen and coworkers (6) examined the release of hydroxycinnamic acids from four prepared methyl hydroxycinnamoyl esters including feruloyl ester in the rat intestinal mucosa. The results showed that rat intestinal mucosa from the jejunum and ileum had the highest feruloyl ester enzymatic esterase activities at 45 and 47 μmol/h, respectively, and from the total GI tract mucosa feruloyl ester esterase activity was 105.4 μmol/h. The group also examined the release of ferulic acid from wheat bran by human fecal extract and compared it to ferulic acid alkali extraction. The fecal extract was able to release 25% compared to the alkali extraction method, which is a significant amount considering that alkali extraction is nearly 100% effective. These results indicate that colonic cinnamoyl esterases may be the major route of release of ferulic acid from wheat bran.

In 2003, Zhao and coworkers (7) tested the percent of intestinal absorption, percent of recovery in urine, and percent of degradation rate of ferulic acid, FAF, and FAXn as individual compounds in single-dose *in situ* administration in the rat. The results demonstrated that over 99% of free ferulic acid was absorbed by the GI tract, while FAF and FAXn were absorbed at 98.7% and 80.0%, respectively. Ferulic acid also had the highest recovery in urine at 72% followed by FAF and FAXn at 54.0% and 20.0%, respectively. Degradation rate percentages were calculated by total ferulic acid recovered in the urine divided by the total ferulic acid absorbed. The free ferulic acid also had the lowest degradation rate percentage (27.4%), followed by FAF (44.7%), and FAXn (60.0%) (7).

In 2002, Adam et al. (8) examined ferulic acid absorption and metabolism by *in situ* intestinal perfusion from 10 to 50 nmol/min in rats. They found that there was a significant correlation ($r = 0.9968$) between dose and absorption in the tested concentration range. Additionally, the group examined the bioavailability of ferulic acid in ferulic acid supplemented diets compared to the bioavailability of ferulic acid in complex cereal matrix diets including wheat flour and wheat bran using the rat model. The bioavailability of ferulic acid in the supplemented feed was over 99%, and the bioavailability of ferulic acid in the whole-wheat flour and wheat bran was 79% and 62%, respectively. The degradation rate percentages were 47.4%, 58.1%, and 75.4% for the ferulic acid supplemented diet, wheat bran, and whole-wheat flour, respectively. The researchers concluded that the bioavailability of ferulic acid is strongly interfered with complex cereal matrices demonstrated by the lower yields compared to that of the ferulic acid supplemented feed (8). It is interesting that the degradation in the ferulic acid supplemented feed was 47.4%, which was much higher than in the single-dose *in situ* individual compound experiment that had a degradation rate of 27.4% (7). It is possible that the supplemented ferulic acid may have had a strong interaction with other compounds in the food matrix that extended absorption from the gut to the large intestine where ferulic acid could have been chemically altered by microbial fermentation.

In another study by Zhao and coworkers (9), the absorption of ferulic acid as free ferulic acid, FAF, and FAXn were tested in the rat model. The results showed that free ferulic acid was almost completely absorbed before reaching the cecum. Approximately 40% of FAF was absorbed in the small intestine and about 57% was absorbed in the cecum; while only a trace of FAXn was absorbed in the small intestine, about 44% was absorbed in the cecum, and 23% was absorbed in the colon. The results indicate that fermentation may be the major releaser of ferulic acid in the gut considering that more than 99% of ferulic acid is found naturally in the bound glycoside form.

In 2004, Zhao and coworkers (10) examined the absorption of free ferulic acid and FAF in the rat stomach. The results demonstrated that approximately 74% of free ferulic and 19% of FAF were absorbed following 25 min of *in situ* exposure, suggesting that FAF may serve to protect the GI tract from oxidative damage longer than the free form, while the free form could provide antioxidant protection inside the body.

13.2.1.4 Caspase-3 Protein and Apoptosis

Caspase-3 protein is expressed by cells in the process of programmed cell death (apoptosis). Apoptosis is a critical cellular activity responsible for several important functions that include maintaining the homeostatic cell number of an organism. A cell without properly functioning apoptosis may divide without either daughter cell dying in a continuing process, and this may lead to the development of cancer. Apoptosis is a series of cellular events that leads to death by disassembly of cellular components. Apoptosis differs from cellular necrosis in that the disassembled cell components are packaged neatly and engulfed by surrounding cells such as macrophages, whereas cell necrosis mainly involves cell membrane rupture and spillage of internal contents from the cell onto surrounding cells, which may have harmful effects considering the release of organelles such as peroxisomes and lysosomes and other free enzymes. Caspase-3 is expressed posttranslational following the initiation of apoptosis when it is hydrolyzed from inactive procaspase-3, and as an activated protease, it begins to cleave internal cell proteins leading to the eventual death of the cell. Measuring cellular caspase-3 expression is one method of determining apoptosis as a mechanism of antiproliferation or cell death from an effective cell growth inhibitor.

13.2.1.5 Cell Cycle Analysis

Eukaryotic cells proliferate by means of progressing through the cell cycle followed by cytokinesis, dividing into two identical daughter cells. The process of the cell cycle is described in phases (G_1, S, G_2, and M) and has distinct checkpoints that must be met to continue through the cycle in normal cell cycle progression. In some cases, checkpoints may be skipped leading to uncontrolled cell division and possibly cancer. It is possible that a compound or compounds that are not produced by the cell and introduced by other means may interfere with or suppress progression of the cell cycle, which may prevent or slow the proliferation of tumor cells.

13.2.2 Effects of Wheat Bran Extract on IEC-6 Cell Proliferation

Extracts from the bran of two wheat varieties "Wichita" and "Lakin" were tested for their effect on IEC-6 cells for antiproliferation and recovery. Wichita is a hard red winter wheat while Lakin is a hard white winter wheat. Extracts from both bran varieties were evaluated for antiproliferation and recovery on IEC-6 cells.

13.2.2.1 Antiproliferation and Antiproliferation/Recovery

Wichita and Lakin bran samples were extracted with 50% acetone and dried and then resuspended in 50% DMSO which was used for the *in vitro* anti-proliferative capacity assays using the IEC-6 cell line. IEC-6 cells were treated with Wichita extract at 3 and 6 mg bran equivalents/mL and with Lakin extract at 6.5 and 5 mg bran equivalents/mL. The results for the antiproliferation/recovery experiments were similar for both wheat bran extract studies. During the treatment phase, the cells grew slower compared to the control and were significantly lower on days 2 and 3 for the Wichita extract (Figs. 13.1 and 13.2) and on day 3 for the Lakin bran extract (Figs. 13.3 and 13.4). Following the removal of treatments, the cells proliferated similar to the control in both experiments. The significant proliferation inhibition of IEC-6 cells treated with Lakin bran extract at 6.5 mg/mL final concentration is not shown. At 6.5 mg/mL, the cells did not recover following the removal of the treatment. These results show that it may be possible that an undefined "moderate" level of consumption of wheat bran could slow the growth of intestinal cells *in vivo*. The

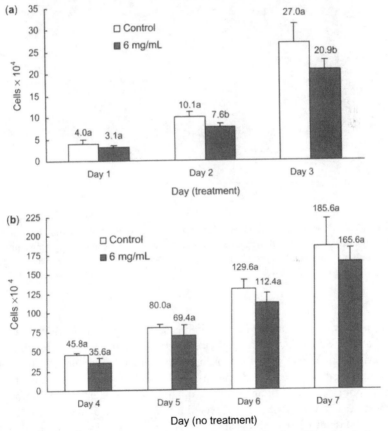

Figure 13.1   Effects of Wichita wheat bran extract on IEC-6 cell proliferation during treatment followed by treatment removal. IEC-6 cells treated with Wichita wheat bran extract at 6 mg bran equivalents/mL. Figure (**a**) represents cell proliferation during 3 days of treatment with the extract. Figure (**b**) represents cell proliferation for the 4 following days of treatment as control cultures. Triplicate measurements were taken. Bars with different letters on the same day are significantly different ($P < 0.05$).

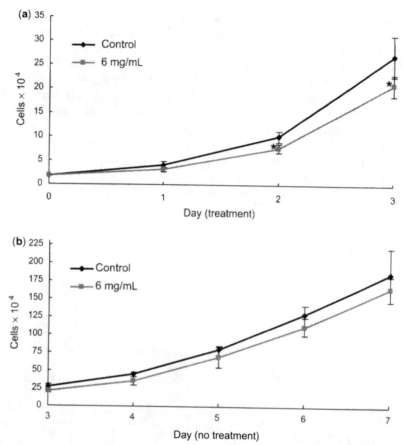

Figure 13.2 Growth curve effect of Wichita wheat bran extract on IEC-6 cells followed by treatment removal. IEC-6 cells treated with Wichita wheat bran extract at 6 mg bran equivalents/mL. Figure (**a**) represents cell proliferation during 3 days of treatment with the extract. Figure (**b**)

resulting slowed growth may lead to a decreased absorption of calories through the small intestine contributing to weight loss without causing irreversible cell damage. However, a moderate level of consumption cannot be estimated based on the methods used in these experiments. It would seem likely that moderate consumption of wheat bran would be qualitatively higher than the term would suggest. Also, it would seem that irreversible intestinal cell damage caused by wheat bran would require laborious consumption for an individual with a normal intestine.

13.2.3 Ferulic Acid and IEC-6 Cell Proliferation

Ferulic acid is the predominating phenolic acid in wheat primarily located in the bran of the kernel. Ferulic acid has a relatively strong antioxidant capacity that is similar to

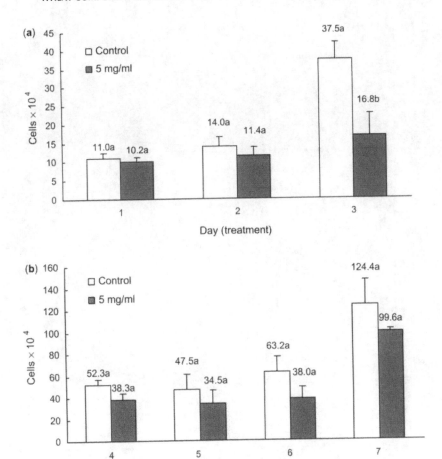

Figure 13.3   Effect of Lakin wheat bran extract on IEC-6 cell proliferation during treatment followed by treatment removal. IEC-6 cell treated with Lakin wheat bran extract at 5 mg bran equivalents/mL. Figure (**a**) represents cell proliferation during 3 days of treatment with the extract. Figure (**b**) represents cell proliferation for the 4 following days of treatment as control cultures. Triplicate measurements were taken. Bars with different letters on the same day are significantly different ($P < 0.05$).

other phenolic acids found in wheat. Therefore, it was tested for antiproliferation, antiproliferation/recovery, caspase-3 expression using western analysis, and cell cycle phase distribution using flow cytometry.

13.2.3.1   Antiproliferation and Antiproliferation/Recovery   Ferulic acid was examined for antiproliferation effect on IEC-6 cells. IEC-6 cells were treated with 1 and 2 mM ferulic acid and counted for 4 days. Results showed no difference in proliferation among the treated and control cells following 1 day of exposure; however, 2 mM ferulic acid treatment significantly inhibited the growth of IEC-6

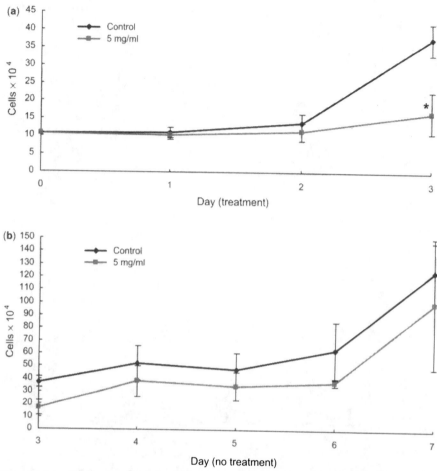

Figure 13.4 Growth curve effect of Lakin wheat bran extract on IEC-6 cells followed by treatment removal. IEC-6 cells treated with Lakin wheat bran extract at 5 mg bran equivalents/ mL. Figure (a) represents cell proliferation during 3 days of treatment with the extract. Figure (b) represents cell proliferation for the 4 following days of treatment as control cultures. Values with '*' are significantly different ($P < 0.05$).

cells, beginning at day 2 and continued through day 4 (Fig. 13.5). At 1 mM concentration, ferulic acid significantly inhibited cell growth on days 3 and 4, which is shown on the line growth curve (Fig. 13.6). There was also a significant dose-dependent cell growth inhibition between the 2 ferulic acid concentrations with stronger inhibition associated with the higher concentration. The exposure to ferulic acid also had a significant effect on cellular morphology (Image 13.3a and b). The cells treated with ferulic acid have jagged attachment borders on the culture flask, whereas the attachment borders of the control cells are smooth. Also, the tight junctions between cells treated with ferulic acid are not as clear as those of the control. These observations may in part be explained by cell death and detachment from the group.

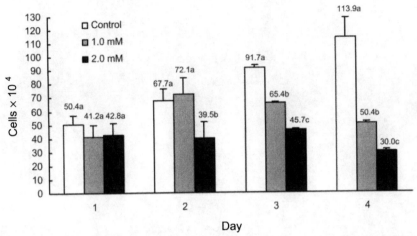

Figure 13.5   Effect of ferulic acid on IEC-6 cell proliferation. Cells were treated with 1.0 and 2.0 mM ferulic acid for 4 days. Cells were counted daily in triplicate. Bars with different letters on the same day are significantly different ($P < 0.05$).

For antiproliferation/recovery assays, IEC-6 cells were treated with ferulic acid at 1 mM and counted for 3 days, and then cultured in control media and counted for 4 more days. The results for three treatment days were similar to the growth inhibition study with the ferulic acid significantly inhibiting cell proliferation (Fig. 13.7a). Following the removal of treatment on day 4 through day 7, cell growth rates returned to the same level as the control cultures (Fig. 13.7b).

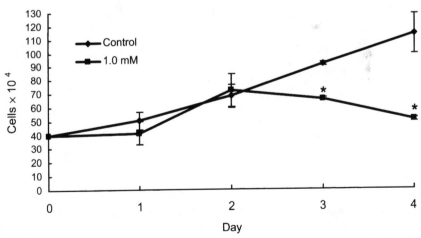

Figure 13.6   Growth curve effect of ferulic acid on IEC-6. IEC-6 cells were treated with 1.0 mM ferulic acid and counted for 4 days. Measurements were taken in triplicate. Values with "*" are significantly different ($P < 0.05$).

(a)

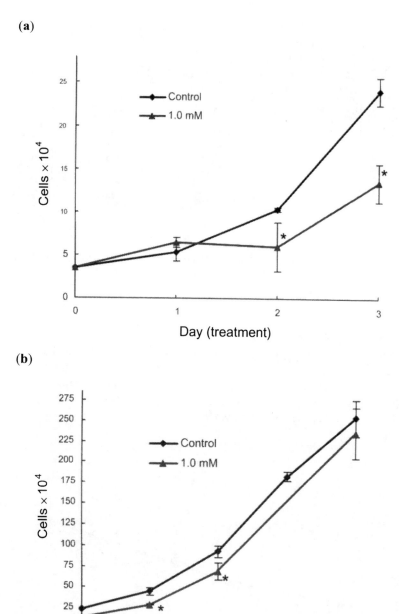

(b)

Figure 13.7   Growth curve effect of ferulic acid on IEC-6 cell proliferation followed by treatment removal. Figure (**a**) represents the growth curve of IEC-6 cells treated with 1.0 mM ferulic acid for 3 days. Figure (**b**) represents the growth curve of IEC-6 for the 4 following days of treatment as control cultures. Measurements were taken in triplicate. Values with "*" are significantly different ($P < 0.05$).

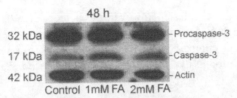

Figure 13.8   Effect of ferulic acid on caspase-3 expression in IEC-6 cells. Caspase-3, actin, and procaspase-3 expression examined by western analysis of IEC-6 cells treated with 1.0 and 2.0 mM ferulic acid for 48 h.

13.2.3.2   Caspase-3   Expression   Apoptosis was examined in IEC-6 cells by caspase-3 expression using western analysis. IEC-6 cells were treated with 1 and 2 mM ferulic acid and were examined following 48 and 96 h of treatment (Fig. 13.8). Densitometry readings of the western analyses determined that caspase-3 expression was significantly higher in the ferulic acid-treated cells compared to the control cells using β-actin as the internal control and may be dose dependent (Fig. 13.9). Duplicated data that demonstrated similar findings are not shown. These results suggest that ferulic acid may increase apoptosis in IEC-6 normal intestinal epithelial cells by increasing the expression of caspase-3 and that ferulic acid may increase apoptosis in a dose-dependent fashion.

13.2.3.3   Effect of Ferulic Acid on Cell Cycle   Ferulic acid was examined for its effects on cell cycle using flow cytometry. IEC-6 cells were treated with 0.5, 1, and 2 mM ferulic acid for 48 and 96 h. The results of the cell cycle analysis are shown in Fig. 13.10. Following both 48 and 96 h of treatment with ferulic acid, there was a dose-dependent trend of an increasing percent of cells in the S phase and a decreasing

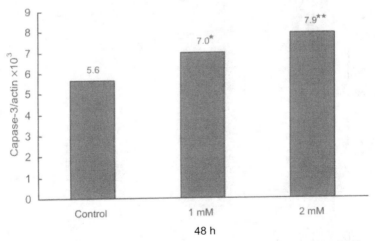

Figure 13.9   Relative density of caspase-3/actin in IEC-6 cells. Caspase-3 and actin relative density measurements following western analysis of IEC-6 cells treated with 1.0 and 2.0 mM ferulic acid for 48 h.

percent of cells in the G_1 phase. However, the percent of cells in S phase slightly dropped from 48 to 96 h of treatment. After 96 h of treatment, percentages of cells in G_2/M phase increased compared to 48 h for all treatments (Fig. 13.10b). These results are similar to those found by Janicke and others (11), who examined the effect of

(a)

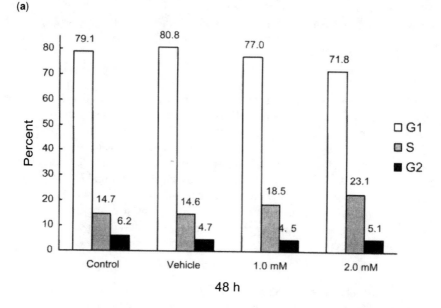

(b)

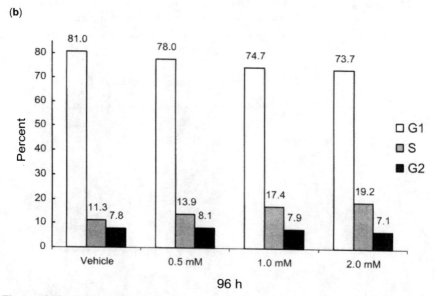

Figure 13.10   Cell cycle analysis of IEC-6 cells treated with ferulic acid. Flow cytometry analysis of IEC-6 cells treated with 0.5, 1.0, and 2.0 mM ferulic acid. Figure (**a**) represents treatment for 48 h. Figure (**b**) represents treatment for 96 h.

ferulic acid on the cell cycle of human colon cancer Caco-2 cells. Caco-2 cells were treated with 1.5 mM ferulic acid for 3 days and analyzed for length of phase and proportion of cells in each phase. The results showed that treatment caused an increase in the length of time in S phase and number of cells in the S phase, and also the percent of cells in the G_2 phase was significantly higher than control on days 2 and 3. The percent of cells in G_1 phase was significantly lower than control for all 3 days of the study (11).

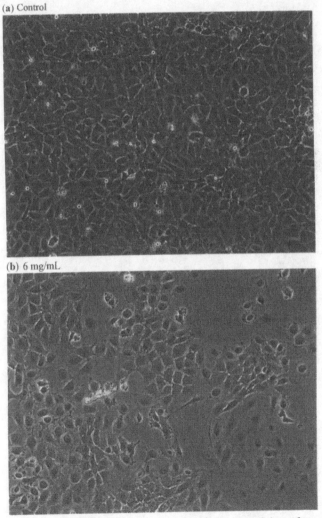

Image 13.1 IEC-6 cells treated with Wichita bran extract following 3 days of treatment. Image (**a**) is a picture of control cells. Image (**b**) is a picture of cells treated with Wichita bran extract at 6 mg bran equivalents/mL.

13.3   DISCUSSION

Very few studies have examined the effects of wheat components on the normal intestinal cell *in vitro*. In the current study, we examined the effect of extracts from two different varieties of wheat on the growth of IEC-6 normal intestinal epithelial cells. Extracts from both varieties were able to significantly inhibit cell growth, and the inhibition was reversible upon removal of the treatment. In the present study, ferulic acid, which is the primary phenolic compound in wheat, significantly and dose dependently inhibited IEC-6 proliferation, and like the wheat bran extracts, effects were reversible upon removal of the treatment. Ferulic acid also dose dependently

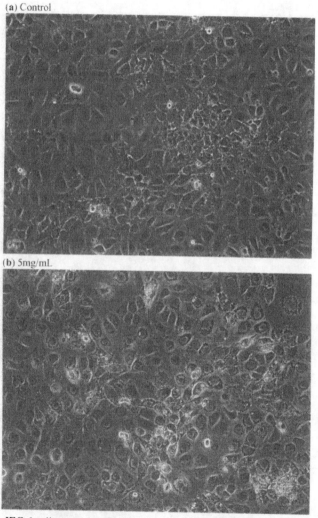

Image 13.2   IEC-6 cells treated with Lakin bran extract following 3 days of treatment. Image **(a)** is a picture of control cells. Image **(b)** is a picture of cells treated with Lakin bran extract at 5 mg bran equivalents/mL.

increased the expression of caspase-3 and increased the S phase cell cycle arrest with a concomitant decrease in G_1 phase suggesting that part of the reasons for antiproliferation may be explained by an increase in apoptosis and cell cycle phase arrest.

One previous study investigated the effect of ferulic acid on ECV304 cell (12), which is an endothelial cell from the human umbilical vein. Ferulic acid was examined for effects on cell proliferation, nitric oxide production, extracellular signal-related kinase (ERK1/2) regulation, and cell cycle. Antiproliferation tests showed that there were significant reductions in proliferation from 20, 40, and 80 μM ferulic acid compared to control. Effects of ferulic acid showed an increased nitric oxide production that may have led to the down regulation and inhibition of ERK1/2 phosphorylation, which is involved in cell proliferation events. The inhibition of ERK1/2 may have been

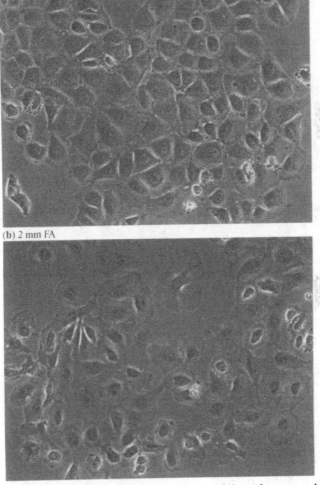

(a) Control

(b) 2 mm FA

Image 13.3 IEC-6 cells treated with ferulic acid following 3 days of treatment. Image (**a**) is a picture of control cells. Image (**b**) is a picture of cells treated with 2 mM ferulic acid.

partially responsible for antiproliferative effects and was supported by the restoration of cell proliferation by a nitric oxide synthase inhibitor. Ferulic acid treatment at 80 μM on the cell cycle was shown to significantly inhibit cells from entering S phase from G_1 phase, which could also explain a reduction in proliferation. The ferulic acid treatment effects on cell cycle were in contrast to our findings on intestinal epithelial IEC-6 cell cycle that showed an increase in S phase arrest with a decrease in G_1 (12). Although both experiments involved epithelial cells and had similar results regarding antiproliferation, the difference in cell cycle analysis may demonstrate that different mechanisms of proliferation regulation may exist even between similar cells.

13.4 CONCLUSION

Several mechanisms of body weight regulation have been proposed for whole grains including wheat. Some of these include their high volume and low density saturation promotion, the prolonging of GI emptying time leading to a slower nutrient absorption rate that may delay the return of hunger, and satiety from hormonal effects released by the mechanics of chewing and swallowing, among others. It is also possible that other factors may influence body weight gain from the consumption of wheat discussed in this chapter. Both wheat bran extracts and ferulic acid from our study displayed antiproliferation effects against the normal intestinal IEC-6 cell line. Although antiproliferation may raise an issue of consumption safety, the growth inhibitory effects were halted with removal of the treatment. It may be possible that the presence of wheat in the GI tract may effectively and safely reduce the number of nutrient-absorbing cells leading to a decrease in nutrient uptake and a reduction in total body weight.

ACKNOWLEDGMENT

The author would like to thank Dr. Jian-Ying Wang at the University of Maryland in the Departments of Surgery and Pathology, Baltimore, MD, for his technical assistance to perform the *in vitro* studies.

REFERENCES

1. Quaroni, A.; Wands, J.; Trelstad, R. L.; Isselbacher, K. J. Epithelial cell cultures from rat small intestine, Characterization by morphologic and immunologic criteria. *J. Cell Biol.* **1979**, *80*, 248–265.
2. Kim, K.-H.; Tsao, R.; Yang, R.; Cui, S. W. Phenolic acid profiles and antioxidant activities of wheat bran extracts and the effect of hydrolysis conditions. *Food Chem.* **2006**, *95*, 466–473.
3. Moore, J.; Yin, J.-J.; Yu, L. L. Novel fluorometric assay for hydroxyl radical scavenging capacity (HOSC) estimation. *J. Agric. Food Chem.* **2006**, *54*, 617–626.
4. Kroon, P. A.; Faulds, C. B.; Ryden, P.; Robertson, J. A.; Williamson, G. Release of covalently bound ferulic acid from fiber in the human colon. *J. Agric. Food Chem.* **1997**, *45*, 661–667.
5. Day, A. J.; DuPont, S.; Ridley, S.; Rhodes, M.; Rhodes, M. J. C.; Morgan, M. R. A.; Williamson, G. Deglycosylation of flavonoid and isoflavonoid glycosides by human small intestine and liver β-glucosidase activity. *FEBS Lett.* **1998**, *436*, 71–75.

6. Andreasen, M. F.; Kroon, P. A.; Williamson, G.; Garcia-Conesa, M.-T. Esterase activity able to hydrolyze dietary antioxidant hydroxycinnamates is distributed along the intestine of mammals. *J. Agric. Food Chem.* **2001**, *49*, 5679–5684.

7. Zhao, Z.; Egashira, Y.; Sanada, H. Ferulic acid sugar esters are recovered in rat plasma and urine mainly as the sulfoglucuronide of ferulic acid. *J. Nutr.* **2003**, *133*, 1355–1361.

8. Adam, A.; Crespy, V.; Levrat-Verny, M.-A.; Leenhardt, F.; Leuillet, M.; Demigne, C.; Remesy, C. The bioavailability of ferulic acid is governed primarily by the food matrix rather than its metabolism in intestine and liver in rats. *J. Nutr.* **2002**, *132*, 1962–1968.

9. Zhao, Z.; Egashira, Y.; Sanada, H. Digestion and absorption of ferulic acid sugar esters in rat gastrointestinal tract. *J. Agric. Food Chem.* **2003**, *51*, 5534–5539.

10. Zhao, Z.; Egashira, Y.; Sanada, H. Ferulic acid is quickly absorbed from rat stomach as the free form then conjugated mainly in the liver. *J. Nutr.* **2004**, *134*, 3083–3088.

11. Janicke, B.; Onning, G.; Oredsson, S. M. Differential effects of ferulic acid and *p*-coumaric acid of S phase distribution and length of S phase in the human colonic cell line Caco-2. *J. Agric. Food Chem.* **2005**, *53*, 6658–6665.

12. Hou, Y. Z.; Yang, J.; Zhao, G. R.; Yuan, Y. J. Ferulic acid inhibits endothelial cell proliferation through NO down-regulating ERK1/2 pathway. *J. Cell Biochem.* **2004**, *93*, 1203–1209.

CHAPTER 14

WHEAT ANTIOXIDANTS AND CHOLESTEROL METABOLISM

Huiping Zhou
Elaine Studer
Junjun Zhang

14.1 INTRODUCTION

Cholesterol is an essential component of mammalian cell membranes and plays an important role in maintaining normal cellular function (1). However, excessive accumulation of cholesterol is toxic and responsible for various diseases including cardiovascular disease, a major health care problem in developed countries (2). Numerous studies indicate that oxidative damage is a major contributor to the development of cardiovascular diseases, such as atherosclerosis (3). It has been well recognized that proper nutrition is important for the prevention of cardiovascular diseases, as many cardiovascular diseases are secondary to inappropriate diet. Epidemiological studies also have shown that dietary antioxidants can reduce the risk of cardiovascular disease, although the underlying mechanisms remain elusive (4–6). The awareness of the potential role of dietary nutrition in maintaining a healthy life has driven the rapid development of various functional foods designed to improve the health of millions of consumers. Wheat is a globally important agricultural commodity and food ingredient, and contains considerable beneficial nutritional components. Wheat and wheat-based food ingredients can ideally serve as the basis for the development of these functional foods. Our recent studies have shown that wheat antioxidants have cholesterol-lowering activities through the modulation of cholesterol metabolism in liver. This chapter will summarize the wheat antioxidants and their potential effects on hepatic cholesterol homeostasis.

14.2 WHEAT ANTIOXIDANTS

Wheat is an important agricultural crop across the world. Previous studies indicated that different wheat varieties contain specific profiles of antioxidant properties (7,8). Even for the same variety and growing conditions, the interaction between environmental factors and genotype may alter phytochemical compositions and antioxidant properties of wheat grain and bran (9,10). The major components of antioxidants present in wheat grain are phenolic acids, carotenoids, and tocopherols. Early studies indicated that wheat antioxidants mainly concentrate in the aleuronic fraction of bran, while some are located in the pericarp, nuclear envelope, and germ (11).

14.2.1   Phenolic Acids

Several phenolic acids including ferulic, vanillic, 4-coumaric, caffeic, chlorogenic, gentisic, syringic, and 4-hydroxybenzoic acids have been detected in wheat grain and fractions (8,10). All of these phenolic acids are derivatives of either benzoic or cinnamic acids. In a study on Trego wheat, ferulic acid was found to be the dominant phenolic acid in wheat bran fractions and accounted for around 59–60% of the total phenolic acids on a per weight basis, along with significant levels of syringic, *p*-hydroxybenzoic, vanillic, and coumaric acids at a concentration range of 4–33 μg/g (10). Recently, Zhou et al., also evaluated the bran samples of seven wheat varieties produced in four different countries and found that ferulic acid prevailed in all of the tested samples with a concentration range of 99–231 μg/g and accounted for about 46–67% of total phenolic acids on a per weight basis (12).

14.2.2   Carotenoids

In addition to phenolic acids, significant levels of carotenoids including lutein, zeaxanthin, and cryptoxanthin are also present in wheat whole grain and brans (8,12). The carotenoid profile differs significantly among different varieties. The total carotenoid level ranges from 0.12 to 0.68 μmol per 100 g of bran sample in a study involving seven varieties from four countries (12). Lutein was the primary carotenoid in four wheat bran samples and zeaxanthin was the major carotenoid in the other three samples, implying that zeaxanthin is also important in the measurement of total carotenoid contents in wheat and wheat bran. This study differed from the observation that lutein was the predominant carotenoid in wheat grain of 11 wheat varieties (13) and eight durum wheat varieties (14). The discrepancy may be partially explained by the difference in wheat variety and environmental growing conditions. Cryptoxanthin was found in both wheat bran and grain samples at considerable levels.

14.2.3   Tocopherols

Tocopherols may play an important role in reducing the risk of cardiovascular disease according to evidence from both epidemiological and clinical studies (15). There are four tocopherol isomers, including α-, β-, γ-, and δ-tocopherols in nature. In 2004, the total tocopherol and tocopherol composition in the bran samples of seven wheat varieties were examined and evaluated for the first time. Three of the four tocopherol isomers, α-, γ-, and δ-tocopherols, were detected in all the tested samples. The total tocopherol content varied greatly among the seven samples with a range of 0.92–6.9 μmol per 100 gram bran, while the tocopherol composition exhibited 16.6-,

30.6-, and 3.7-fold differences in their α-, γ-, and δ-tocopherol contents among the seven samples. Basically, different bran samples from different countries or regions contain specific tocopherol profiles and dominant components. Canadian durum wheat bran possesses the highest level of α-tocopherol and total tocopherol among all samples. Bran samples from the United States may contain more γ- or δ-tocopherols. The significant variation suggests the potential influence of wheat variety and growing condition on tocopherol production in wheat bran and grain (12).

14.3   WHEAT ANTIOXIDANT PROPERTIES

Oxidative damage is a major contributor to the development of cardiovascular diseases, such as atherosclerosis (6). Increased production of oxidants such as superoxide, hydrogen peroxide, or nitric oxide may damage tissues either by direct oxidation of key biological molecules or by alteration of transcriptional factors (16). It has been demonstrated that native low density lipoprotein (LDL) itself is not atherogenic, but the oxidized LDL is highly atherogenic and toxic to vascular cells (17). The oxidative modification of LDL plays a pivotal role in the initiation of atherosclerosis. Inhibition of LDL oxidation therefore limits the cytotoxicity of LDL. One potential mechanism proposed in the beneficial actions of antioxidants in biological or food systems is that they can directly quench free radicals to terminate the radical chain reaction and prevent free radical mediated LDL oxidation (18,19).

Free radical scavenging activities of wheat grain or bran extracts against 2, 2-diphenyl-1-picrylhydrazyl radical (DPPH$^{\bullet}$), radical cation ABTS$^{\bullet+}$, peroxide radical anion ($O^{\bullet2-}$), hydroxyl radical (OH$^{\bullet}$), and oxygen radical (ORAC) have been evaluated in various studies by spectrophotometric and electron spin resonance (ESR) spectrometry methods. Different wheat varieties or the same variety under different growing conditions or from different regions varied greatly in their free radical scavenging properties (8,12,20). A recent study done by Yu et al. (7) shows that three hard winter wheat varieties from Akron, Trego, and Platte have different profiles for quenching free radicals. Akron had the greatest capacity to quench DPPH radicals ($ED_{50} = 15.4$ mg of grain/mL), while Platte showed the highest activity against ABTS$^{\bullet+}$ (1.91 0.06 μmol of trolox equiv (TE)/g of grain). The ED_{50} values are comparable to that of 20 and 23.2–27.4 mg of grain/mL detected in the Swiss red wheat grain (11) and eight Maryland soft wheat samples (21). However, the scavenging activity of wheat extracts against ABTS$^{\bullet+}$ (1.08–1.91 μmol TE/g of grain) was much lower than that observed in different studies (22). Furthermore, there was no significant correlation between total phenolic content and radical scavenging capacities for DPPH$^{\bullet}$ and ABTS$^{\bullet+}$, which contrasts with what was reported by Adom et al., who found that total antioxidant activity was strongly correlated with the phenolic content of wheat extracts (8). The discrepancies between these studies may be explained by the wheat varieties, different extraction solvents, and experimental conditions for scavenging activity assays. A most recent study done by Moore et al. (24) confirmed the potential influences of growing conditions on the antioxidant properties of hard winter wheat, suggesting that it is possible to produce wheat with specific antioxidant properties by optimizing the growing conditions.

14.4 CHOLESTEROL HOMEOSTASIS

The liver plays a central role in the regulation of cholesterol homeostasis in the body (1). Disruption of cholesterol homeostasis has been associated with pathogenesis of various diseases such as hyperlipidemia and atherosclerosis.

Under normal physiological conditions, lipid input equals lipid output from the body and cholesterol homeostasis is maintained. There are two major input pathways for cholesterol. One is the receptor-mediated endocytosis of cholesterol-carrying lipoproteins. Another is the *de novo* biosynthesis of cholesterol from acetate. Hepatic uptake of LDL cholesterol occurs via receptor-mediated endocytosis by LDL receptor (LDLR). High density lipoprotein (HDL) cholesterol is taken up by the scavenger receptor (SR) B1. After uptake into the liver, cholesterol can be stored as cholesterol esters by activation of acyl-CoA-cholesterol acyl transferase (ACAT) that catalyzes the esterification of cholesterol. Cholesterol can also be incorporated into very low density lipoproteins (VLDL) and directly secreted into the blood. The liver is a major site for *de novo* biosynthesis of cholesterol, which involves dozens of distinct enzymatic reactions. The 3-hydroxy-3-methylglutaryl-CoA reductase (HMG-CoA-R) is the rate-determining enzyme of the cholesterol biosynthetic pathway.

In the liver, there are two major output pathways for cholesterol. The hepatocyte disposes of excess cholesterol by secreting free cholesterol directly into the canaliculus and by biosynthesis of bile acids; each constitutes approximately 50% of the total output in man (25). The biosynthesis of bile acid from free cholesterol occurs through two major synthetic pathways: the classic or neutral pathway and the alternative or acidic pathway. The classic pathway is initiated by a microsomal cholesterol 7α-hydroxylase (CYP7A1), the rate-limiting enzyme in this pathway. The acidic pathway is initiated by a mitochondrial cholesterol 27-hydroxylase (CYP27A1), the rate-limiting enzyme in this pathway. Under normal physiological conditions, the classic pathway is the predominant pathway and the acidic pathway may contribute little to the overall bile acid synthesis (26). Conversion of hepatic cholesterol to bile acids represents the major regulatory pathway by which the body eliminates excess cholesterol (27).

Cholesterol homeostasis is maintained through the coordinated regulation of cholesterol synthesis, degradation, and secretion. The sterol regulatory element-binding proteins (SREBPs) are a family of membrane-bound transcription factors regulating a number of biochemical pathways of lipid metabolism (28–30). SREBPs are synthesized as inactive precursors bound to the endoplasmic reticulum (ER) and must be released from the membrane by a two-step proteolytic process, transported to the nucleus where they act as transcription factors by binding to sterol responsive elements (SRE). The SREBP family comprises of three subtypes: SREBP1a, SREBP1c, and SREBP2, which control the expression of more than 30 genes required for the biosynthesis of cholesterol, fatty acids, triacylglycerols, and phospholipids, and for cholesterol transport and glucose/insulin metabolism. Studies using transgenic and knockout mice indicate that SREBPs play related but distinct roles: SREBP1 preferentially regulates fatty acid and glucose metabolism, while SREBP2 preferentially activates the LDLR gene and various genes required for cholesterol synthesis (28–30).

14.5   EFFECTS OF WHEAT ANTIOXIDANTS ON CHOLESTEROL METABOLISM

It has been widely accepted that diet can significantly alter overall human health and quality of life. Wheat is a globally important agricultural commodity and food ingredient, which contains considerable nutritional components. Growing evidence indicates that whole-grain intake reduces the risk of diabetes and cardiovascular diseases (31,32). Wheat antioxidants and the insoluble fiber have been considered as the possible primary contributors for these beneficial effects. However, little is known about whether wheat antioxidants may directly regulate cholesterol biosynthesis and metabolism other than simply acting as antioxidative agents.

Recently, we examined the potential effects of wheat antioxidants on the key enzymes involved in cholesterol biosynthesis and metabolism. Primary rat hepatocytes were treated with wheat antioxidants at a final concentration of 0.12 mg/mL,

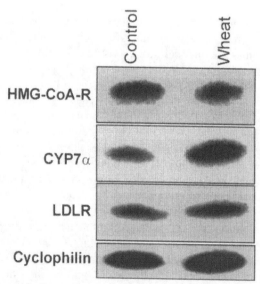

Figure 14.1   Effects of wheat antioxidants on mRNA levels of genes involved in cholesterol and bile acid metabolism in rat primary hepatocytes. Rat primary hepatocytes were treated with wheat antioxidants for 24 h. Total RNA was isolated. The mRNA levels of CYP7A1, HMG-CoA-R, and LDL-R were determined by RPA analysis. Relative amounts of mRNA were determined by quantifying the densities of the radioactive bands using Image J software and normalized to rat cyclophilin mRNA as loading control.

equivalent to 2 mg wheat grain/mL medium, for 24 h; the mRNA levels of HMG-CoA-R, CYP7A1, and LDLR were examined using a ribonuclease protection assay (RPA). As shown in Fig. 14.1, wheat antioxidants significantly reduced HMG-CoA-R but increased CYP7A1 mRNA expression, while showing no significant effect on LDLR mRNA expression. We further investigated the effects of wheat antioxidants on the mRNA stabilities of HMG-CoA-R and CYP7A1, and found that wheat antioxidants significantly increased the degradation of HMG-CoA-R mRNA, but increased the CYP7A1 mRNA stability (data not shown). Consistently, wheat antioxidants also increased bile acid synthesis. Taken together, these data suggest that regulation of the key genes involved in cholesterol biosynthesis and metabolism may represent one of the major cellular/molecular mechanisms by which wheat antioxidants reduce the risk of cardiovascular diseases.

14.6 SUMMARY

Epidemiological studies have strongly suggested that consumption of wheat and wheat-based products may reduce the risk of cardiovascular disease (4,5,31,33,34). The results summarized in this chapter clearly indicate that wheat antioxidants are able to not only reduce the oxidative stress, but also regulate cholesterol metabolism in liver. In addition, our studies demonstrate the potential of a common food staple in disease prevention and health promotion, and warrant further multidisciplinary investigation of the wheat antioxidants in human health and the approaches needed to enhance their bioavailability in wheat-based functional foods.

ACKNOWLEDGMENTS

This work is supported by grants from the National Institutes of Health (R21 AI068432, R01 AI057189, P01 DK38030), GlaxoSmithKline research fund and A.D. Williams fund.

REFERENCES

1. Hylemon, P.; Pandak, W.; Vlahcevic, Z. Regulation of hepatic cholesterol homestasis. In:Arias, I. M., Boyer, J. L., Chisari, F. V., Fausto, N., Schachter, D., Shafritz, D. A. (Eds.). *Liver: Biology and Pathobiology.*, Philadelphia, **2001**. pp. 231–247.
2. Chung, C. P.; Avalos, I.; Raggi, P.; Stein, C. M. Atherosclerosis and inflammation: insights from rheumatoid arthritis. *Clin. Rheumatol.* **2007**, *26*, 1228–1233.
3. Chung, C. P.; Avalos, I.; Stein, C. M. Oxidative stress, microvascular dysfunction, and scleroderma: an association with potential therapeutic implications, a commentary on Postocclusive reactive hyperemia inversely correlates with urinary 15-F2t-isoprostane levels in systemic sclerosis. *Free Radic. Biol. Med.* **2006**, *40*, 1698–1699.
4. Willcox, J. K.; Ash, S. L.; Catignani, G. L. Antioxidants and prevention of chronic disease. *Crit. Rev. Food Sci. Nutr.* **2004**, *44*, 275–295.

5. Halliwell, B.; Gutteridge, J. M.; Cross, C. E. Free radicals, antioxidants, and human disease: where are we now? *J. Lab. Clin. Med.* **1992**, *119*, 598–620.
6. Minhajuddin, M.; Beg, Z. H.; Iqbal, J. Hypolipidemic and antioxidant properties of tocotrienol rich fraction isolated from rice bran oil in experimentally induced hyperlipidemic rats. *Food Chem. Toxicol.* **2005**, *43*, 747–753.
7. Yu, L.; Haley, S.; Perret, J.; Harris, M.; Wilson, J.; Qian, M. Free radical scavenging properties of wheat extracts. *J. Agric. Food Chem.* **2002**, *50*, 619–1624.
8. Adom, K. K.; Sorrells, M. E.; Liu, R. H. Phytochemical profiles and antioxidant activity of wheat varieties. *J. Agric. Food Chem.* **2003**, *51*, 7825–7834.
9. Yu, L.; Perret, J.; Harris, M.; Wilson, J.; Haley, S. Antioxidant properties of bran extracts from "Akron" wheat grown at different locations. *J. Agric. Food Chem.* **2003**, *51*, 1566–1570.
10. Zhou, K.; Yu, L.;Antioxidant properties of bran extracts from Trego wheat grown at different locations. *J. Agric. Food Chem.* **2004**, *52*, 1112–1117.
11. Zhou, K.; Laux, J. J.; Yu, L. Comparison of Swiss red wheat grain and fractions for their antioxidant properties. *J. Agric. Food Chem.* **2004**, *52*, 1118–1123.
12. Zhou, K.; Su, L.; Yu, L. L. Phytochemicals and antioxidant properties in wheat bran. *J. Agric. Food Chem.* **2004**, *52*, 6108–6114.
13. Yu, L.; Perret, J.; Davy, D.; Wilson, J.; Melby, C. L. Antioxidant properties of cereal products. *J. Food Sci.* **2002**, *67*, 2600–2603.
14. Hentschel, V.; Kranl, K.; Hollmann, J.; Lindhauer, M. G.; Bohm, V.; Bitsch, R. Spectrophotometric determination of yellow pigment content and evaluation of carotenoids by high-performance liquid chromatography in durum wheat grain. *J. Agric. Food Chem.* **2002**, *50*, 6663–6668.
15. Pham, D. Q.; Plakogiannis, R. Vitamin E supplementation in cardiovascular disease and cancer prevention: Part 1. *Ann. Pharmacother.* **2005**, *39*, 1870–1878.
16. Bowers, R.; Cool, C.; Murphy, R. C.; Tuder, R. M.; Hopken, M. W.; Flores, S. C.; Voelkel, N. F. Oxidative stress in severe pulmonary hypertension. *Am. J. Respir. Crit. Care Med.* **2004**, *169*, 764–769.
17. Rota, S.; McWilliam, N. A.; Baglin, T. P.; Byrne, C. D. Atherogenic lipoproteins support assembly of the prothrombinase complex and thrombin generation: modulation by oxidation and vitamin E. *Blood* **1998**, *91*, 508–515.
18. Thurnham, D. I.; Functional foods: cholesterol-lowering benefits of plant sterols. *Br. J. Nutr.* **1999**, *82*, 255–256.
19. Chopra, M.; Thurnham, D. I. Antioxidants and lipoprotein metabolism. *Proc. Nutr. Soc.* **1999**, *58*, 663–671.
20. Zhou, K.; Yin, J. J.; Yu, L. L. Phenolic acid, tocopherol and carotenoid compositions, and antioxidant functions of hard red winter wheat bran. *J. Agric. Food Chem.* **2005**, *53*, 3916–3922.
21. Moore, J.; Hao, Z.; Zhou, K.; Luther, M.; Costa, J.; Yu, L. L. Carotenoid, tocopherol, phenolic acid, and antioxidant properties of Maryland-grown soft wheat. *J. Agric. Food Chem.* **2005**, *53*, 6649–6657.
22. Zielinski, H.; Kozlowska, H. Antioxidant activity and total phenolics in selected cereal grains and their different morphological fractions. *J. Agric. Food Chem.* **2000**, *48*, 2008–2016.
23. Adom, K. K.; Liu, R. H. Antioxidant activity of grains. *J. Agric. Food Chem.* **2002**, *50*, 6182–6187.
24. Moore, J.; Liu, J. G.; Zhou, K.; Yu, L. L. Effects of genotype and environment on the antioxidant properties of hard winter wheat bran. *J. Agric. Food Chem.* **2006**, *54*, 5313–5322.
25. Hylemon, P. B.; Pandak, W. M.; Vlahcevic, Z. R. Regulation of hepatoc cholesterol homeostasis. In: Arias, I. M.; Boyer, J. L.; Chisari, F. V.; Fausto, C. N.; Schachter, D.; Shafritz, D. A. (Eds.). *The Liver: Biology and Pathobiology*, Philadelphia, PA, **2001**. pp. 231–247.
26. Chiang, J. Y. Regulation of bile acid synthesis: pathways, nuclear receptors, and mechanisms. *J. Hepatol.* **2004**, *40*, 539–551.
27. Romero, A. L.; West, K. L.; Zern, T.; Fernandez, M. L. The seeds from *Plantago ovata* lower plasma lipids by altering hepatic and bile acid metabolism in guinea pigs. *J. Nutr.* **2002**, *132*, 1194–1198.
28. Horton, J. D.; Goldstein, J. L.; Brown, M. S. SREBPs: transcriptional mediators of lipid homeostasis. *Cold Spring Harb. Symp. Quant. Biol.* **2002**, *67*, 491–498.
29. Horton, J. D.; Goldstein, J. L.; Brown, M. S. SREBPs: activators of the complete program of cholesterol and fatty acid synthesis in the liver. *J. Clin. Invest.* **2002**, *109*, 1125–1131.
30. Eberle, D.; Hegarty, B.; Bossard, P.; Ferre, P.; Foufelle, F. SREBP transcription factors: master regulators of lipid homeostasis. *Biochimie.* **2004**, *86*, 839–848.

31. Jensen, M. K.; Koh-Banerjee, P.; Franz, M.; Sampson, L.; Gronbaek, M.; Rimm, E. B. Whole grains, bran, and germ in relation to homocysteine and markers of glycemic control, lipids, and inflammation 1. *Am. J. Clin. Nutr.* **2006**, *83*, 275–283.

32. Jensen, M. K.; Koh-Banerjee, P.; Hu, F. B.; Franz, M.; Sampson, L.; Gronbaek, M.; Rimm, E. B. Intakes of whole grains, bran, and germ and the risk of coronary heart disease in men. *Am. J. Clin. Nutr.* **2004**, *80*, 1492–1499.

33. Gorinstein, S.; Bartnikowska, E.; Kulasek, G.; Zemser, M.; Trakhtenberg, S. Dietary persimmon improves lipid metabolism in rats fed diets containing cholesterol. *J. Nutr.* **1998**, *128*, 2023–2027.

34. Sabovic, M.; Lavre, S.; Keber, I. Supplementation of wheat fibre can improve risk profile in patients with dysmetabolic cardiovascular syndrome. *Eur. J. Cardiovasc. Prev. Rehabil.* **2004**, *11*, 144–148.

CHAPTER 15

WHEAT ANTIOXIDANT BIOAVAILABILITY

Yutaka Konishi

15.1 INTRODUCTION

In 2003, WHO reported a causal link between the regular ingestion of polyphenols and a reduction in cardiovascular diseases (1). Consumption of fruits and vegetables, as well as grains, has been associated with reduced risk of chronic diseases (2,3). Indeed, an increased consumption of fruits and vegetables containing high levels of antioxidant is generally regarded to be beneficial to human health. However, much less attention has been focused on grain consumption, although epidemiological studies also show the importance and health benefits of this food source. Furthermore, nutritional guidelines put whole grains or grain-based products at the base of the food guide pyramid (4). The most abundant types of polyphenols in the human diet are the flavonoids and phenolic acids (PAs). Although many studies have been carried out to investigate the physiological role of flavonoids, PAs have not been extensively studied and are not considered to be of great nutritional interest. PAs are present in many foods including grains, vegetables, and fruits. Consumption of PAs may vary considerably according to dietary habits, especially in terms of the quantity of coffee consumed. It was reported that persons who drink several cups of coffee per day, may ingest as much as 500–800 mg PAs/day (5), caffeic acid (CA) intake alone running at 206 mg/day (5). By contrast, the intake of flavonoids and isoflavonoids is only estimated to be 100–150 mg/day (5). The total polyphenol intake commonly reaches 1 g/day (5).

Grains contain unique phytochemicals that complement those in fruits and vegetables. Most phytochemicals in fruits and vegetables are free or soluble conjugated molecules in the form of glycosides. However, grain phytochemicals may exist in free, soluble conjugated, and insoluble bound forms. PAs such as ferulic acid (FA) and diferulates are predominantly found in grains but are not present in significant quantities in fruits and vegetables (4). The FA content of wheat grain is 0.8–2 g/kg dry wt, which may represent up to 90% of total polyphenols (5). Wheat is also a rich source of other PAs, such as syringic acid (SRA), *p*-coumaric acid (PCA), synapic acid (SPA), vanillic acid (VA), CA, and *p*-hydroxybenzoic acid (PBA) (6,7). The physiological impact of PAs, as well as flavonoids, on human health depends on their availability for intestinal absorption and subsequent interaction with target tissues. However, the absorption efficiency of flavonoids in humans is generally low, and the biological activity and health effect of polyphenols, especially flavonoids, must be reevaluated in terms of their bioavailability and bioactive compounds *in vivo* (8). The physiological significance of microbial metabolites of poorly absorbed flavonoids has now been firmly established. Specifically, these compounds have been detected in large quantities within urine and serum (8).

Hitherto, the affinity of polyphenols for the biomembrane was thought to govern their absorption characteristics (i.e., transcellular passive diffusion was presumed to be the main mechanism). Recently the transporter-mediated active absorption system for PAs has been elucidated, and the diversity of the absorption characteristics responsible for PAs has been demonstrated (9–16). This transport system is also responsible for microbial metabolites of poorly absorbed flavonoids (17,18). Recent findings highlight the profound significance of this transport system on the health benefits of dietary constituents. This chapter attempts to shed new light on the absorption characteristics and bioavailability of polyphenols, especially PAs found in wheat. The review focuses on the transport system and the concept of novel dietary nutrients, known as "metabonutrients." First, the principles related to the absorption and bioavailability of dietary PAs will be discussed. Thereafter, an overview of the antioxidants found in wheat will be given.

15.2 ABSORPTION CHARACTERISTICS OF FLUORESCEIN *IN VITRO*

Fluorescein (FC) (Table 15.1) is a fluorescent marker dye customarily applied to the evaluation of paracellular permeability of epithelial cell monolayers (19). Detailed analysis of the transepithelial transport of FC in Caco-2 cells has demonstrated that FC is absorbed by the monocarboxylic acid transporter (MCT). This conclusion was drawn by careful analysis of several possible transport mechanisms, such as H^+-driven polarized transport in the apical-to-basolateral direction, paracellular permeability-independent saturable transport (Michaelis constant, 16.2 mM; maximum velocity, 220.4 nmol/min mg protein), and from competitive inhibition of FC transport by benzoic acid (BA), a known substrate of MCT (K_i value, 1.2 ± 0.1 mM) (19). In the absence of H^+ gradient between the apical and basolateral sides, an inverse correlation between FC transport and paracellular permeability has been reported (19). Importantly, FC is absorbed by the paracellular pathway only in the absence of H^+ gradient (19). Because most of the earlier transport experiments were done in the absence of H^+ gradient, this specific FC transport might have been overlooked. The key components of a substrate for MCTs are thought to be a monoanionic carboxyl group and a nonpolar side chain or aromatic hydrophobic moiety (20). FC, comprising an aromatic bulky portion and carboxyl group that is dissociated under physiological conditions of the gastrointestinal (GI) tract, fulfills the structural criteria for an MCT substrate. We have recently developed a novel means of readily identifying compounds that can be transported by the MCT in Caco-2 cells (21). This technique measures the inhibition or competitive inhibition of dietary substances on the transport of FC ((21), Table 15.1).

TABLE 15.1 Structures and Effects of Various Phenolic Acids and their Derivatives on Fluorescein Transport Across Caco-2 Cell Monolayers

								Relative permeation,
No.	Compound name	Type	R2	R3	R4	R5	R6	% of control
1	Cinnamic acid	A	H	H	H	H	H	11.9 ± 1.8*
2	p-Coumaric acid	A	H	H	OH	H	H	84.1 ± 2.2*
3	m-Coumaric acid	A	H	OH	H	H	H	43.7 ± 2.1*
4	o-Coumaric acid	A	OH	H	H	H	H	133.3 ± 18.4
5	p-Methoxycinnamic acid	A	H	H	OMe	H	H	73.2 ± 7.6*
6	m-Methoxycinnamic acid	A	H	OMe	H	H	H	20.2 ± 3.8*
7	o-Methoxycinnamic acid	A	OMe	H	H	H	H	22.2 ± 2.9*
8	Caffeic acid	A	H	OH	OH	H	H	115.8 ± 12.1
9	Isoferulic acid	A	H	OH	OMe	H	H	33.9 ± 5.9*
10	Ferulic acid	A	H	OMe	OH	H	H	51.3 ± 2.4*
11	3,4-Dimethoxycinnamic acid	A	H	OMe	OMe	H	H	15.9 ± 1.5*
12	Sinapic acid	A	H	OMe	OH	OMe	H	74.9 ± 2.9*
13	Benzoic acid	B	H	H	H	H	H	5.8 ± 1.7*
14	p-Hydroxybenzoic acid	B	H	H	OH	H	H	34.5 ± 7.1*
15	m-Hydroxybenzoic acid	B	H	OH	H	H	H	97.1 ± 7.6
16	Salicylic acid	B	OH	H	H	H	H	22.8 ± 9.9*
17	p-Anisic acid	B	H	H	OMe	H	H	51.4 ± 9.4*
18	m-Anisic acid	B	H	OMe	H	H	H	18.2 ± 1.1*
19	o-Anisic acid	B	OMe	H	H	H	H	34.2 ± 3.2*
20	2,3-Dihydroxybenzoic acid	B	OH	OH	H	H	H	105.2 ± 2.4
21	Gentisic acid	B	OH	H	H	OH	H	102.7 ± 5.0
22	2,6-Dihydroxybenzoic acid	B	OH	H	H	H	OH	124.6 ± 14.7
23	Protocatechuic acid	B	H	OH	OH	H	H	115.3 ± 8.3
24	Isovanillic acid	B	H	OH	OMe	H	H	46.4 ± 5.7*
25	Vanillic acid	B	H	OMe	OH	H	H	55.2 ± 2.4*
26	Veratric acid	B	H	OMe	OMe	H	H	20.0 ± 4.3*
27	2,3,4-Trihydroxybenzoic acid	B	OH	OH	OH	H	H	110.4 ± 5.5
28	2,4,6-Trihydroxybenzoic acid	B	OH	H	OH	H	OH	105.8 ± 7.8
29	Gallic acid	B	H	OH	OH	OH	H	109.5 ± 5.0
30	Syringic acid	B	H	OMe	OH	OMe	H	104.8 ± 10.2

The amount of fluorescein transported was measured by incubating Caco-2 cells at 37°C for 40 min in the absence or presence of a compound. Each value is the mean $\pm$ SD of three or more experiments. (Adapted from Reference 21.)
*$P < 0.01$.

Other than passive diffusion, there has been a paucity of studies on possible mechanisms of absorption of PAs. Because PAs fulfill the structural criteria for an MCT substrate (20), we have applied our method to examine the mechanism of intestinal absorption of these compounds. FA, the major PA in wheat, competitively inhibits the transport of FC (K_i value, 2.99 $\pm$ 0.08 mM) (19). PAs are derivatives of BA and cinnamic acid (CNA). We have demonstrated that the hydroxylation or methoxylation of BA and CNA influences the inhibitory effects on the transport of FC. The investigation revealed three major findings: (i) an increase in hydroxylation decreases the affinity for MCT, (ii) metahydroxylation plays a particular role in decreasing the affinity for MCT, (iii) carbon chain length between the aromatic ring and the carboxylic acid group has little effect on the affinity for MCT ((21), Table 15.1). From the inhibitory effect on FC transport (Table 15.1), the affinity of each PA for MCT is considered to increase in the following order: PCA < FA < CNA. Indeed, the affinity of each PA for MCT matches the order of the K_i values (i.e., CNA, 2.0 $\pm$ 0.4 mM; FA, 2.99 $\pm$ 0.08 mM; PCA, 27.2 $\pm$ 4.3 mM). These data also confirmed the method for evaluating an MCT substrate using FC to be reliable (19,21). Thus, for the first time, it was possible to readily determine the affinity of each PA for MCT.

15.3   ABSORPTION CHARACTERISTICS OF PHENOLIC ACID *IN VITRO*

We have used an HPLC–electrochemical detector (ECD), fitted with a coulometric detection system (ESA, Boston, MA), to analyze samples and identify the constituents in two dimensions (chromatographic and voltammetric). Resolution was achieved by arranging several coulometric detectors in series set to different electrode detector potentials (22). Purity of the peaks was assessed using peak area ratio accuracies for the adjacent oxidation channels (lower or upper) to the dominant oxidation channel (more than 70% ratio accuracy). The voltammetric response of the analyte across these channels was unique for each compound. This provides for high sensitivity and accuracy without requiring any sample pretreatment (i.e., detection limit of <0.5 pmol for each PA tested). The technique allowed us to measure the kinetic velocity analysis for the permeation of PA, which hitherto had not been possible (9–13).

15.3.1   FA and PCA

We investigated the absorption characteristics of two PAs, FA and PCA, found in wheat grain. We clarified that both FA and PCA exhibit H^+-driven polarized saturable transport in the apical-to-basolateral direction (Michaelis constant for FA and PCA, 16.2 and 17.5 mM, respectively; maximum velocity for FA and PCA, 220.4 and 82.7 nmol/min mg of protein, respectively) by measuring transepithelial transport in Caco-2 cells. The preferable distribution of apically loaded FA and PCA into the basolateral side in the presence of H^+ gradient was determined. Significant inhibition of FA and PCA transport occurred in the presence of specific MCT substrates ((9,10),

TABLE 15.2 Absorption Characteristics of Each PA

FA, PCA	MCT-mediated active transport
	H^+-driven polarized transport in apical-to-basolateral direction
	Saturable transport
	Inhibition by MCT substrate
	J_{ap-bl} of 1 mM of FA, 9.79 nmol/min mg of protein
	J_{ap-bl} of 1 mM of PCA, 3.73 nmol/min mg of protein
CA	Paracellular passive diffusion with some affinity for MCT
	H^+-driven polarized transport in apical-to-basolateral direction
	Nonsaturable transport
	Inversely correlated with paracellular permeability
	Inhibition by MCT substrate
	J_{ap-bl} of 1 mM of CA, 0.46 nmol/min mg of protein
CLA, GA, and RA	Paracellular passive diffusion.
	Nonpolarized and nonsaturable transport.
	Inversely correlated with paracellular permeability.
	J_{ap-bl} of 5 mM of CLA, 0.16 nmol/min mg of protein
	J_{ap-bl} of 5 mM of GA, 0.16 nmol/min mg of protein
	J_{ap-bl} of 5 mM of RA, 0.13 nmol/min mg of protein
AC	Transcellular passive diffusion.
	Nonpolarized and nonsaturable transport.
	Constant J_{ap-bl} irrespective of the paracellular permeability.
	H^+-driven intracellular accumulation.
	J_{ap-bl} of 0.1 mM of AC, 0.91 nmol/min mg of protein.

Abbreviations: FA, ferulic acid; PCA, *p*-coumaric acid; CA, caffeic acid; CLA, chlorogenic acid; GA, gallic acid; RA, rosmarinic acid; AC, artepillin C. From References (9–13).

Table 15.2). Furthermore, the affinity of FA and PCA for MCT was consistent with that predicted from the inhibition of FC transport mentioned earlier (9,10,21).

15.3.2 CA, CLA, GA, and RA

The absorption characteristics of CA and chlorogenic acid (CLA) have also been investigated. These PAs are particularly abundant in fruits and vegetables, but not in wheat grain. Both CA and CLA exhibit nonsaturable transport in Caco-2 cells, whereas CA also showed H^+-driven polarized transport in the apical-to-basolateral direction, which was inhibited by substrates of MCT. Furthermore, almost all of the apically loaded CA and CLA were retained on the apical side, and the transepithelial flux of CA and CLA was inversely correlated with paracellular permeability ((11), Table 15.2). These results demonstrate that absorption of both CA and CLA is mainly via paracellular diffusion, although CA was absorbed to a lesser extent by MCT ((11), Table 15.2).

Furthermore, we examined the absorption characteristics of gallic acid (GA), which is found in tea leaves (constituting up to about 5% of the dry weight of tea leaf) (23). In Caco-2 cells, the permeation of GA was nonpolarized and independent of pH. Permeation increased linearly with increasing concentration of GA. At a concentration of 5 mM, the initial permeation rate of GA in the apical-to-basolateral direction, $J_{ap \to bl}$, was the same as that for CLA (0.16 nmol/min mg of protein; apical pH, 6.0; basolateral pH, 7.4), indicating the transepithelial transport of GA to be via the paracellular pathway ((10,11), Table 15.2).

Rosmarinic acid (RA), an ester of CA and 3,4-dihydroxyphenyllactic acid, is commonly distributed in species of *Boraginaceae* and the subfamily *Nepetoideae* of *Lamiaceae*, as well as other higher plant families (24). RA has been reported to have a number of biological activities *in vitro*, including antiviral, antibacterial, antioxidant, anti-inflammatory, and antiallergic activities. Orally administered RA exhibits antiallergic and anticarcinogenic effects *in vivo* (24). In Caco-2 cells, permeation of RA was nonpolarized, independent of pH and linearly increased with concentration (up to 30 mM). At a concentration of 5 mM, the initial permeation rate of RA in the apical-to-basolateral direction, $J_{ap \to bl}$, was almost the same as that of CLA and GA (0.13 nmol/min mg of protein, apical pH, 6.0; basolateral pH, 7.4), and was inversely correlated with the paracellular permeability. These results indicate that the transepithelial transport of RA occurs via the paracellular pathway ((10–12), Table 15.2). The affinity of each PA for MCT was found to decrease in the following order: FA, PCA > CA > CLA, GA, RA. Hydroxylation of CNA or BA decreased the affinity for MCT. These results also indicate that the method for evaluating an MCT substrate using FC is both robust and accurate (21). Esterification of PAs, observed for CLA and RA, is thought to prevent interaction with the MCT molecule (11,12).

15.3.3 Artepillin C (AC)

Artepillin C (AC) is a highly hydrophobic PA with two prenyl groups and is one of the principal PAs found in Brazilian propolis. AC exhibits various biological activities, such as antibacterial, antiviral, and anticarcinogenic properties (25). The permeation rate of AC (0.1 mM) across Caco-2 cell monolayers in the basolateral-to-apical direction, $J_{bl \to ap}$, in the presence of H^+ gradient (apical pH, 6.0; basolateral pH, 7.4) was 0.14 nmol/min mg of protein, whereas $J_{bl \to ap}$ in the absence of H^+ gradient (apical pH, 7.4; basolateral pH, 7.4) was 1.14 nmol/min mg of protein ((13), Fig. 15.1). The latter value is almost the same as the permeation rate in the apical-to-basolateral direction, $J_{ap \to bl}$, both in the presence and absence of H^+ gradient (0.91 and 0.83 nmol/min mg of protein) ((13), Fig. 15.1). In the presence of H^+ gradient, $J_{ap \to bl}$ was almost constant, irrespective of NaN_3 or BA. However, $J_{bl \to ap}$ dramatically increased upon addition of NaN_3 or BA, specifically to the apical side (13). In both the presence and absence of H^+ gradient, $J_{ap \to bl}$ also appeared to be constant irrespective of the paracellular permeability ((13), Fig. 15.1). After AC was loaded apically in the presence of H^+ gradient, the intracellular concentration of AC increased with time. This accumulation was inhibited by apically loaded NaN_3 ((13), Fig. 15.1). From these data, AC appears to be absorbed mainly via transcellular passive

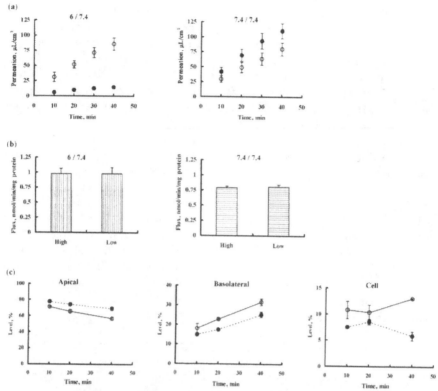

Figure 15.1 Transport characteristics and changes of distribution of AC in Caco-2 cell monolayers. (**a**) Permeation of AC (0.1 mM) from the apical to the basolateral side ($\circ$) and from the basolateral to the apical side ($\bullet$) was measured at 37°C in the presence (left panel) and absence (right panel) of a proton gradient (apical pH, 6.0 or 7.4; basolateral pH, 7.4). Each point is the mean $\pm$ SD of three experiments.(**b**) AC (0.1 mM) was loaded in the apical side and the flux from the apical side to the basolateral side was measured at 37°C in the presence (left panel) and absence (right panel) of a proton gradient (apical pH, 6.0 or 7.4; basolateral pH, 7.4). Each point is the mean $\pm$ SD of three experiments. TER values are indicated as follows: (left panel), high, 1110 $\pm$ 67; low, 273 $\pm$ 8; (right panel), high, 965 $\pm$ 36; low, 259 $\pm$ 13.(**c**) AC (0.1 mM) was loaded in the apical side and changes of distribution with incubation time were measured at 37°C in the presence ($\circ$) and absence ($\bullet$) of 5 mM NaN$_3$ (apical pH, 6.0; basolateral pH, 7.4). Each point is the mean $\pm$ SD of three or more experiments. (Adapted from Reference (13).)

diffusion, although a considerable amount of AC is taken up intracellularly by MCT on the apical side and not transported out across the basolateral membrane. This also suggests that different subtypes of MCT are involved, as is the case for glucose and fructose ((13), Table 15.2, Fig. 15.2).

In conclusion, owing to the affinity for MCT and the liphophilicity of PAs per se, each PA exhibits a range of intestinal absorption pathways: MCT-mediated absorption, partial MCT-mediated absorption, paracellular diffusion, and transcellular passive diffusion (Table 15.2).

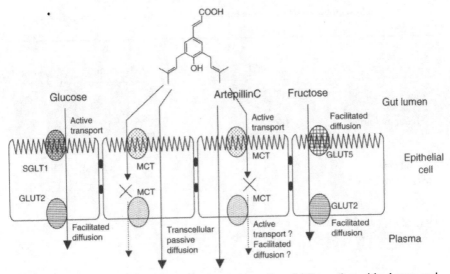

Figure 15.2 Proposed pathways for intestinal absorption of AC together with glucose and fructose. (Adapted from Reference (13).)

15.4 ABSORPTION EFFICIENCY AND BIOAVAILABILITY OF PHENOLIC ACID IN RATS

The absorption efficiency and bioavailability of PCA, CA, GA, RA, and AC *in vivo* have been examined. Because each PA exhibits different absorption characteristics, we have examined the oral administration of individual PAs (100 μmol/kg) in rats under identical experimental conditions. The pharmacokinetic parameters, obtained by measuring the serum concentration of each intact PA in the portal vein and in the abdominal artery, are listed in Table 15.3 (14–16). Comparison of the intact levels of PCA and GA in the portal vein shows that PCA has a high C_{max} (165.7 μmol/L) and AUC (2991.3 μmol min/L), and a low t_{max} (10 min) and $t_{1/2}$ (15.9 min). In contrast, t_{max} (60 min) and $t_{1/2}$ (42.6 min) for GA is high, while C_{max} (0.71 μmol/L) and AUC (42.6 μmol min/L) are extremely low. The concentration of intact PCA was about 230-fold greater than that of GA in the portal vein, and the relative bioavailability of PCA against GA was estimated to be 70-fold greater (the ration of AUC) (14). These findings indicate that PCA is absorbed efficiently, whereas the absorption of GA is relatively slow. The absorption characteristics observed *in vivo* are different between PCA and GA, as is the case *in vitro*. In the abdominal artery, t_{max} and C_{max} of intact PCA were 10 min and 99.3 ± 22.4 μmol/L, respectively, whereas the pharmacokinetic parameters of intact GA could not be determined because only a trace amount of intact GA was detected. These differences in the absorption efficiency and bioavailability observed in PCA and GA *in vivo* are consistent with the differences in the absorption characteristics and affinity for MCT *in vitro* (10,14). A simultaneous excess dose of BA (10 mmol/kg) and PCA (100 μmol/kg) given to rats was found to inhibit

TABLE 15.3 Pharmacokinetic Parameters for Each Intact PA After Oral Administration of 100 μmol/kg

	C_{max}, μmol/L	t_{max}, min	$AUC_{0-1.5h}$, μmol min L	$t_{1/2}$, min
PCA				
Portal vein	165.7	10	2991.3	15.9
Abdominal artery	99.32	10	1703.7	11.6
GA				
Portal vein	0.71	60	42.6	24.1
Abdominal artery	—	—	—	—
CA				
Portal vein	11.24	10	585.0	34.8
Abdominal artery	2.27	10	109.7	34.3
RA				
Portal vein	1.36	10	60.4	56.9
Abdominal artery	0.46	5	16.6	63.9
AC				
Portal vein	19.67	5	182.6	86.6
Abdominal artery	0.55	5	7.72	23.2

Abbreviations: PCA, p-coumaric acid; CA, caffeic acid; GA, gallic acid; RA, rosmarinic acid; AC, artepillin C; C_{max}, maximum serum concentration; t_{max}, time to reach the C_{max}; AUC, area under the serum concentration–time curve; $t_{1/2}$, elimination half-life; —, not measured. From References 14–16.

the absorption of intact PCA, whereas mannitol (10 mmol/kg), a compound that is paracellularly absorbed, did not. This indicates that PCA shares the common transport carrier with BA *in vivo*, as already shown by the *in vitro* studies (10,14). Similarly, comparing the values of AUC, t_{max} and C_{max} for other PAs, the absorption efficiency increases in the order GA = RA < CA < AC < PCA, while the bioavailability increases in the order: GA = RA = AC < CA < PCA ((14–16), Table 15.3). Except in the case of AC, the absorption characteristics and the affinity for MCT *in vitro* display a good correlation with the absorption efficiency and bioavailability *in vivo*. These results highlight the fact that MCT-mediated absorption and distribution operates in the body, and contributes to exhibiting the health effects of various PAs in humans.

In spite of the relatively high absorption efficiency of AC, its bioavailability is quite low ((16), Table 15.3). Xenobiotic compounds with an affinity for biomembranes could potentially permeate across the epithelium of enterocytes to cause harm. However, specific elimination mechanisms to preserve homeostasis, such as glucuronidation or sulfation followed by efflux into the GI tract or further degradation in the liver, can lessen the harmful effects of these compounds. In fact, the ratio of $AUC_{abdominal}$ to AUC_{portal} for intact AC and PCA was 0.04 and 0.57, respectively, which suggests that intact AC is much more susceptible than intact PCA to hepatic elimination ((14,16), Table 15.3). This is consistent with an *in vivo* mechanism for eliminating AC, or other PAs with a high affinity for biomembranes.

PAs transported and distributed in the peripheral tissues by MCT are also susceptible to glucuronidation or sulfation, although the extent of conjugation varies depending on the precise chemical structure. The methoxy and hydroxyl groups on aromatic moiety of FA and VA are presumably good targets for conjugation enzymes, such as UDP-glucuronosyltransferase and sulfotransferase (26). These enzymes are also effective in conjugating catechol groups such as the 3,4-dihydroxy structure of CA (27). The ratio of each intact PA in the portal vein to that in the abdominal artery decrease in the following order: PCA (0.57) > GA = RA (0.27) > CA (0.19) > AC (0.04) ((14–16), Table 15.3). These findings indicate that the bioavailability of each PA is also affected by the susceptibility to conjugation enzymes in addition to the affinity for MCT and the liphophilicity per se.

15.5 ABSORPTION CHARACTERISTICS OF COLONIC METABOLITES OF POORLY ABSORBED POLYPHENOLS *IN VITRO*

Recent studies have indicated that in order to fully assess the health benefits of dietary polyphenols, the biological properties of both the ingested parent compound and its microbial metabolites must be investigated. Indeed, the absorption efficiency of intact polyphenols (i.e., catechin, quercetin, and CA) is known to be low. In general, the phenolic degradation products formed by gut microflora can be classified as hydroxylated phenylpropionic acids or hydroxylated phenylacetic acids, depending on the structural characteristics of the parent polyphenol (28). For instance, it have been reported that the hydroxylated phenylpropionic acids, such as *m*-hydroxyphenylpropionic acid (mHPP) and 3,4-dihydroxyphenylpropionic acid (DHPP), can be formed from CA, catechin, procyanidin or naringin, while quercetin was metabolized to give hydroxylated phenylacetic acids, such as 4-hydroxy-3-methoxyphenylacetic acid (HMPA), *m*-hydroxyphenylacetic acid (mHPA), and 3,4-dihydroxyphenylacetic acid (DHPA) (17,18,28–30). Currently there is a paucity of data on the absorption and distribution of these compounds within the body. We have attempted to assess the biological effects of microbial metabolites of ingested polyphenols. By way of an example, the absorption characteristics of the metabolites of CA and quercetin have been examined in the Caco-2 cell system (9–13).

m-Coumaric acid (MCA) and mHPP display the same characteristics as those observed in the case of FA and PCA, that is, H^+-driven polarized saturable transport in the apical-to-basolateral direction (Michaelis constant for MCA and mHPP of 32.5 and 12.9 mM, respectively; maximum velocity for MCA and mHPP, 204.3 and 91.2 nmol/min mg of protein, respectively). Preferred distribution of apically loaded MCA and mHPP into the basolateral side in the presence of H^+ gradient was observed. Furthermore, significant inhibition of MCA and mHPP transport was seen in the presence of specific MCT substrates ((17), Table 15.4). The $J_{ap \to bl}$ of 1 mM for MCA and mHPP in the presence of H^+ gradient (7.23 nmol/min mg of protein for MCA and 7.47 nmol/min mg of protein for mHPP) is similar to that for FA

TABLE 15.4   **Absorption Characteristics of Microbial Metabolites of Polyphenols**

MCA, mHPP	MCT-mediated active transport H^+-driven polarized transport in apical-to-basolateral direction. Saturable transport. Inhibition by MCT substrate. $J_{ap \to bl}$ of 1 mM of MCA, 7.23 nmol/min mg of protein
HMPA, mHPA	$J_{ap \to bl}$ of 1 mM of mHPP, 7.47 nmol/min mg of protein MCT-mediated active transport H^+-driven polarized transport in apical-to-basolateral direction. Nonsaturable transport. Constant $J_{ap \to bl}$ irrespective of the paracellular permeability. Inhibition by MCT substrate. $J_{ap \to bl}$ of 1 mM of HMPA, 1.83 nmol/min mg of protein.
DHPP	$J_{ap \to bl}$ of 1 mM of mHPA, 1.17 nmol/min mg of protein. Paracellular passive diffusion with some affinity for MCT H^+-driven polarized transport in apical-to-basolateral direction. Nonsaturable transport. Inversely correlated with paracellular permeability. Inhibition by MCT substrate.
DHPA	$J_{ap \to bl}$ of 5 mM of DHPP, 2.27 nmol/min mg of protein. Paracellular passive diffusion. Nonpolarized and nonsaturable transport. Inversely correlated with paracellular permeability. $J_{ap \to bl}$ of 5 mM of DHPA, 0.54 nmol/min mg of protein

Abbreviations: MCA, *m*-coumaric acid; mHPP, *m*-hydroxyphenylpropionic acid; HMPA, 4-hydroxy-3-methoxyphenylacetic acid; mHPA, *m*-hydroxyphenylacetic acid; DHPP, 3,4-dihydroxyphenylpropionic acid; DHPA, 3,4-dihydroxyphenylacetic acid. From references (17 and 18).

(9.79 nmol/min mg of protein), but greater than that for PCA (3.73 nmol/min mg of protein) ((9,10), Table 15.4).

In contrast, HMPA and mHPA exhibit different profiles from those of MCA and mHPP. HMPA and mHPA show H^+-driven polarized nonsaturable transport in the apical-to-basolateral direction (18). In the presence of H^+-gradient, the amount of apically loaded HMPA and mHPA transported out into the basolateral side increases in the following order: CA (1.57%) < mHPA (2.67%) < HMPA (5.51%) < MCA (13.74%) and mHPP (16.90%). The $J_{ap \to bl}$ of 1 mM for HMPA, mHPA, and CA in the presence of H^+ gradient are 1.83, 1.17, and 0.46 nmol/min mg of protein, respectively. Both HMPA and mHPA transport in the apical-to-basolateral direction is inhibited by typical MCT substrates, but is constant irrespective of paracellular permeability. This is different from the transport characteristics of CA ((11,17,18), Table 15.4).

On the contrary, the absorption characteristics of DHPP are identical to those of CA, that is, H^+-driven polarized transport in the apical-to-basolateral direction, nonsaturable transport, but inversely correlated with paracellular permeability (11,17). Furthermore, the $J_{ap \to bl}$ of DHPP and CA (at 5 mM; 2.27 nmol/min mg of protein for DHPP; 2.69 nmol/min mg of protein for CA) and the amount of apically loaded DHPP and CA transported out into the basolateral side (CA, 1.57%; DHPP, 1.71%) in the presence of H^+ gradient also support the above conclusion ((11,17), Table 15.4).

Although $J_{ap \to bl}$ of 5 mM DHPA in the presence of H^+ gradient (0.54 nmol/min mg of protein) is greater than that of GA, RA, and CLA (0.13–0.16 nmol/min mg of protein), DHPA shows nonpolarized and nonsaturable transport in the apical-to-basolateral direction, and is inversely correlated with paracellular permeability as observed for GA, RA, and CLA. This clearly indicates that DHPA is paracellularly transported, as is the case with GA, RA, and CLA ((11,12,18), Table 15.4). Furthermore, the affinity of microbial metabolites, such as mHPP or MCA, matches the level of inhibition of FC transport. These observations verify the method of assessing the affinity for MCT using FC and the general predictive rules based on structural characteristics of an MCT substrate (11,21).

In summary, the absorption characteristics of various PAs and microbial metabolites of polyphenols can be classified into five groups: (a) MCT-mediated active transport (FA, PCA, MCA, and mHPP); (b) MCT-mediated active transport with lower affinity for MCT than (a) (HMPA and mHPA); (c) paracellular passive diffusion with some affinity for MCT (CA and DHPP); (d) paracellular passive diffusion with no affinity for MCT (GA, RA, CLA, and DHPA); and (e) transcellular passive diffusion (AC) (Fig. 15.3). The knowledge gained from these studies can be used to accurately predict the absorption and bioavailability profiles of other PAs or microbial metabolites of polyphenols. These studies exemplify the physiological

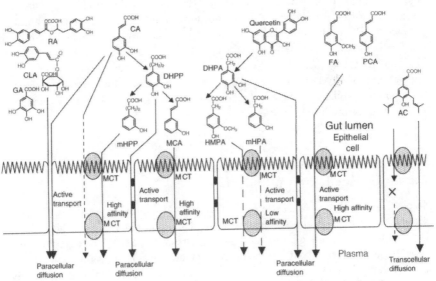

Figure 15.3 Proposed pathways for intestinal absorption of ingested polyphenols.

significance of MCT-mediated absorption to give a fuller understanding of the health benefits of dietary polyphenols.

15.6   CURRENT KNOWLEDGE AND STATUS OF THE MCT-MEDIATED TRANSPORT SYSTEM

15.6.1   Gastric Absorption

Recent studies have demonstrated that gastric absorption occurs for some flavonoids such as quercetin, daizein, and anthocyanins (31–33). Thus, the significance of gastric absorption in terms of the health effects of dietary phytochemicals is becoming an increasingly important issue. Various PAs with different affinities for MCT (i.e., PCA, FA, CA, GA, and CLA) were administered (2.25 μmol) into rat stomach *in situ* and then the plasma concentration of the PA in the portal vein and the abdominal artery was measured (26). The plasma concentration of FA peaked 5 min after administration in the stomach. At 5 min after administration, the plasma concentration of each PA increased in the following order: GA = CLA < CA < PCA = FA. This order matches their respective affinity for MCT, indicating that the MCT-mediated transport system operates in the gastric absorption of PAs, similar to the intestinal absorption mentioned earlier (14–16,26). Furthermore, the ratio of the proportion of each intact PA to each total PA in the abdominal artery to that in the portal vein decrease in the following order: PCA (0.78) > CA (0.44) > FA (0.29) (26). These results suggest that CA and FA are subject to conjugation by glucuronosyltransferases and/or sulfotransferases. Garrait et al. have also demonstrated that CNA and PCA are absorbed across the epithelium via MCT in all GI organs, stomach, duodenum, jejunum, ileum, cecum and colon, although the stomach is a minor absorption site compared to all the other digestive organs (34). The pK_a values for PAs suggest that both protonated and dissociated forms are likely to exist in the stomach under physiological conditions. Dissociated PA, as the monocarboxylic anion, will be recognized and transported by MCT, whereas the protonated PA is subject to transcellular passive diffusion by means of affinity for the biomembrane. Further studies to characterize both transport systems are required to fully assess the health effects of dietary phenolics. In particular, it will be important to elucidate the relative contribution of the two transport systems to the gastric absorption of dietary PAs.

15.6.2   MCT Subtype Responsible for Transport of PAs and Microbial Metabolites of Polyphenols

To date, 14 isoforms of MCT have been identified (35), although only MCT1–MCT4 are characterized in terms of their substrate and inhibitor kinetics. Each MCT isoform is likely to have a unique biological role, which is related to the different tissue and subcellular distribution. MCT1, the best characterized isoform, is found in almost all tissues of the human body (i.e., heart, skeletal muscle, small intestine, colon, liver, brain, spinal cord, testis, ovary, placenta, and adrenal gland) (36), which implies a physiological significance for humans. Recent analysis using MCT1 expressing Caco-2 cells and oocytes has demonstrated that MCT1 does not participate in the transport of PA (37). This is consistent with the observation that lactic acid, a typical substrate for MCT1–MCT4, has no effect on the MCT-mediated transport of PAs such as FA and PCA (9,10). These findings indicate the possible involvement of a novel MCT subtype other than MCT1–MCT4 in the transport of PAs and microbial metabolites of polyphenols. Various subtypes of MCT, such as MCT1, MCT3, MCT4, MCT5, and MCT6, are expressed in Caco-2 cells (38). Furthermore, it has recently been reported that different subtypes of MCTs participate in the absorption of PA in Caco-2 cells: MCT on the apical side appears to be different from MCT on the basolateral side in terms of affinity for AC ((13), Fig. 15.2). Further studies are required to establish the MCT subtype responsible for gastric and intestinal absorption of PA and microbial metabolites of polyphenols. Thereafter, tissue distribution and subcellular localization of the relevant MCT subtype will have to be determined.

15.6.3   Concept of Metabonutrients

The most abundant types of polyphenol in the human diet are the flavonoids and PAs. The MCT-mediated transport system is involved in the absorption of both PAs and the microbial metabolites of flavonoids that are poorly absorbed and reach the colon. Thus, the MCT-mediated transport system plays an important physiological role in humans. The antioxidant activity of PAs and the microbial metabolites of flavonoids are generally lower than that of flavonoids. However, the high antioxidant activity of flavonoids can result in cytotoxic effects. The high affinity of flavonoids for biomembranes allows them to readily permeate across the epithelium of enterocytes. Several mechanisms are employed to reduce cytotoxicity resulting from an influx of flavonoids, such as conjugation by sulfation or glucuronidation, energy-dependent efflux by specific efflux pumps (e.g., multidrug resistance associated protein 2 or p-glycoprotein), and hepatic metabolism/elimination involving cytochrome-P450 enzymes. To preserve homeostasis preferably, it is more reasonable for living things to absorb and utilize dietary constituents with lower antioxidant activity efficiently.

Short chain fatty acids (SCFA), such as acetate, propionate and butyrate, are the most abundant organic anions in the colonic lumen. SCFA are also microbial metabolites of undigested carbohydrates and proteins, as well as poorly absorbed polyphenols mentioned above. Butyrate is the principal energy source for colonic epithelial cells (39), exerting a number of biological effects to induce cell differentiation and regulate growth and proliferation of the colonic mucosal epithelia (40,41). The mechanism of absorption of butyrate is still not fully resolved. MCT1 is expressed at the protein level predominantly in the apical membrane of colonic epithelial cells, and can recognize and transport butyrate (38), although the nonelectrogenic $SCFA^-/HCO3^-$ antiporter is also implicated in its absorption (42). By extension, during the transition from normality to malignancy in human colon, MCT1 expression is significantly reduced. By contrast, the expression of MCT1 is upregulated by its substrate, butyrate, and contributes to the ability of butyrate to induce cellular metabolism and for the maintenance of colonic tissue homoeostasis (43,44).

In summary, SCFA and hydroxylated phenylpropionic acids or hydroxylated phenylacetic acids are also colonic microbial metabolites of poorly absorbed food materials that exert various biological activities. These microbial metabolites are absorbed and distributed in the body by the MCT-mediated transport system to exert their effects on human health. It may be useful to classify these microbial metabolites as metabonutrients because this nomenclature distinguishes them from classical nutrients, such as sugars, lipids, amino acids, minerals, and vitamins (18). Until recently, research concerning the physiological function of food has focused on the biologically active constituents of the food itself. In future the metabonutrients, which are generated by colonic microflora from ingested food materials, should also be investigated for their health benefits.

15.7   OVERVIEW OF ABSORPTION AND BIOAVAILABILITY OF WHEAT ANTIOXIDANTS: FUTURE STUDIES

15.7.1   SRA, SPA, VA, and PBA

Until recently, there has been a paucity of information concerning the absorption characteristics or pharmacokinetic parameters of SRA, SPA, VA, and PBA. Furthermore, the bioavailability of these compounds is unknown. On the basis of the deduced novel principle of absorption and bioavailability for PAs, however, we can now estimate their absorption efficiency and bioavailability. All these compounds fulfill the structural criteria for an MCT substrate, that is, monoanionic carboxyl group and an aromatic hydrophobic moiety. Inhibition of FC transport by each PA increases in the following order: SRA (104.8%) < SPA (74.9%) < VA (55.2%) < PBA (34.5%) ((21), Table 15.1). Compared to the inhibition of FC transport by CA (115.8%), PCA (84.1%) and FA (51.3%), the affinity for MCT of each PA increases in the following order: CA < SRA < PCA, SPA < VA = FA < PBA. Therefore, SPA, VA, and PBA are presumed to be actively absorbed by the MCT-mediated transport system. Absorption of SRA is thought to be mainly via paracellular diffusion although it is absorbed to a lesser extent by MCT, as is the case with CA (11). Susceptibility to conjugation enzymes, such as glucuronosyltransferase and sulfotransferase, also affects their bioavailability. VA, having a methoxy group in addition to the hydroxyl group on the aromatic ring, is likely to be a good target for conjugation (26). The affinities of VA and FA for MCT, together with their susceptibility to conjugation, suggest that the bioavailability of these two compounds should be similar. Although the absorption efficiency of SRA is similar to that of CA, susceptibility to conjugation is different. SRA lacks a catechol moiety and is thus less likely to be subject to conjugation (27). These findings suggest a greater bioavailability of SRA over CA. It is believed that the bioavailability of PBA, SPA, and SRA is similar to that of PCA, whereas the bio-availability of VA is similar to that of FA. Germano et al. reported that after rats were orally treated with hydrolyzed extract of *Trichilia emetica* root including various PAs, such as free CA, SRA, GA, VA, and PCA, the absorption of free VA was relatively fast

TABLE 15.5 Relative Bioavailability of Eight Phenolics in Hamsters that were Fed 40 μmol Total Phenolics of Oat Bran Phenol-rich Powder

Relative	Oral dose[a], μmol	Plasma C_{max}[b], μmol/L	Plasma C_{max}/oral dose, (μmol/L)/μmol	Apparent relative bioavailability[c]
p-Coumaric acid	0.32	1.55 ± 0.91	4.84 ± 2.84	100
Sinapic acid	0.06	0.26 ± 0.38	4.30 ± 6.30	89.5
Syringic aicd	0.10	0.38 ± 0.25	3.80 ± 2.50	78.5
p-Hydroxybenzoic acid	0.03	0.10 ± 0.04	3.33 ± 1.33	68.8
Ferulic acid	0.50	1.20 ± 1.08	2.40 ± 2.60	49.5
Vanillic acid	0.13	0.15 ± 0.05	1.20 ± 0.38	23.8
Avenanthramide A	0.63	0.04 ± 0.03	0.06 ± 0.05	1.3
Avenanthramide B	0.49	0.03 ± 0.02	0.06 ± 0.04	1.3

[a]Oral dose is the absolute amount of each phenolic compound fed to each hamster.
[b]Values are means ± SD, $n = 5$.
[c]The ratio of plasma C_{max}/oral dose for p-coumaric acid was arbitrarily set at 100. (Adapted from Reference (46).)

and efficient (45). Chen et al. used hamsters fed oat bran phenol-rich powder to determine the apparent bioavailability of total SPA, SRA, PBA, VA, FA, and PCA, including free and conjugated forms of each PAs, by calculating the ratio of plasma C_{max}/oral dose ((46), Table 15.5). The anticipated absorption and bioavailability of these PAs, based on our novel FC-assay protocol, are broadly consistent with these two reports. However, the absorption and bioavailability obtained from these studies is influenced by a number of factors (i.e., effects of various PAs dosed simultaneously, and substantial differences of bioavailability between intact and total PAs). To determine the absorption and bioavailability of these PAs accurately, it is desirable to perform both *in vitro* and *in vivo* studies. This has been done for PCA and CA (10,11,14,15).

15.7.2 Free, Soluble Conjugate, and Insoluble Bound PAs in Wheat

The major portion of PAs in grains, such as wheat, rice and corn, are characteristically esterified or etherified with arabinose or galactose residues in the pectic or hemi-cellulosis components of cell wall and exist in an insoluble bound form (85% in corn, 75% in wheat, and 62% in rice) (4). Free, soluble conjugated and insoluble bound FAs in wheat are present in the ratio 0.1:1:100 (4). Moreover, bound phenolics of both PAs and flavonoids have been found to be the major contributors to the total antioxidant activity: 90% in wheat, 87% in corn, and 71% in rice (4). Hence, to fully understand the health effects of dietary PAs in wheat-derived foodstuffs, the absorption and bioavailability of soluble conjugated and insoluble bound PAs, in addition to free PA, should be considered.

The majority of bound FA in an insoluble fiber, such as wheat bran, is thought to avoid stomach or intestinal digestion/absorption to reach the colon. Once in the colon, colonic microbial enzymes, such as xylanase and esterase, solubilize and release feruloylated oligosaccharides (i.e., FAA; 5-o-feruloyl-arabinofuranose, FAXn;

feruloyl-arabinoxylane) or free FA (47,48). During transit through the GI tract to the colon, a portion of bound FA is attacked by an esterase from the mucosa to release feruloylated oligosaccharides or free FA (47,48). Ingested free PAs, together with the newly liberated PAs, are absorbed across the GI epithelium into the bloodstream and distributed into various tissues. Soluble conjugated PAs, such as FAA, are substrates for mucosal and microbial esterases to generate free PA for absorption (49,50). The potential mechanism for absorption and distribution of PAs in wheat foodstuffs are presented in Fig. 15.4.

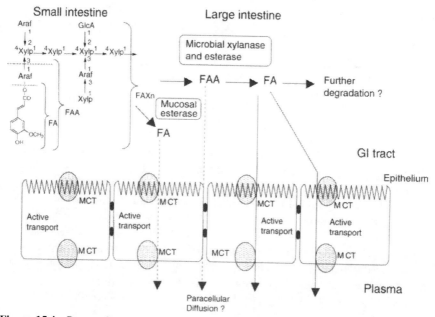

Figure 15.4   Proposed routes for absorption and metabolism of insoluble fibers such as wheat bran.

TABLE 15.6   Factors Affecting Absorption and Bioavailability of Wheat Derived Foodstuffs

A. absorption across GI epithelium of free PA
   affinity for MCT and the biomembrane

B. metabolism of free PA in tissues
   affinity for MCT and susceptibility to conjugating enzymes

C. form of ingested wheat PA
   content of free, soluble conjugated and insoluble bound PAs

D. release of free PA
   enzymatic attack on soluble conjugated and insoluble bound PAs and its site through the GI tract.

In summary, various forms of wheat PAs (i.e., free, soluble conjugated, and insoluble bound forms) are converted to free PAs by mucosal or colonic microbial enzymatic hydrolysis. Free PAs are then absorbed into the blood circulation and further distributed and metabolized into tissues according to the susceptibility to conjugating enzymes (Fig. 15.4). Various factors affect absorption and bioavailability of wheat PAs at each stage, which have previously never been systematically recognized (Table 15.6). These include (a) absorption across the GI epithelium of free PA, affinity for MCT and the biomembrane; (b) metabolism of free PA in tissues, affinity for MCT and susceptibility for conjugating enzymes; (c) form of ingested wheat PA, content of free, soluble conjugated and insoluble bound PAs; and (d) release of free PA, enzymatic attack on soluble conjugated forms and insoluble bound PAs and its site through the GI tract. In addition to these factors, the effect of the ingested food matrix with wheat PAs has also been the focus of recent research (51). Further detailed studies to investigate the effects of each factor on the absorption and bioavailability of wheat PAs will help to assess the health benefits of whole grains such as wheat and rice.

ABBREVIATIONS

PA	phenolic acid
MCT	monocarboxylic acid transporter
CA	caffeic acid
FA	ferulic acid
SRA	syringic acid
PCA	*p*-coumaric acid
SPA	synapic acid
VA	vanillic acid
PBA	*p*-hydroxybenzoic acid
BA	benzoic acid
GI	gastrointestinal
CNA	cinnamic acid
ECD	electrochemical detector
CLA	chlorogenic acid
GA	gallic acid
RA	rosmarinic acid
AC	artepillin C
mHPP	*m*-hydroxyphenylpropionic acid
DHPP	3,4-dihydroxyphenylpropionic acid
HMPA	4-hydroxy-3-methoxyphenylacetic acid
mHPA	*m*-hydroxyphenylacetic acid
DHPA	3,4-dihydroxyphenylacetic acid
MCA	*m*-coumaric acid
SCFA	short chain fatty acid
FAA	5-*o*-feruloyl-arabinofuranose
FAXn	feruloyl-arabinoxylane

REFERENCES

1. WHO/FAO.Diet, nutrition and the prevention of chronic diseases.World Health Organization,Geneva, **2003**.
2. Temple, N. J. *Nutr. Res.* **2000**, *20*, 449–459.
3. Willet, W. C. *Science* **1994**, *254*, 532–537.
4. Adom, K. K.; Liu, R. H. *J. Agric. Food Chem.* **2002**, *50*, 6182–6187.
5. Manach, C.; Saclbert, A.; Morand, C.; Remesy, C.; Jimenez, L. *Am. J. Clin. Nutr.* **2004**, *79*, 727–747.
6. Sosulski, F.; Krygier, K.; Hogge, L. *J. Agric. Food Chem.* **1982**, *30*, 337–340.
7. Hatcher, D. W.; Kruger, J. E. *Cereal Chem.* **1997**, *74*, 337–343.
8. Scalbert, A.; Johnson, I. T.; Saltmarsh, M. *Am. J. Clin. Nutr.* **2005**, *81*, 215S–217S.
9. Konishi, Y.; Shimizu, M. *Biosci. Biotechnol. Biochem.* **2003**, *67*, 856–862.
10. Konishi, Y.; Kobayashi, S.; Shimizu, M. *Biosci. Biotechnol. Biochem.* **2003**, *67*, 2317–2324.
11. Konishi, Y.; Kobayashi, S. *J. Agric. Food Chem.* **2004**, *52*, 2518–2526.
12. Konishi, Y.; Kobayashi, S. *Biosci. Biotechnol. Biochem.* **2005**, *69*, 583–591.
13. Konishi, Y. *Biochim. Biophys. Acta* **2005**, *1713*, 138–144.
14. Konishi, Y.; Hitomi, Y.; Yoshioka, E. *J. Agric. Food Chem.* **2004**, *52*, 2527–2532.
15. Konishi, Y.; Hitomi, Y.; Yoshida, M.; Yoshioka, E. *J. Agric. Food Chem.* **2005**, *53*, 4740–4746.
16. Konishi, Y.; Hitomi, Y.; Yoshida, M.; Yoshioka, E. *J. Agric. Food Chem.* **2005**, *53*, 9928–9933.
17. Konishi, Y.; Kobayashi S. *J. Agric. Food Chem.* **2004**, *52*, 6418–6424.
18. Konishi, Y. *J. Agric. Food Chem.* **2005**, *53*, 601–607.
19. Konishi, Y.; Hagiwara, K.; Shimizu, M. *Biosci. Biotechnol. Biochem.* **2002**, *66*, 2449–2457.
20. Rahman, B.; Schneider, H. P.; Broer, A.; Deitmer, J. W.; Broer, S. *Biochemistry* **1999**, *38*, 11577–11584.
21. Konishi, Y.; Kubo, K.; Shimizu, M. *Biosci. Biotechnol. Biochem.* **2003**, *67*, 2014–2017.
22. Guo, C.; Cao, G.; Sofic, E.; Prior, R. L. *J. Agric. Food Chem.* **1997**, *45*, 1787–1796.
23. Harbowy, M. E.; Ballentine, D. A. *Crit. Rev. Plant Sci.* **1997**, *16*, 415–480.
24. Petersen, M.; Simonds, M. S. *Phytochemistry* **2003**, *62*, 121–125.
25. Burdock, G. A. *Food Chem. Toxicol.* **1998**, *36*, 347–363.
26. Konishi, Y.; Zhao, Z.; Shimizu, S. M. *J. Agric. Food Chem.* **2006**, *54*, 7539–7543.
27. Nardini, M.; Natella, F.; Sccaccini, C.; Ghiseli, A. *J. Nutr. Biochem.* **2006**, *17*, 14–22.
28. Rechner, A. R.; Smith, M. A.; Kuhnle, G.; Gibson, G. R.; Debnam, E. S.; Srai, S. K. S.; Moore, K. P.; Rice-Evans, C. A. *Free Radical Biol. Med.* **2004**, *36*, 221–225.
29. Rechner, A. R.; Kuhnle, G.; Bremner, P.; Hubbard, G. P.; Moore, K. P.; Rice-Evans, C. A. *Free Radical Biol. Med.* **2002**, *33*, 220–235.
30. Gonthier, M. P.; Donovan, J. L.; Texier, O.; Felgines, C.; Remesy, C.; Scalbert, A. *Free Radical Biol. Med.* **2003**, *35*, 837–844.
31. Crespy, V.; Morand, C.; Besson, C.; Manach, C.; Demigne, C.; Remesy, C. *J. Agric. Food Chem.* **2002**, *50*, 618–621.
32. Piskula, M. K.; Yamakoshi, J.; Iwai, Y. *FEBS Lett.* **1999**, *447*, 287–291.
33. Passamonti, S.; Vrhovsek, U.; Vanzo, A.; Mattivi, F. *FEBS Lett.* **2003**, *544*, 210–213.
34. Garrait, G.; Jarrige, J. F.; Blanquet, S.; Beyssac, E.; Cardot, J. M.; Alric, M. *J. Agric. Food Chem.* **2006**, *54*, 2944–2950.
35. Halestrap, A. P.; Meredith, D. *Pflugers Arch.* **2004**, *447*, 619–628.
36. Lin, R. Y.; Vera, J. C.; Chaganti, R. S. K.; Golde, D. *J. Biol. Chem.* **1998**, *273*, 28959–28969.
37. Watanabe, H.; Yashiro, T.; Tohjo, Y.; Konishi, Y. *Biosci. Biotechnol. Biochem.* **2006**, *70*, 1928–1933.
38. Hadjiagapiu, C.; Schmidt, L.; Dudeja, P. K.; Layden, T. J.; Ramaswamy, K. *Am. J. Physiol. Gastrointest. Liver Physiol.* **2000**, *279*, G775–G780.
39. Cummings J. H. *Scand. J. Gastroenterol.* **1984**, *19*, 89–99.
40. Roediger, W. E. W. *Gut.* **1980**, *21*, 793–798.
41. Scheppach, W.; Bartram, H. P.; Richter, A.; Richter, F.; Liepold, H.; Dusel, G.; Hofstetter, G.; Ruthlein, J.; Kasper, H. *J. Parent Enter. Nutr.* **1992**, *16*, 43–48.
42. Stein, J.; Schroder, O.; Milovic, V.; Caspary, W. F. *Gastroenterology* **1995**, *108*, 673–679.
43. Lambert, D. W.; Wood, I. S.; Ellis, A.; Shirazi-Beechey, S. P. *Br. J. Cancer* **2002**, *86*, 1262–1269.

44. Cuff, M. A.; Shirazi-Beechey, S. P. *Biochem. Soc. Trans.* **2004**, *32*, 1100–1102.

45. Germano, M. P.; Angelo, V. D.; Biasini, T.; Sanogo, R .,De Pasquale, R.; Catania, S. *J. Ethnopharmacal.* **2006**, *105*, 368–373.

46. Chen, C. Y.; Milbury, P. E.; Kwak, H. K.; Collins, F. W.; Samuel, P.; Blumberg, J. B. *J. Nutr.* **2004**, *134*, 1459–1466.

47. Kroon, P. A.; Faulds, C. B.; Ryden, P.; Robertson, J. A.; Williamson, G. *J. Agric. Food Chem.* **1997**, *45*, 661–667.

48. Andreasen, M. F.; Kroon, P. A.; Williamson, G.; Garcia-Conesa, M. T. *Free Radical Biol. Med.* **2001**, *31*, 304–314.

49. Zhao, Z.; Egashira, Y.; Sanada, H. *J. Nutr.* **2003**, *133*, 1355–1361.

50. Zhao, Z.; Egashira, Y.; Sanada, H. *J. Agric. Food Chem.* **2003**, *51*, 5534–5539.

51. Adam, A.; Crespy, V.; Levrat-Verny, M. A.; Leenhardt, F.; Leuillet, M.; Demigne, C.; Remesy, C. *J. Nutr.* **2002**, *132*, 1962–1968.

CHAPTER 16

WHEAT LIGNANS: PROMISING CANCER PREVENTIVE AGENTS[1]

Weiqun Wang
Allan Ayella
Yu Jiang
Ping Ouyang
Hongyan Qu

16.1 INTRODUCTION

Adequate dietary fiber intakes, particularly fiber from cereal grains, have been associated with a low risk of colorectal cancer (1). Wheat bran appears to protect against tumor development more consistently than other sources of plant fiber in a number of experimental colon cancer studies (2–4). Several hypotheses have been established to explain the link between wheat bran and cancer prevention, including the increase of overall GI transit time, dilution of carcinogenic compounds, release of short-chain fatty acids, and promotion of tumor suppressor signaling, among others (5–8). However, the experimental evidence by using different fiber sources or different doses on colon cancer prevention is controversial (9,10). Especially, a study conducted in the Takemoto laboratory showed that antitumor activities of wheat bran from various wheat cultivars were significantly different even when fiber content was equal (11).

Lignans are a group of the phytochemicals that are composed of phenylpropane dimer linked with a 1,4-diarylbutane structure by β-β bonds. Figure 16.1 shows the chemical structure of a few prominent lignans. Lignans not only present abundantly in flaxseed but also present in various grains such as wheat (12). In wheat, lignans are located in the pericarp and aleurone layers with the highest concentration in wheat bran (13). The main lignan in wheat bran is SDG. Table 16.1 lists the contents of SDG in flaxseed, wheat bran, and some other selected plant foods.

As one group of secondary metabolites, lignans are synthesized via phenyl-propanoid pathway. However, the completed biosynthetic pathways to the lignans in wheat are not clear. The Lewis laboratory from the Washington State University

---

[1] This work is supported in part by a USDA Cooperative Project (KS 680-0199184), Agricultural Experiment Station, Kansas State University (contribution No. 07-77-B).

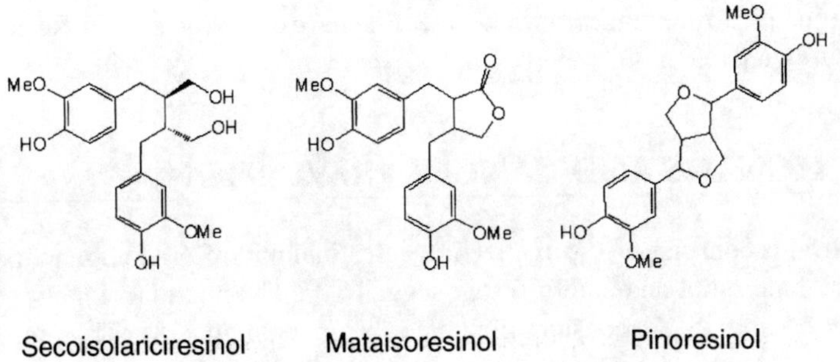

Secoisolariciresinol   Mataisoresinol   Pinoresinol

Figure 16.1   Chemical structures of the prominent lignans: secoisolariciresinol, mataisoresinol, and pinoresinol.

TABLE 16.1   Level of Lignans in Selected Plant Foods

Food	Total lignans, μg/g
Whole flaxseed	636–2213[a]
Flaxseed meal	675[b]
Flaxseed flour	527[b]
Wheat bran	0.5–83[a]
Oat bran	6.5[b]
Soy bean	8.6[b]

[a]The contents of lignans in 10 flaxseed samples (unpublished data) and 4 wheat cultivars were measured by HPLC analysis as described in our previous publication (Qu et al. 2005).
[b]Adapted from Reference (14).

reported a series of studies of the lignan biosynthesis in *Forsythia intermedia* (15–17). As shown in Fig. 16.2, the biosynthetic pathways to secoisolariciresinol occur via coupling of two coniferyl alcohol molecules to afford pinoresinol. Then pinoresinol undergoes sequential reduction by pinoresinol–lariciresinol reductase (PLR) to generate lariciresinol and secoisolariciresinol (15,16). Our ongoing studies are trying to

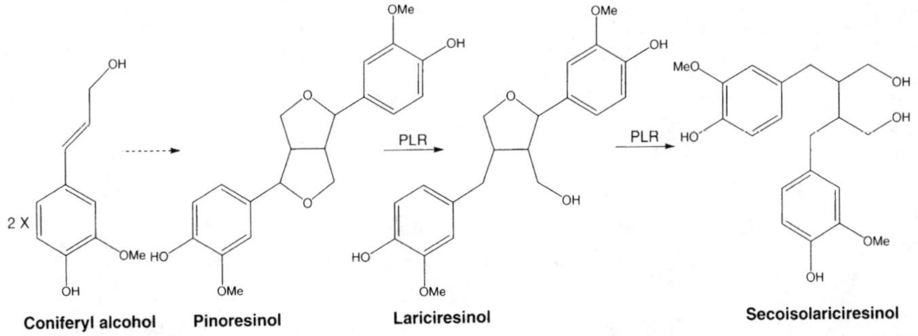

Coniferyl alcohol   Pinoresinol   Lariciresinol   Secoisolariciresinol

Figure 16.2   Schematic overview of the lignan biosynthesis pathway. Pinoresinol–lariciresinol reductase (PLR) is the last enzyme directly downstream of the biosynthetic pathways from coniferyl alcohol to secoisolariciresinol (modified from Refrence 16).

develop genetically engineered wheat lines that have overexpressed PLR and thus may potentially generate a high level of SDG.

16.2 LIGNANS AND CANCER PREVENTION

When SDG is consumed, it is oxidized by intestinal microflora to lignan metabolites including enterodiol and enterolactone (Fig. 16.3). The pharmacokinetics of lignan metabolites after SDG consumption have been reported (18). Therefore, dietary

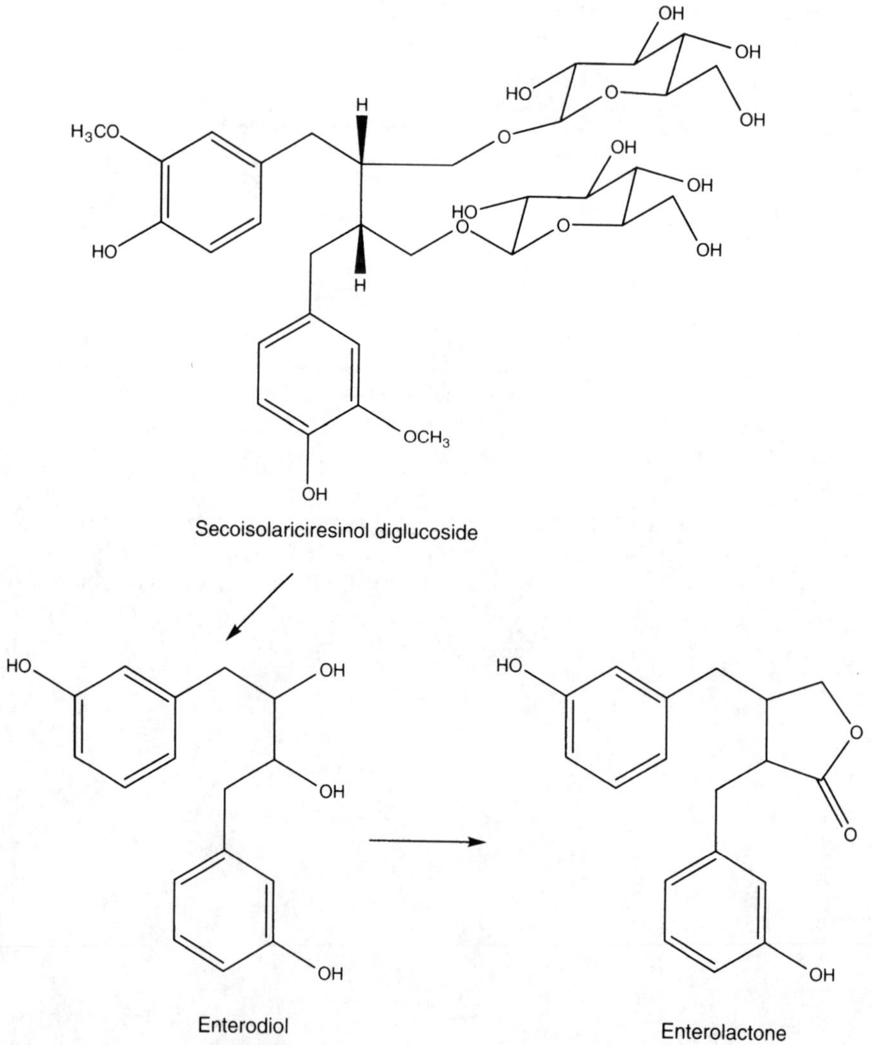

Secoisolariciresinol diglucoside

Enterodiol

Enterolactone

Figure 16.3 Chemicals structures of the secoisolariciresinol diglucoside and its mammalian metabolites: enterodiol and enterolactone.

lignans such as SDG may exert their biological effects through their metabolites. Compelling data from epidemiological, clinical, and experimental studies, along with *in vitro* mechanistic studies, have suggested that lignans may be promising for cancer prevention.

16.2.1   Epidemiological and Clinical Studies

Although some controversial data exist, many epidemiological studies have suggested an inverse relationship between lignan consumption and various cancer risks. Adlercreutz and his team in the University of Helsinki have intensively studied the relationship between plasma or urine levels of lignan metabolites and cancer risk since 1981. They found that a lignan-low diet was related to an increased breast cancer risk in case-control and prospective studies (19). A review paper summarized a conclusion that the most support for a cancer preventive role of dietary lignans was observed for premenopausal breast cancer (20). Furthermore, a case-control study conducted in the Netherlands recently demonstrated that a substantial reduction of colorectal adenoma risk was associated with a high plasma level of lignan metabolites (21). Another case-control study from Sweden further supported a reverse correlation of serum levels of lignan metabolites with prostate cancer risk (22). However, the conflicting results have been reported. A case-control study nested within a prospective cohort study by the New York University Women's Health Study did not find a protective role of circulating lignan enterolactone against breast cancer development (23). A prospective Zutphen Elderly cohort study conducted in the Netherlands did not find an association of a total lignan intake with cancer risk (24). Arts and Hollman (25) reviewed three prospective nested case-control studies and three case-control studies and found an inverse association between lignans, and breast cancer, and observed prostate cancer only in case-control studies but not in prospective studies. The conflicting data have been suggested, at least in part, due to inadequate databases used in dietary lignan estimation (26). It is interesting to note that a clinical intervention study by the Thompson laboratory has shown a potential reduction of breast cancer growth in patients supplemented with dietary flaxseed, the richest source of dietary lignans (27).

16.2.2   Experimental Animal Studies

In comparison with epidemiological studies, more consistent results regarding cancer-preventive role of dietary lignans have been demonstrated in animal models. For example, dietary flaxseed at 10% or purified SDG at an equivalent dose during sucking significantly suppressed later 9,10-dimethyl-1,2-benzanthracene-induced mammary tumorigenesis in rats (28). A study conducted in athymic mice carrying LNCaP human prostate cancer xenografts showed that dietary lignan 7-hydroxymatairesinol at 0.3% inhibited the tumor growth significantly (29). A similar model using athymic mice but with MCF-7 human breast cancer xenografts demonstrated dietary flaxseed at 10% attenuated soy protein-stimulated tumor growth (30). Recently, the Thompson laboratory reported that dietary flaxseed at 10% inhibited metastasis but

TABLE 16.2   Inhibition of Azoxymethane (AOM)-induced Formation of Aberrant Crypt Foci (ACF) in F344 Rats by Dietary Secoisolariciresinol Diglucoside (SDG)

Treatment	No. rats with ACF	Total no. ACF/colon
Control diet	6/6	264 ± 43
0.01% SDG	6/6	$196 \pm 51*$

Male F344 rats at 6 weeks old were fed either control or experimental diet containing SDG at 0.01% for 4 weeks. One day after the first dietary treatment, rats were given s.c. injection with 15 mg/kg B.W. of AOM once per week for 2 weeks. At the end of 4-week dietary treatment, rats were terminated under ether euthanasia and the entire colon was resected for 0.2% methylene blue staining. The number of ACF per colon was scored under a light microscope. Values are means $\pm$ SD, $*P < 0.05$ versus the controls ($n = 6$).

not the recurrence of estrogen receptor negative human breast cancer cells after excision in nude mice (31).

We also conducted an animal study to assess the effects of dietary lignan SDG on azoxymethane-induced aberrant crypt foci (ACF) formation in rat colons. ACF are morphologically altered crypts, alone or in cluster, first identified by microscopic examination of methylene blue-stained whole-mount preparations of colonic mucosa from azoxymethane-treated rodents (32). Studies demonstrated that ACF in humans are important precursors to human colon cancer (33). As shown in Table 16.2, we found that 0.01% SDG significantly reduced the formation of ACF.

16.3   PLAUSIBLE MECHANISMS OF LIGNANS FOR CANCER PREVENTION

As a group of the phytoestrogens, lignans may act through estrogen receptor-mediated mechanisms. Webb and McCullough (20) have reviewed this potential mechanism in details. They also discussed a potential interaction of the phytoestrogenic activity with other growth hormones as well as antioxidant and antiproliferative activities. In fact, our previous studies and others have shown that enterolactone is a strong antioxidant against human LDL oxidation (34,35). We also found that enterolactone was capable of inhibiting colon cancer cell growth and inducing detoxification enzyme activity (36,37). Furthermore, the chemopreventive mechanisms of lignans and lignan metabolites via antiinflammatory and immunosuppressive activities have been intensively reviewed by Saleem et al. (38).

We assessed the cancer preventive mechanisms of two prominent lignan metabolites on a human colonic cancer cell line SW480. Treatment of SW480 cells with enterolactone and enterodiol, alone or in combination, at 0–40 μM resulted in a dose- and time-dependent decrease in cell numbers (37). While the cytotoxicity as measured by tyrpan blue staining was not significantly changed, DNA flow cytometric analysis indicated that treatments induced cell cycle arrest at S phase (Fig. 16.4). Furthermore, apoptosis analysis by TUNEL assay showed an increased percentage of apoptotic cells in the treated cells (Fig. 16.5). These results suggest that inhibition of cancer cell growth by lignan metabolites appears to be mediated with cytostatic and apoptotic mechanisms.

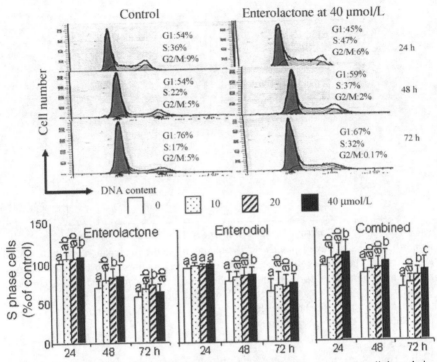

Figure 16.4   Induction of cell cycle arrest at S phase by enterolactone, enterodiol, or their combination. The human colon cancer SW480 cells were cocultured with the indicated lignan metabolites at 0–40 μmol/L for 24–72 h and the cell cycle was measured by DNA flow cytometric analysis. Values are mean ± SD ($n = 5$–6). Means within a treatment without a common letter differ, $P \leq 0.05$. (Adapted from Reference 3.).

Taken together, the observations, direct *in vitro* cell culture studies along with *in vivo* animal experiment, as well as the epidemiological correlation between lignan intake and antitumor activities, suggest that dietary lignans may attribute to the observed cancer prevention. However, most studies so far have been performed by using lignan-rich foods such as flaxseed. We recently conducted a study by measuring lignan contents in four wheat cultivars (e.g., "Madison," "Ernie," "Betty" and "Arapahoe") and tried to link them with antitumorigenesis in APC^{Min} mouse model. APC^{Min} mice carry truncated adenomatous polyposis coli (APC) and thus spontaneously develop multiple intestinal neoplasia (Min). Drankhan et al. (11) applied this model into tumor prevention by feeding wheat bran and found a significant difference in antitumor activity among various wheat cultivars when the fiber contents were equal. As shown in Table 16.3, we compared the SDG contents in four wheat cultivars with antitumor activities in APC^{Min} mice and found a significant correlation between SDG contents and antitumor activities, suggesting lignans may attribute to the observed colon cancer prevention by wheat bran or whole grain products.

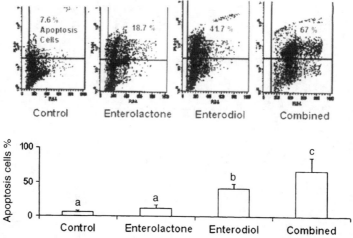

Figure 16.5 Increase of apoptotic cells by enterodiol and the combination of enterodiol and enterolactone. The human colon cancer SW480 cells were cocultured with the indicated lignan metabolites at 40 μmol/L for 72 h and the cellular apoptosis was measured by the terminal deoxynucleotidyl transferase-mediated dUTP nick end labeling assay. Values are mean ± SD ($n = 4$). Means with different alphabetic letters are significantly different, $P \leq 0.05$. (Adapted from Reference 37 with permission.)

TABLE 16.3   Correlation between the Contents of Secoisolariciresinol Diglucoside (SDG) in Various Wheat Cultivars and Antitumor Activities in APC[Min] Mice

Wheat cultivars	SDG, μg/g[a] Means ± SE ($n = 2$–3)	Antitumor activities[b], % of inhibition
Madison	82.9 ± 16.0	58.6 ± 1.8
Ernie	52.2 ± 22.4	36.1 ± 3.5
Betty	42.7 ± 0.1	23.7 ± 4.5
Arapahoe	Undetectable	27.2 ± 2.6
Correlation coefficient	$r = 0.73$ ($P < 0.02$)	

[a]The contents of lignans in four wheat cultivars were measured by HPLC analysis as described in our previous publication (37).
[b]The antitumor activities assessed in APC[Min] mice were adapted from Drankhan et al. (2003). Briefly, female APC[Min] mice at 5 weeks old (Jackson Lab, Bar Harbour, ME) were fed wheat bran at 45% in basal diet ($n = 10$ per group). The antitumor activities were calculated based on a formula as follows: (total tumor numbers in mice fed basal diet – total tumor numbers in mice fed wheat bran diet)/total tumor numbers in mice fed basal diet.

REFERENCES

1. Marlett, J. A.; McBurney, M. I.; Slavin, J. L. American Dietetic Association. Position of the American Dietetic Association: health implications of dietary fiber. *J. Am. Diet Assoc.* **2002**, *102*, 993–1000.
2. Hu, Y.; Martin, J.; Le Leu, R.; Young, G. P. The colonic response to genotoxic carcinogens in the rat: regulation by dietary fibre. *Carcinogenesis* **2002**, *23*, 1131–1137.

3. Alabaster, O.; Tang, Z.; Shivapurkar, N. Dietary fiber and the chemopreventive model of colon carcinogenesis. *Mutat. Res.* **1996**, *350*, 185–197.

4. Slvin, J.; Jacobs, D.; Marquart, L. Whole grain consumption and chronic disease: protective mechanisms. *Nutr. Cancer* **1997**, *27*, 14–21.

5. Eastwood, M. A. The physiological effect of dietary fiber: an update. *Annu. Rev. Nutr.* **1992**, *12*, 19–35.

6. Weisburger, J. H.; Reddy, B. S.; Rose, D. P.; Cohen, L. A.; Kendall, M. E.; Wynder, E. L. Protective mechanisms of dietary fibers in nutritional carcinogenesis. *Adv. Exp. Med. Biol.* **1993**, *348*, 85–97.

7. Velazquez, O. C.; Lederer, H. M.; Rombeau, J. L. Butyrate and the colonocytes: implications for neoplasia. *Dig. Dis. Sic.* **1996**, *41*, 727–739.

8. Nguyen, K. A.; Cao, Y.; Chen, J. R.; Townsend, C. M. Jr.; Ko, T. C. Dietary fiber enhances a tumor suppressor signaling pathway in the gut. *Ann. Surg.* **2006**, *243*, 619–625.

9. Pajari, A. M.; Oikarinen, S.; Grasten, S.; Mutanen, M. Diets enriched with cereal brans or inulin modulate protein kinase C activity and isozyme expression in rat colonic mucosa. *Br. J. Nutr.* **2000**, *84*, 635–643.

10. Otani, T.; Iwasaki, M.; Ishihara, J.; Sasazuki, S.; Inoue, M.; Tsugane, S. Japan Public Health Center-Based Prospective Study Group. Dietary fiber intake and subsequent risk of colorectal cancer: the Japan Public Health Center-based prospective study. *Int. J. Cancer* **2006**, *119*, 1475–1480.

11. Drankhan, K.; Carter, J.; Madl, R.; Klopfenstein, C.; Padula, F.; Lu, Y.; Warren, T.; Schmitz, N.; Takemoto, D. J. Anti-tumor activity of wheats with high orthophenolic content. *Nutr. Cancer* **2003**, *47*, 188–194.

12. MacRae, W. D.; Towers, G. H. N. Biological activities of lignans. *Phytochemistry* **1984**, *23*, 1207–1220.

13. Adlercreutz, H.; Mazur, W. Phyto-oestrogens and Western diseases. *Ann. Med.* **1997**, *29*, 95–120.

14. Reinli, K.; Block, G. Phytoestrogen content of foods–a codium of literature values. *Nutr. Cancer* **1996**, *26*, 123–148.

15. Dinkova-Kostova, A. T.; Gang, D. R.; Davin, L. B.; Bedgar, D. L.; Chu, A.; Lewis, N. G. (+)-Pinoresinol/ (+)-lariciresinol reductase from Forsythia intermedia. *J. Biol. Chem.* **1996**, *271*, 29473–29482.

16. Min, T.; Kasahara, H.; Bedgar, D. L.; Youn, B.; Lawrence, P. K.; Gang, D. R.; Halls, S. C.; Park, H.; Hilsenbeck, J. L.; Davin, L. B.; Lewis, N. G.; Kang, C. Crystal structures of pinoresinil-lariciresinol and phenylcoumaran benzylic ether reductases and their relationship to isoflavone reductases. *J. Biol. Chem.* **2003**, *278*, 50714–50723.

17. Youn, B.; Moinuddin, S. G.; Davin, L. B.; Lewis, N. G.; Kang, C. Crystal structures of apo-form and binary/ternary complexes of Podophyllum secoisolariciresinol dehydrogenase, an enzyme involved in formation of health-protecting and plant defense lignans. *J. Biol. Chem.* **2005**, *280*, 12917–12926.

18. Kuijsten, A.; Arts, I. C.; Vree, T. B.; Hollman, P. C. Pharmacokinetics of enterolignans in healthy men and women consuming a single dose of secoisolariciresinol diglucoside. *J. Nutr.* **2005**, *135*, 795–801.

19. Adlercreutz, H. Phytoestrogens and breast cancer. *J. Steroid Biochem. Mol. Biol.* **2002**, *83*, 113–118.

20. Webb, A. L.; McCullough, M. L. Dietary lignans: potential role in cancer prevention. *Nutr. Cancer* **2005**, *51*, 117–131.

21. Kuijsten, A.; Arts, I. C.; Vree, T. B.; Hollman, P. C.; van't Veer, P.; Kampman, E. Plasma enterolignans are associated with lower colorectal adenoma risk. *Cancer Epidemiol. Biomarkers Prev.* **2006**, *15*, 1132–1136.

22. Hedelin, M.; Klint, A.; Chang, E. T.; Bellocco, R.; Johansson, J. E.; Andersson, S. O.; Heinonen, S. M.; Adlercreutz, H.; Adami, H. O.; Gronberg, H.; Balter, K. A. Dietary phytoestrogen, serum enterolactone and risk of prostate cancer: the cancer prostate Sweden study (Sweden). *Cancer Causes Control* **2006**, *17*, 169–180.

23. Zeleniuch-Jacquotte, A.; Adlercreutz, H.; Shore, R. E.; Koenig, K. L.; Kato, I.; Arslan, A. A.; Toniolo, P. Circulating enterolactone and risk of breast cancer: a prospective study in New York. *Br. J. Cancer* **2004**, *91*, 99–105.

24. Milder, I. E.; Feskens, E. J.; Arts, I. C.; Bueno-de-Mesquita, H. B.; Hollman, P. C.; Kromhout, D. Intakes of 4 dietary lignans and cause-specific and all-cause mortality in the Zutphen Elderly Study. *Am. J. Clin. Nutr.* **2006**, *84*, 400–405.

25. Arts, I. C.; Hollman, P. C. Polyphenols and disease risk in epidemiologic studies. *Am. J. Clin. Nutr.* **2005**, *81*, 317S–325S.

26. Thompson, L. U.; Boucher, B. A.; Liu, Z.; Cotterchio, M.; Kreiger, N. *Nutr. Cancer* **2006**, *54*, 184–201.

27. Thompson, L. U.; Chen, J. M.; Li, T.; Strasser-Weippl, K.; Goss, P. E. Dietary flaxseed alters tumor biological markers in postmenopausal breast cancer. *Clin. Cancer Res.* **2005**, *11*, 3828–3835.
28. Chen, J.; Tan, K. P.; Ward, W. E.; Thompson, L. U. Exposure to flaxseed or its purified lignan during suckling inhibits chemically induced rat mammary tumorigenesis. *Exp. Biol. Med.* **2003**, *228*, 951–958.
29. Bylund, A.; Saarinen, N.; Zhang, J. X.; Bergh, A.; Widmark, A.; Johansson, A.; Lundin, E.; Adlercreutz, H.; Hallmans, G.; Stattin, P.; Makela, S. Anticancer effects of a plant lignan 7-hydroxymatairesinol on a prostate cancer model *in vivo. Exp. Biol. Med.* **2005**, *230*, 217–223.
30. Saarinen, N. M.; Power, K.; Chen, J.; Thompson, L. U. Flaxseed attenuates the tumor growth stimulating effect of soy protein in ovariectomized athymic mice with MCF-7 human breast cancer xenografts. *Int. J. Cancer* **2006**, *119*, 925–931.
31. Chen, J.; Wang, L.; Thompson, L. U. Flaxseed and its components reduce metastasis after surgical excision of solid human breast tumor in nude mice. *Cancer Lett.* **2006**, *234*, 168–175.
32. Bird, R. P. Role of aberrant crypt foci in understanding the pathogenesis of colon cancer. *Cancer Lett.* **1995**, *93*, 55–71.
33. Takayama, T.; Katsuki, S.; Takahashi, Y.; Ohi, M.; Nojiri, S.; Sakamaki, S.; Kato, J.; Kogawa, K.; Miyake, H.; Niitsu, Y. Aberrant crypt foci of the colon as precursors of adenoma and cancer. *N. Engl. J. Med.* **1998**, *339*, 1277–1284.
34. Wang, W.; Goodman, M. T. Antioxidant properties of dietary phenolic agents in a human LDL-oxidation ex vivo model: interaction of protein binding activity. *Nutr. Res.* **1999**, *19*, 191–202.
35. Lu, H.; Liu, G. T. Effect of dibenzo(α,χ)cycloctene lignans isolated from Fructus Schizandrae on lipid peroxidation and antioxidative enzyme activity. *Chem. Biol. Interact* **1991**, *78*, 77–84.
36. Wang, W.; Liu, L. Q.; Higuchi, C. M.; Chen, H. Induction of NADPH:quinone reductase by dietary phytoestrogens in human colonic Colo205 cells. *Biochem. Pharmacol.* **1998**, *56*, 189–195.
37. Qu, H.; Madl, R. L.; Takemoto, D. J.; Baybutt, R. C.; Wang, W. Lignans are involved in the antitumor activity of wheat bran in colon cancer SW480 cells. *J. Nutr.* **2005**, *135*, 598–602.
38. Saleem, M.; Kim, H. J.; Ali, M. S.; Lee, Y. S. An update on bioactive plant lignans. *Nat. Prod. Rep.* **2005**, *22*, 696–716.

INDEX